DATA MINING AND BUSINESS ANALYTICS WITH *R*

DATA MINING AND BUSINESS ANALYTICS WITH *R*

Johannes Ledolter
Department of Management Sciences
Tippie College of Business
University of Iowa
Iowa City, Iowa

Published by John Wiley & Sons, Inc., Hoboken, New Jersey
Published simultaneously in Canada

For general information on our other products and services or for technical support, please contact our Customer Care Department within the United States at (800) 762-2974, outside the United States at (317) 572-3993 or fax (317) 572-4002.

Wiley also publishes its books in a variety of electronic formats. Some content that appears in print may not be available in electronic formats. For more information about Wiley products, visit our web site at www.wiley.com.

Library of Congress Cataloging-in-Publication Data:

Ledolter, Johannes.
Data mining and business analytics with R / Johannes Ledolter, University of Iowa.
pages cm
Includes bibliographical references and index.
ISBN 978-1-118-44714-7 (cloth)
1. Data mining. 2. R (Computer program language) 3. Commercial statistics. I. Title.
QA76.9.D343L44 2013
006.3′12–dc23

2013000330

Printed in the United States of America

CONTENTS

PREFACE

This book is about useful methods for data mining and business analytics. It is written for readers who want to apply these methods so that they can learn about their processes and solve their problems. My objective is to provide a thorough discussion of the most useful data-mining tools that goes beyond the typical "black box" description, and to show why these tools work.

Powerful, accurate, and flexible computing software is needed for data mining, and Excel is of little use. Although excellent data-mining software is offered by various commercial vendors, proprietary products are usually expensive. In this text, I use the R Statistical Software, which is powerful and free. But the use of R comes with start-up costs. R requires the user to write out instructions, and the writing of program instructions will be unfamiliar to most spreadsheet users. This is why I provide R sample programs in the text and on the webpage that is associated with this book. These sample programs should smooth the transition to this very general and powerful computer environment and help keep the start-up costs to using R small.

The text combines explanations of the statistical foundation of data mining with useful software so that the tools can be readily applied and put to use. There are certainly better books that give a deeper description of the methods, and there are also numerous texts that give a more complete guide to computing with R. This book tries to strike a compromise that does justice to both theory and practice, at a level that can be understood by the MBA student interested in quantitative methods. This book can be used in courses on data mining in quantitative MBA programs and in upper-level undergraduate and graduate programs that deal with the analysis and interpretation of large data sets. Students in business, the social and natural sciences, medicine, and engineering should benefit from this book. The majority of the topics can be covered in a one semester course. But not every covered topic will be useful for all audiences, and for some audiences, the coverage of certain topics will be either too advanced or too basic. By omitting some topics and by expanding on others, one can make this book work for many different audiences.

Certain data-mining applications require an enormous amount of effort to just collect the relevant information, and in such cases, the data preparation takes a lot more time than the eventual modeling. In other applications, the data collection effort is minimal, but often one has to worry about the efficient storage and retrieval of high volume information (i.e., the "data warehousing"). Although it is very important to know how to acquire, store, merge, and best arrange the information, this text does not cover these aspects very deeply. This book concentrates on the modeling aspects of data mining.

The data sets and the R-code for all examples can be found on the webpage that accompanies this book (**http://www.biz.uiowa.edu/faculty/jledolter/DataMining**). Supplementary material for this book can also be found by entering ISBN 9781118447147 at booksupport.wiley.com. You can copy and paste the code into your own R session and rerun all analyses. You can experiment with the software by making changes and additions, and you can adapt the R templates to the analysis of your own data sets. Exercises and several large practice data sets are given at the end of this book. The exercises will help instructors when assigning homework problems, and they will give the reader the opportunity to practice the techniques that are discussed in this book. Instructions on how to best use these data sets are given in Appendix A.

This is a first edition. Although I have tried to be very careful in my writing and in the analyses of the illustrative data sets, I am certain that much can be improved. I would very much appreciate any feedback you may have, and I encourage you to write to me at johannes-ledolter@uiowa.edu. Corrections and comments will be posted on the book's webpage.

ACKNOWLEDGMENTS

I got interested in developing materials for an MBA-level text on Data Mining when I visited the University of Chicago Booth School of Business in 2011. The outstanding University of Chicago lecture materials for the course on Data Mining (BUS41201) taught by Professor Matt Taddy provided the spark to put this text together, and several examples and R-templates from Professor Taddy's notes have influenced my presentation. Chapter 19 on the analysis of text data draws heavily on his recent research. Professor Taddy's contributions are most gratefully acknowledged.

Writing a text is a time-consuming task. I could not have done this without the support and constant encouragement of my wife, Lea Vandervelde. Lea, a law professor at the University of Iowa, conducts historical research on the freedom suits of Missouri slaves. She knows first-hand how important and difficult it is to construct data sets for the mining of text data.

CHAPTER 1

Introduction

Today's statistics applications involve enormous data sets: many cases (rows of a data spreadsheet, with a row representing the information on a studied case) and many variables (columns of the spreadsheet, with a column representing the outcomes on a certain characteristic across the studied cases). A case may be a certain item such as a purchase transaction, or a subject such as a customer or a country, or an object such as a car or a manufactured product. The information that we collect varies across the cases, and the explanation of this variability is central to the tools that we study in this book. Many variables are typically collected on each case, but usually only a few of them turn out to be useful. The majority of the collected variables may be irrelevant and represent just noise. It is important to find those variables that matter and those that do not.

Here are a few types of data sets that one encounters in data mining. In marketing applications, we observe the purchase decisions, made over many time periods, of thousands of individuals who select among several products under a variety of price and advertising conditions. Social network data contains information on the presence of links among thousands or millions of subjects; in addition, such data includes demographic characteristics of the subjects (such as gender, age, income, race, and education) that may have an effect on whether subjects are "linked" or not. Google has extensive information on 100 million users, and Facebook has data on even more. The recommender systems developed by firms such as Netflix and Amazon use available demographic information and the detailed purchase/rental histories from millions of customers. Medical data sets contain the outcomes of thousands of performed procedures, and include information on their characteristics such as the type of procedure and its outcome, and the location where and the time when the procedure has been performed.

While traditional statistics applications focus on relatively small data sets, data mining involves very large and sometimes enormous quantities of information. One talks about megabytes and terabytes of information. A megabyte represents a million bytes, with a byte being the number of bits needed to encode a single character of text. A typical English book in plain text format (500 pages with 2000

Data Mining and Business Analytics with R, First Edition. Johannes Ledolter.

characters per page) amounts to about 1 MB. A terabyte is a million megabytes, and an exabyte is a million terabytes.

Data mining attempts to extract useful information from such large data sets. Data mining explores and analyzes large quantities of data in order to discover meaningful patterns. The *scale* of a typical data mining application, with its large number of cases and many variables, exceeds that of a standard statistical investigation. The analysis of millions of cases and thousands of variables also puts pressure on the *speed* that is needed to accomplish the search and modeling steps of the typical data mining application. This is why researchers refer to data mining as statistics at scale and speed. The large scale (lots of available data) and the requirements on speed (solutions are needed quickly) create a large demand for automation. Data mining uses a combination of pattern-recognition rules, statistical rules, as well as rules drawn from machine learning (an area of computer science).

Data mining has wide applicability, with applications in intelligence and security analysis, genetics, the social and natural sciences, and business. Studying which buyers are more likely to buy, respond to an advertisement, declare bankruptcy, commit fraud, or abandon subscription services are of vital importance to business.

Many data mining problems deal with categorical outcome data (e.g., no/yes outcomes), and this is what makes machine learning methods, which have their origins in the analysis of categorical data, so useful. Statistics, on the other hand, has its origins in the analysis of continuous data. This makes statistics especially useful for correlation-type analyses where one sifts through a large number of correlations to find the largest ones.

The analysis of large data sets requires an efficient way of storing the data so that it can be accessed easily for calculations. Issues of data warehousing and how to best organize the data are certainly very important, but they are not emphasized in this book. The book focuses on the analysis tools and targets their statistical foundation.

Because of the often enormous quantities of data (number of cases/replicates), the role of traditional statistical concepts such as confidence intervals and statistical significance tests is greatly reduced. With large data sets, almost any small difference becomes significant. It is the problem of overfitting models (i.e., using more explanatory variables than are actually needed to predict a certain phenomenon) that becomes of central importance. Parsimonious representations are important as simpler models tend to give more insight into a problem. Large models overfitted on training data sets usually turn out to be extremely poor predictors in new situations as unneeded predictor variables increase the prediction error variance. Furthermore, overparameterized models are of little use if it is difficult to collect data on predictor variables in the future. Methods that help avoid such overfitting are needed, and they are covered in this book. The partitioning of the data into training and evaluation (test) data sets is central to most data mining methods. One must always check whether the relationships found in the training data set will hold up in the future.

Many data mining tools deal with problems for which there is no designated response that one wants to predict. It is common to refer to such analysis as *unsupervised learning*. Cluster analysis is one example where one uses feature (variable) data on numerous objects to group the objects (i.e., the cases) into a smaller number of groups (also called *clusters*). Dimension reduction applications are other examples for such type of problems; here one tries to reduce the many features on an object to a manageable few. Association rules also fall into this category of problems; here one studies whether the occurrence of one feature is related to the occurrence of others. Who would not want to know whether the sales of chips are being "lifted" to a higher level by the concurrent sales of beer?

Other data mining tools deal with problems for which there is a designated response, such as the volume of sales (a quantitative response) or whether someone buys a product (a categorical response). One refers to such analysis as *supervised learning*. The predictor variables that help explain (predict) the response can be quantitative (such as the income of the buyer or the price of a product) or categorical (such as the gender and profession of the buyer or the qualitative characteristics of the product such as new or old). Regression methods, regression trees, and nearest neighbor methods are well suited for problems that involve a continuous response. Logistic regression, classification trees, nearest neighbor methods, discriminant analysis (for continuous predictor variables) and naïve Bayes methods (mostly for categorical predictor variables) are well suited for problems that involve a categorical response.

Data mining should be viewed as a *process*. As with all good statistical analyses, one needs to be clear about the purpose of the analysis. Just to "mine data" without a clear purpose, without an appreciation of the subject area, and without a modeling strategy will usually not be successful. The data mining process involves several interrelated steps:

1. Efficient data storage and data preprocessing steps are very critical to the success of the analysis.
2. One needs to select appropriate response variables and decide on the number of variables that should be investigated.
3. The data needs to be screened for outliers, and missing values need to be addressed (with missing values either omitted or appropriately imputed through one of several available methods).
4. Data sets need to be partitioned into training and evaluation data sets. In very large data sets, which cannot be analyzed easily as a whole, data must be sampled for analysis.
5. Before applying sophisticated models and methods, the data need to be visualized and summarized. It is often said that a picture is worth a 1000 words. Basic graphs such as line graphs for time series, bar charts for categorical variables, scatter plots and matrix plots for continuous variables, box plots and histograms (often after stratification on useful covariates), maps for displaying correlation matrices, multidimensional graphs using color, trellis graphs, overlay plots, tree maps for visualizing network data, and geo maps for spatial data are just a few examples of the more useful graphical displays. In constructing good graphs, one needs to be careful about the right scaling, the correct labeling, and issues of stratification and aggregation.
6. Summary of the data involves the typical summary statistics such as mean, percentiles and median, standard deviation, and correlation, as well as more advanced summaries such as principal components.

7. Appropriate methods from the data mining tool bag need to be applied. Depending on the problem, this may involve regression, logistic regression, regression/classification trees, nearest neighbor methods, k-means clustering, and so on.
8. The findings from these models need to be confirmed, typically on an evaluation (test or holdout) data set.
9. Finally, the insights one gains from the analysis need to be implemented. One must act on the findings and spring to action. This is what W.E. Deming had in mind when he talked about process improvement and his Deming (Shewhart) wheel of "plan, do, check, and act" (Ledolter and Burrill, 1999).

Some data mining applications require an enormous amount of effort to just collect the relevant information. For example, an investigation of Pre-Civil War court cases of Missouri slaves seeking their freedom involves tedious study of handwritten court proceedings and Census records, electronic scanning of the records, and the use of character-recognition software to extract the relevant characteristics of the cases and the people involved. The process involves double and triple checking unclear information (such as different spellings, illegible entries, and missing information), selecting the appropriate number of variables, categorizing text information, and deciding on the most appropriate coding of the information. At the end, one will have created a fairly good master list of all available cases and their relevant characteristics. Despite all the diligent work, there will be plenty of missing information, information that is in error, and way too many variables and categories than are ultimately needed to tell the story behind the judicial process of gaining freedom.

Data preparation often takes a lot more time than the eventual modeling. The subsequent modeling is usually only a small component of the overall effort; quite often, relatively simple methods and a few well-constructed graphs can tell the whole story. It is the creation of the master list that is the most challenging task. The steps that are involved in the construction of the master list in such problems depend heavily on the subject area, and one can only give rough guidelines on how to proceed. It is also difficult to make this process automatic. Furthermore, even if some of the "data cleaning" steps can be made automatic, the investigator must constantly check and question any adjustments that are being made. Great care, lots of double and triple checking, and much common sense are needed to create a reliable master list. But without a reliable master list, the findings will be suspect, as we know that wrong data usually lead to wrong conclusions. The old saying "garbage in–garbage out" also applies to data mining.

Fortunately many large business data sets can be created almost automatically. Much of today's business data is collected for transactional purposes, that is, for payment and for shipping. Examples of such data sets are transactions that originate from scanner sales in super markets, telephone records that are collected by mobile telephone providers, and sales and rental histories that are collected by companies such as Amazon and Netflix. In all these cases, the data collection effort is minimal, even though companies have to worry about the efficient storage and retrieval of the information (i.e., the "data warehousing").

Credit card companies collect information on purchases; telecom companies collect information on phone calls such as their timing, length, origin, and destination; retail stores have developed automated ways of collecting information on their sales such as the volume purchased and the price at which products are bought. Supermarkets are now the source of much excellent data on the purchasing behavior of individuals. Electronic scanners keep track of purchases, prices, and the presence of promotions. Loyalty programs of retail chains and frequent-flyer programs make it possible to link the purchases to the individual shopper and his/her demographic characteristics and preferences. Innovative marketing firms combine the customer's purchase decisions with the customer's exposure to different marketing messages. As early as the 1980s, Chicago's IRI (Information Resources Incorporated, now Symphony IRI) contracted with television cable companies to vary the advertisements that were sent to members of their household panels. They knew exactly who was getting which ad and they could track the panel members' purchases at the store. This allowed for a direct way of assessing the effectiveness of marketing interventions; certainly much more direct than the diary-type information that had been collected previously. At present, companies such as Google and Facebook run experiments all the time. They present their members with different ads and they keep track who is clicking on the advertised products and whether the products are actually being bought.

Internet companies have vast information on customer preferences and they use this for targeted advertising; they use recommender systems to direct their ads to areas that are most profitable. Advertising related products that have a good chance of being bought and "cross-selling" of products become more and more important. Data from loyalty programs, from e-Bay auction histories, and from digital footprints of users clicking on Internet webpages are now readily available. Google's "Flu tracker" makes use of the webpage clicks to develop a tool for the early detection of influenza outbreaks; Amazon and Netflix use the information from their shoppers' previous order histories without ever meeting them in person, and they use the information from previous order histories of their users to develop automatic recommender systems. Credit risk calculations, business sentiment analysis, and brand image analysis are becoming more and more important.

Sports teams use data mining techniques to assemble winning teams; see the success stories of the Boston Red Sox and the Oakland Athletics. *Moneyball*, a 2011 biographical sports drama film based on Michael Lewis's 2003 book of the same name, is an account of the Oakland Athletics baseball team's 2002 season and their general manager Billy Beane's attempts to assemble a competitive team through data mining and business analytics.

It is not only business applications of data mining that are important; data mining is also important for applications in the sciences. We have enormous data bases on drugs and their side effects, and on medical procedures and their complication rates. This information can be mined to learn which drugs work and under which conditions they work best; and which medical procedures lead to complications and for which patients.

Business analytics and data mining deal with collecting and analyzing data for better decision making in business. Managers and business students can gain a competitive advantage through business analytics and data mining. Most tools and methods for data mining discussed in this book have been around for a very long time. But several developments have come together over the past few years, making the present period a perfect time to use these methods for solving business problems.

1. More and more data relevant for data mining applications are now being collected.
2. Data is being warehoused and is now readily available for analysis. Much data from numerous sources has already been integrated, and the data is stored in a format that makes the analysis convenient.
3. Computer storage and computer power are getting cheaper every day, and good software is available to carry out the analysis.
4. Companies are interested in "listening" to their customers and they now believe strongly in customer relationship management. They are interested in holding on to good customers and getting rid of bad ones. They embrace tools and methods that give them this information.

This book discusses the modeling tools and the methods of data mining. We assume that one has constructed the relevant master list of cases and that the data is readily available. Our discussion covers the last 10–20% of effort that is needed to extract and model meaningful information from the raw data. A model is a simplified description of the process that may have generated the data. A model may be a mathematical formula, or a computer program. One must remember, however, that no model is perfect, and that all models are merely approximations. But some of these approximations will turn out to be useful and lead to insights. One needs to become a critical user of models. If a model looks too good to be true, then it generally is. Models need to be checked, and we emphasized earlier that models should not be evaluated on the data that had been used to build them. Models are "fine-tuned" to the data of the training set, and it is not obvious whether this good performance carries over to other data sets.

In this book, we use the **R Statistical Software** (Version 15 as of June 2012). It is powerful and free. One may search for the software on the web and download the system. R is similar to Matlab and requires the user to write out simple instructions. The writing of (program) instructions will be unfamiliar to a spreadsheet user, and there will be startup costs to using R. However, the R sample programs in this book and their listing on the book's webpage should help with the transition to this very general and powerful computer environment.

REFERENCE

Ledolter, J. and Burrill, C.: *Statistical Quality Control: Strategies and Tools for Continual Improvement*. New York: John Wiley & Sons, Inc., 1999.

CHAPTER 2

Processing the Information and Getting to Know Your Data

In this chapter we analyze three data sets and illustrate the steps that are needed for preprocessing the data. We consider (i) the 2006 birth data that is used in the book *R in a Nutshell: A Desktop Quick Reference* (Adler, 2009), (ii) data on the contributions to a Midwestern private college (Ledolter and Swersey, 2007), and (iii) the orange juice data set taken from P. Rossi's **bayesm** package for R that was used earlier in Montgomery (1987). The three data sets are of suitable size (427,323 records and 13 variables in the 2006 birth data set; 1230 records and 11 variables in the contribution data set; and 28,947 records and 17 variables in the orange juice data set). The data sets include both continuous and categorical variables, have missing observations, and require preprocessing steps before they can be subjected to the appropriate statistical analysis and modeling. We use these data sets to illustrate how to summarize the available information and how to obtain useful graphical displays. The initial arrangement of the data is often not very convenient for the analysis, and the information has to be rearranged and preprocessed. We show how to do this within R.

All data sets and the R programs for all examples in this book are listed on the webpage that accompanies this book (http://www.biz.uiowa.edu/faculty/jledolter/DataMining). I encourage readers to copy and paste the R programs into their own R sessions and check the results. Having such templates available for the analysis helps speed up the learning curve for R. It is much easier to learn from a sample program than to piece together the R code from first principles. It is the author's experience that even novices catch on quite fast. It may happen that at some time in the future certain R functions and packages become obsolete and are no longer available. Readers should then look for adequate replacements. The R function "help" can be used to get information on new functions and packages.

2.1 EXAMPLE 1: 2006 BIRTH DATA

We consider the 2006 birth data set that is used in the book *R In a Nutshell: A Desktop Quick Reference* (Adler, 2009). The data set *births2006.smpl* consists of

Data Mining and Business Analytics with R, First Edition. Johannes Ledolter.

427,323 records and 13 variables, including the day of birth according to the month and the day of week (DOB_MM, DOB_WK), the birth weight of the baby (DBWT) and the weight gain of the mother during pregnancy (WTGAIN), the sex of the baby and its APGAR score at birth (SEX and APGAR5), whether it was a single or multiple birth (DPLURAL), and the estimated gestation age in weeks (ESTGEST). We list below the information for the first five births.

```
## Install packages from CRAN; use any USA mirror
library(lattice)
library(nutshell)
data(births2006.smpl)
births2006.smpl[1:5,]
```

```
        DOB_MM DOB_WK MAGER TBO_REC WTGAIN SEX APGAR5                   DMEDUC
591430       9      1    25       2     NA   F     NA                     NULL
1827276      2      6    28       2     26   M      9        2 years of college
1705673      2      2    18       2     25   F      9                     NULL
3368269     10      5    21       2      6   M      9                     NULL
2990253      7      7    25       1     36   M     10 2 years of high school
        UPREVIS ESTGEST DMETH_REC  DPLURAL DBWT
591430       10      99   Vaginal 1 Single 3800
1827276      10      37   Vaginal 1 Single 3625
1705673      14      38   Vaginal 1 Single 3650
3368269      22      38   Vaginal 1 Single 3045
2990253      15      40   Vaginal 1 Single 3827
```

```
dim(births2006.smpl)
```

```
[1] 427323     13
```

The following bar chart of the frequencies of births according to the day of week of the birth shows that fewer births take place during the weekend (days 1 = Sunday, 2 = Monday, ..., 7 = Saturday of DOB_WK). This may have to do with the fact that many babies are delivered by cesarean section, and that those deliveries are typically scheduled during the week and not on weekends. To follow up on this hypothesis, we obtain the frequencies in the two-way classification of births according to the day of week and the method of delivery. Excluding births of unknown delivery method, we separate the bar charts of the frequencies for the day of week of delivery according to the method of delivery. While it is also true that vaginal births are less frequent on weekends than on weekdays (doctors prefer to work on weekdays), the reduction in the frequencies of scheduled C-section deliveries from weekdays to weekends (about 50%) exceeds the weekday–weekend reduction of vaginal deliveries (about 25–30%).

```
births.dow=table(births2006.smpl$DOB_WK)
births.dow
```

```
    1     2     3     4     5     6     7
40274 62757 69775 70290 70164 68380 45683
```

```
barchart(births.dow,ylab="Day of Week",col="black")
```

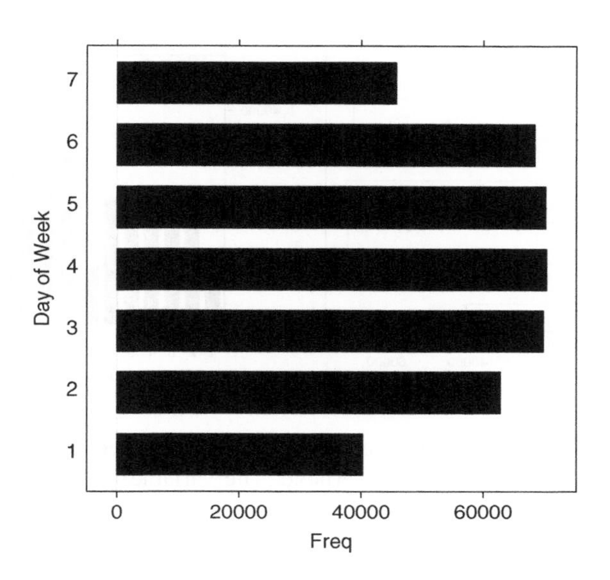


```
dob.dm.tbl=table(WK=births2006.smpl$DOB_WK,
+     MM=births2006.smpl$DMETH_REC)
dob.dm.tbl

   MM
WK  C-section Unknown Vaginal
  1      8836      90   31348
  2     20454     272   42031
  3     22921     247   46607
  4     23103     252   46935
  5     22825     258   47081
  6     23233     289   44858
  7     10696     109   34878

dob.dm.tbl=dob.dm.tbl[,-2]
dob.dm.tbl

   MM
WK  C-section Vaginal
  1      8836   31348
  2     20454   42031
  3     22921   46607
  4     23103   46935
  5     22825   47081
  6     23233   44858
  7     10696   34878

trellis.device()
barchart(dob.dm.tbl,ylab="Day of Week")
barchart(dob.dm.tbl,horizontal=FALSE,groups=FALSE,
+    xlab="Day of Week",col="black")
```

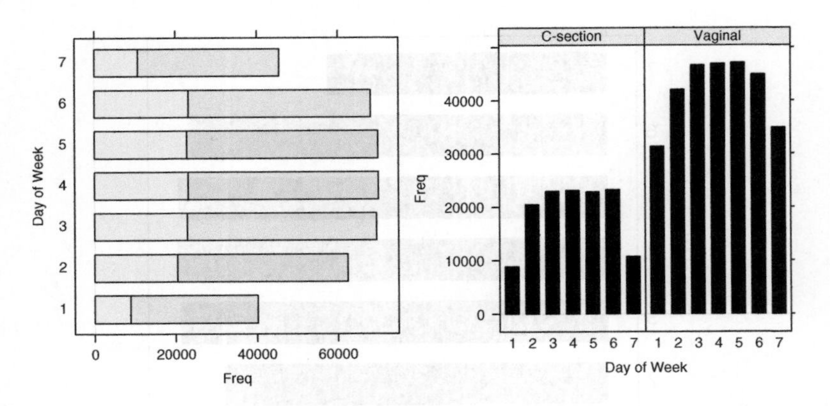

We use lattice (trellis) graphics (and the R package **lattice**) to condition density histograms on the values of a third variable. The variable for multiple births (single births to births with five offsprings (quintuplets) or more) and the method of delivery are our conditioning variables, and we separate histograms of birth weight according to these variables. As expected, birth weight decreases with multiple births, whereas the birth weight is largely unaffected by the method of delivery. Smoothed versions of the histograms, using the lattice command density plot, are also shown. Because of the very small sample sizes for quintuplet and even more births, the density of birth weight for this small group is quite noisy. The dot plot, also part of the **lattice** package, shows quite clearly that there are only few observations in that last group, while most other groups have many observations (which makes the dots on the dot plot "run into each other"); for groups with many observations a histogram would be the preferred graphical method.

```
histogram(~DBWT|DPLURAL,data=births2006.smpl,layout=c(1,5),
+     col="black")
histogram(~DBWT|DMETH_REC,data=births2006.smpl,layout=c(1,3),
+     col="black")
```

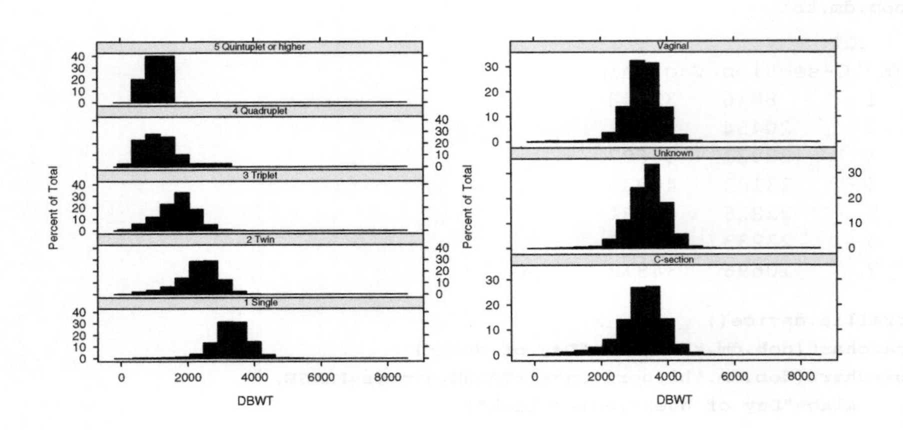

```
densityplot(~DBWT|DPLURAL,data=births2006.smpl,layout=c(1,5),
+    plot.points=FALSE,col="black")
densityplot(~DBWT,groups=DPLURAL,data=births2006.smpl,
+    plot.points=FALSE)
```

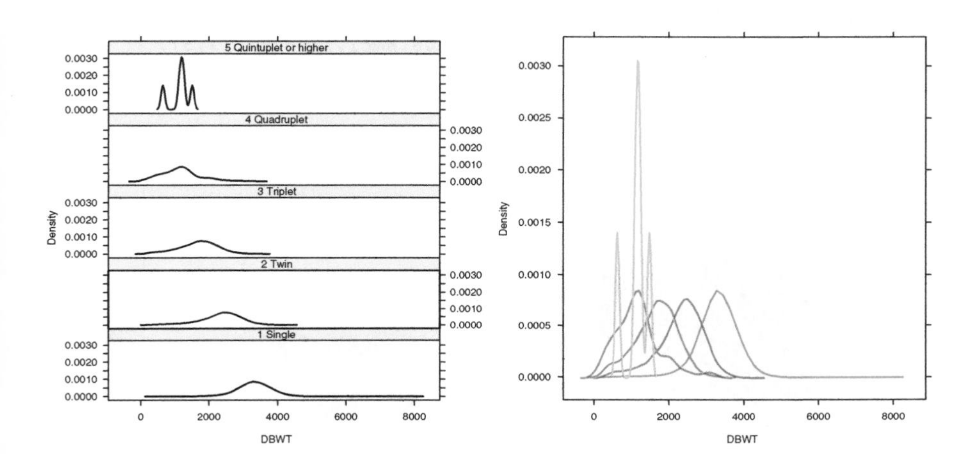


```
dotplot(~DBWT|DPLURAL,data=births2006.smpl,layout=c(1,5),
+    plot.points=FALSE,col="black")
```

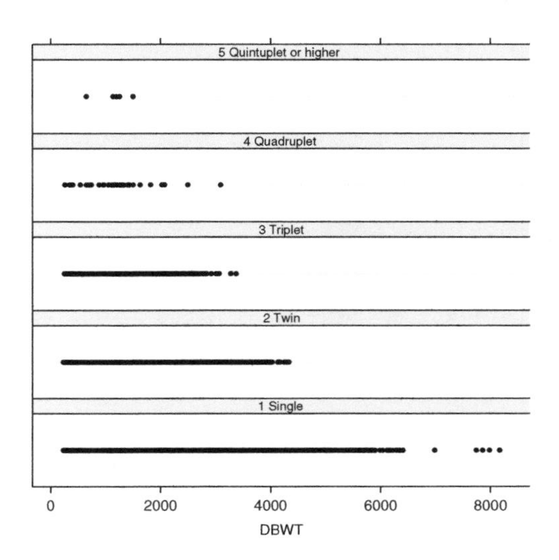

Scatter plots (xyplots in the package **lattice**) are shown for birth weight against weight gain, and the scatter plots are stratified further by multiple births. The last smoothed scatter plot indicates that there is little association between birth weight and weight gain during the course of the pregnancy.

```
xyplot(DBWT~DOB_WK,data=births2006.smpl,col="black")
xyplot(DBWT~DOB_WK|DPLURAL,data=births2006.smpl,layout=c(1,5),
+    col="black")
xyplot(DBWT~WTGAIN,data=births2006.smpl,col="black")
xyplot(DBWT~WTGAIN|DPLURAL,data=births2006.smpl,layout=c(1,5),
+    col="black")
```

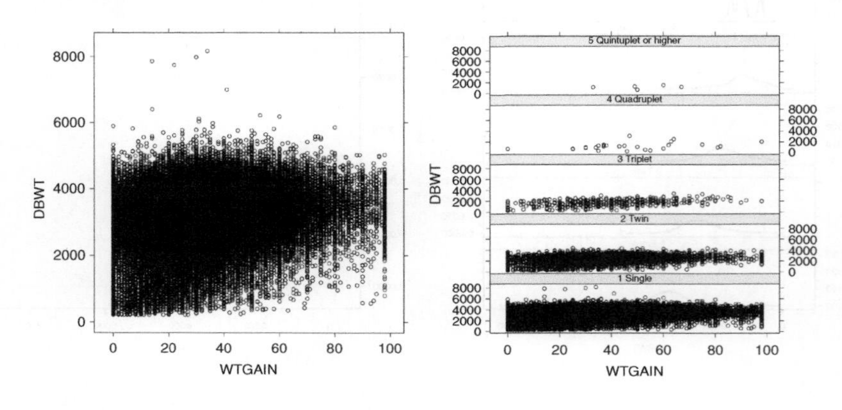


```
smoothScatter(births2006.smpl$WTGAIN,births2006.smpl$DBWT)
```

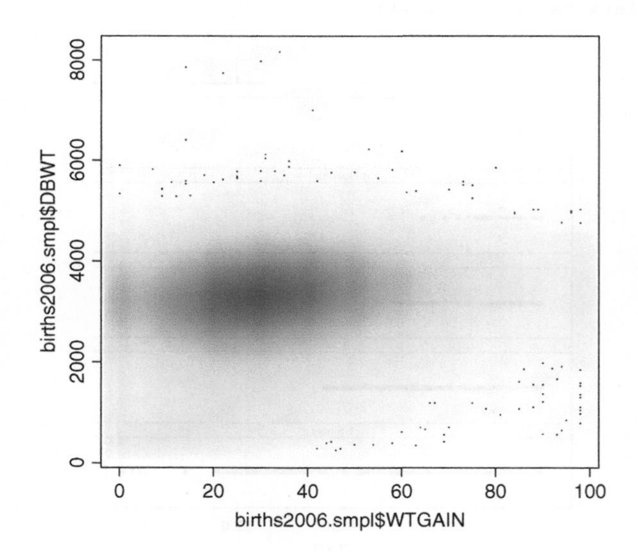

We also illustrate box plots of birth weight against the APGAR score and box plots of birth weight against the day of week of delivery. We would not expect much relationship between the birth weight and the day of week of delivery; there is no reason why babies born on weekends should be heavier or lighter than those born during the week. The APGAR score is an indication of the health status of a newborn, with low scores indicating that the newborn experiences difficulties. The box plot of birth weight against the APGAR score shows a strong relationship. Babies of low birth weight often have low APGAR scores as their health is compromised by the low birth weight and its associated complications.

```
## boxplot is the command for a box plot in the standard graphics
## package
boxplot(DBWT~APGAR5,data=births2006.smpl,ylab="DBWT",
+    xlab="AGPAR5")
boxplot(DBWT~DOB_WK,data=births2006.smpl,ylab="DBWT",
+    xlab="Day of Week")
```

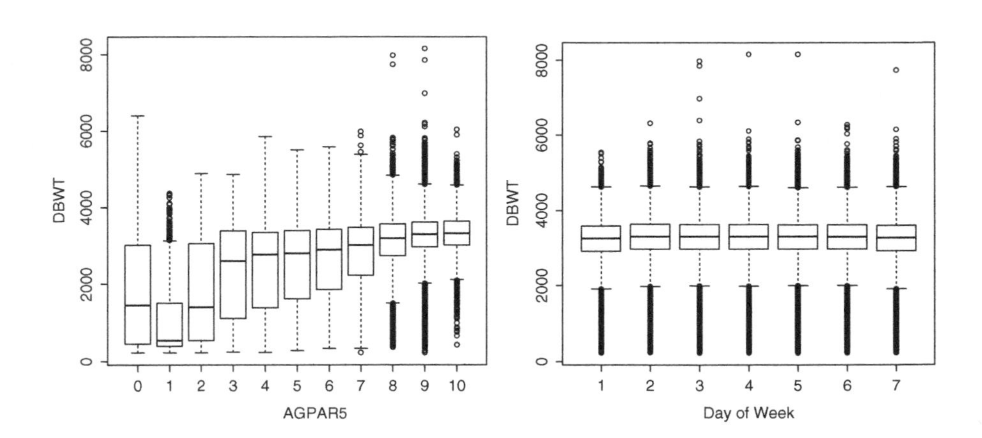


```
## bwplot is the command for a box plot in the lattice graphics
## package. There you need to declare the conditioning variables
## as factors
bwplot(DBWT~factor(APGAR5)|factor(SEX),data=births2006.smpl,
+    xlab="AGPAR5")
bwplot(DBWT~factor(DOB_WK),data=births2006.smpl,
+    xlab="Day of Week")
```

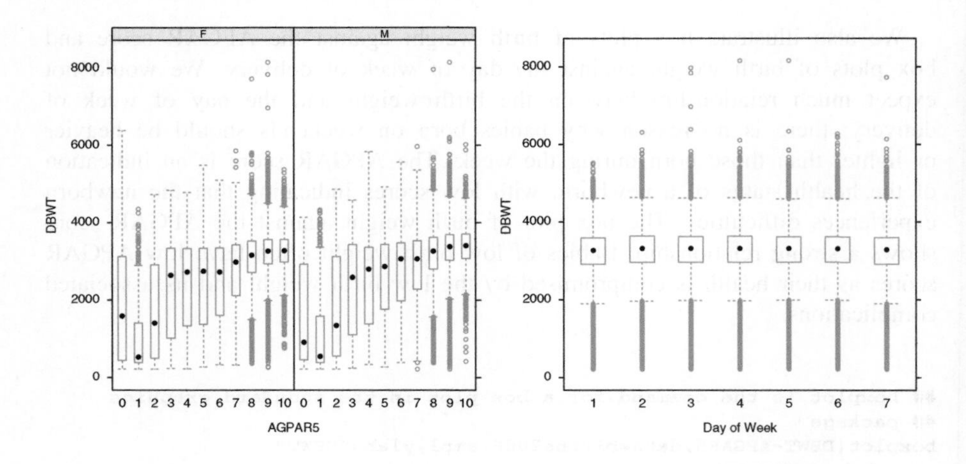

We also calculate the average birth weight as function of multiple births, and we do this for males and females separately. For that we use the tapply function. Note that there are missing observations in the data set and the option `na.rm=TRUE` (remove missing observations from the calculation) is needed to omit the missing observations from the calculation of the mean. The bar plot illustrates graphically how the average birth weight decreases with multiple deliveries. It also illustrates that the average birth weight for males is slightly higher than that for females.

```
fac=factor(births2006.smpl$DPLURAL)
res=births2006.smpl$DBWT
t4=tapply(res,fac,mean,na.rm=TRUE)
t4

        1 Single                      2 Twin                   3 Triplet
        3298.263                    2327.478                    1677.017
    4 Quadruplet 5 Quintuplet or higher
        1196.105                    1142.800

t5=tapply(births2006.smpl$DBWT,INDEX=list(births2006.smpl$DPLURAL,
+    births2006.smpl$SEX),FUN=mean,na.rm=TRUE)
t5

                                  F        M
   1 Single                3242.302 3351.637
   2 Twin                  2279.508 2373.819
   3 Triplet               1697.822 1655.348
   4 Quadruplet            1319.556 1085.000
   5 Quintuplet or higher  1007.667 1345.500

barplot(t4,ylab="DBWT")
barplot(t5,beside=TRUE,ylab="DBWT")
```

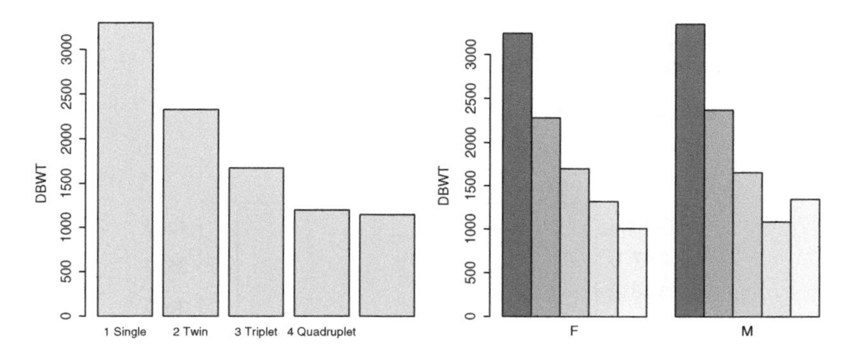

Finally, we illustrate the levelplot and the contourplot of the R package **lattice**. For these plots we first create a cross-classification of weight gain and estimated gestation period by dividing the two continuous variables into 11 nonoverlapping groups. For each of the resulting groups, we compute the average birth weight. An earlier frequency distribution table of estimated gestation period indicates that "99" is used as the code for "unknown". For the subsequent calculations, we omit all records with unknown gestation period (i.e., value 99). The graphs show that the birth weight increases with the estimated gestation period, but that birth weight is little affected by the weight gain. Note that the contour lines are essentially horizontal and that their associated values increase with the estimated gestation period.

```
t5=table(births2006.smpl$ESTGEST)
t5
    12     15     17     18     19     20     21     22     23     24     25
     1      2     18     43     69    116    162    209    288    401    445
    26     27     28     29     30     31     32     33     34     35     36
   461    566    670    703   1000   1243   1975   2652   4840   7954  15874
    37     38     39     40     41     42     43     44     45     46     47
 33310  76794 109046  84890  23794   1931    133     32      6      5      5
    48     51     99
     2      1  57682

new=births2006.smpl[births2006.smpl$ESTGEST != 99,]
t51=table(new$ESTGEST)
t51

    12     15     17     18     19     20     21     22     23     24     25
     1      2     18     43     69    116    162    209    288    401    445
    26     27     28     29     30     31     32     33     34     35     36
   461    566    670    703   1000   1243   1975   2652   4840   7954  15874
    37     38     39     40     41     42     43     44     45     46     47
 33310  76794 109046  84890  23794   1931    133     32      6      5      5
    48     51
     2      1
```

```
t6=tapply(new$DBWT,INDEX=list(cut(new$WTGAIN,breaks=10),
+     cut(new$ESTGEST,breaks=10)),FUN=mean,na.rm=TRUE)
t6
levelplot(t6,scales = list(x = list(rot = 90)))
```

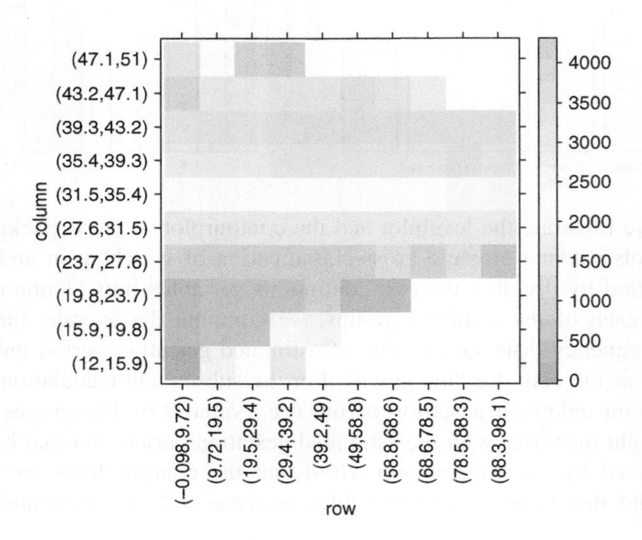


```
contourplot(t6,scales = list(x = list(rot = 90)))
```

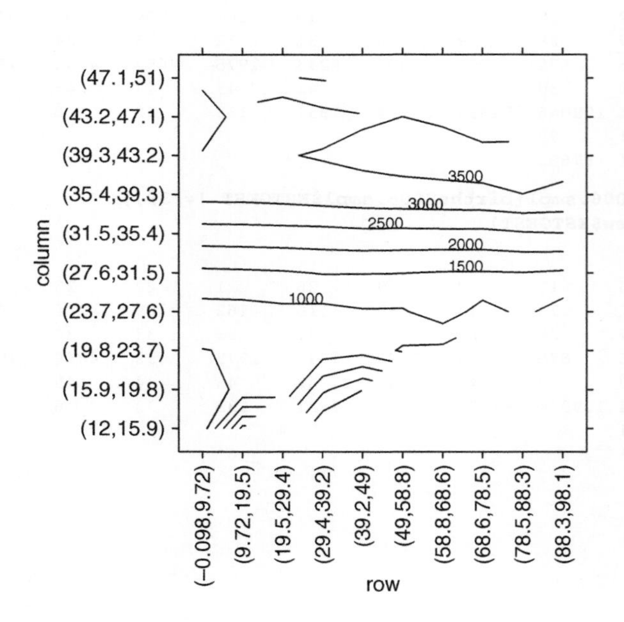

2.1.1 Modeling Issues Investigated in Subsequent Chapters

This discussion, with its many summaries and graphs, has given us a pretty good idea about the data. But what questions would we want to have answered with these data? One may wish to predict the birth weight from characteristics such as the estimated gestation period and the weight gain of the mother; for that, one could use regression and regression trees. Or, one may want to identify births that lead to very low APGAR scores, for which purpose, one could use classification methods.

2.2 EXAMPLE 2: ALUMNI DONATIONS

The file *contribution.csv* (available on our data Web site) summarizes the contributions received by a selective private liberal arts college in the Midwest. The college has a large endowment and, as all private colleges do, keeps detailed records on alumni donations. Here we analyze the contributions of five graduating classes (the cohorts who have graduated in 1957, 1967, 1977, 1987, and 1997). The data set consists of $n = 1230$ living alumni and contains their contributions for the years 2000–2004. In addition, the data set includes several other variables such as gender, marital status, college major, subsequent graduate work, and attendance at fund-raising events, all variables that may play an important role in assessing the success of future capital campaigns. This is a carefully constructed and well-maintained data set; it contains only alumni who graduated from the institution, and not former students who spent time at the institution without graduating. The data set contains no missing observations. The first five records of the file are shown below. Alumni not contributing have the entry "0" in the related column. The 1957 cohort is the smallest group. This is because of smaller class sizes in the past and deaths of older alumni.

```
## Install packages from CRAN; use any USA mirror
library(lattice)
don <- read.csv("C:/DataMining/Data/contribution.csv")
don[1:5,]
```

```
  Gender Class.Year Marital.Status   Major Next.Degree FY04Giving FY03Giving
1      M       1957              M History         LLB       2500       2500
2      M       1957              M Physics          MS       5000       5000
3      F       1957              M   Music        NONE       5000       5000
4      M       1957              M History        NONE          0       5100
5      M       1957              M Biology          MD       1000       1000

FY02Giving FY01Giving FY00Giving AttendenceEvent
1     1400      12060      12000               1
2     5000       5000      10000               1
3     5000       5000      10000               1
4      200        200          0               1
5     1000       1005       1000               1
```

```
table(don$Class.Year)
```

```
1957 1967 1977 1987 1997
127  222  243  277  361
```

```
barchart(table(don$Class.Year),horizontal=FALSE,
+     xlab="Class Year",col="black")
```

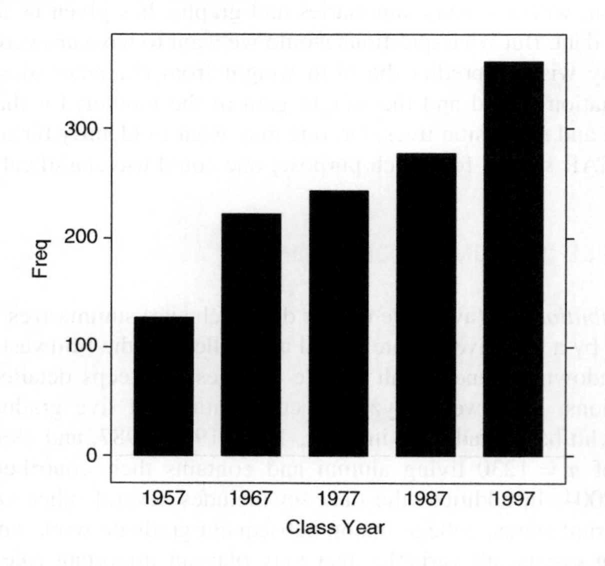

Total contributions for 2000–2004 are calculated for each graduate. Summary statistics (mean, standard deviation, and percentiles) are shown below. More than 30% of the alumni gave nothing; 90% gave $1050 or less; and only 3% gave more than $5000. The largest contribution was $172,000.

The first histogram of total contributions shown below is not very informative as it is influenced by both a sizable number of the alumni who have not contributed at all and a few alumni who have given very large contributions. Omitting contributions that are zero or larger than $1000 provides a more detailed view of contributions in the $1–$1000 range; this histogram is shown to the right of the first one. Box plots of total contributions are also shown. The second box plot omits the information from outliers and shows the three quartiles of the distribution of total contributions (0, 75, and 400).

```
don$TGiving=don$FY00Giving+don$FY01Giving+don$FY02Giving
+     +don$FY03Giving+don$FY04Giving

mean(don$TGiving)

[1] 980.0436

sd(don$TGiving)

[1] 6670.773

quantile(don$TGiving,probs=seq(0,1,0.05))

    0%       5%      10%      15%      20%      25%      30%      35%
   0.0      0.0      0.0      0.0      0.0      0.0      0.0     10.0
   40%      45%      50%      55%      60%      65%      70%      75%
  25.0     50.0     75.0    100.0    150.8    200.0    275.0    400.0
```

```
    80%       85%       90%       95%      100%
  554.2     781.0    1050.0    2277.5 171870.1
```

```
quantile(don$TGiving,probs=seq(0.95,1,0.01))
```

```
      95%        96%        97%        98%        99%       100%
  2277.50    3133.56    5000.00    7000.00   16442.14  171870.06
```

```
hist(don$TGiving)
hist(don$TGiving[don$TGiving!=0][don$TGiving[don$TGiving!=0]<=1000])
```

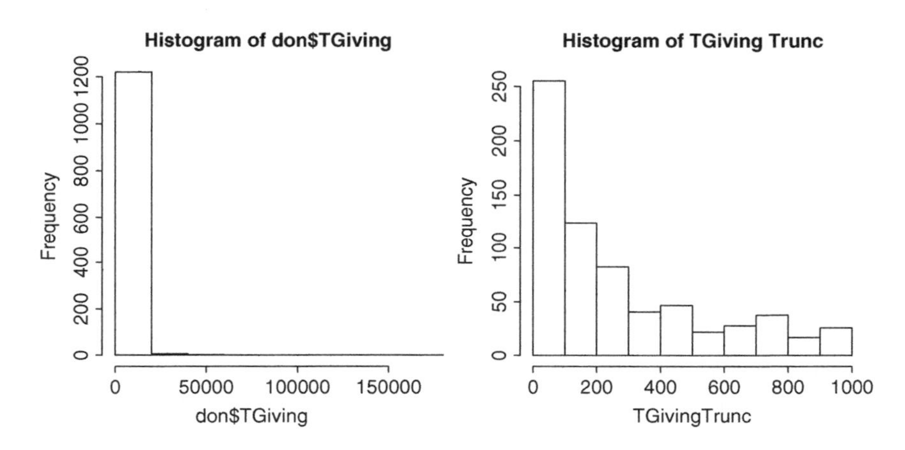


```
boxplot(don$TGiving,horizontal=TRUE,xlab="Total Contribution")
boxplot(don$TGiving,outline=FALSE,horizontal=TRUE,
+     xlab="Total Contribution")
```

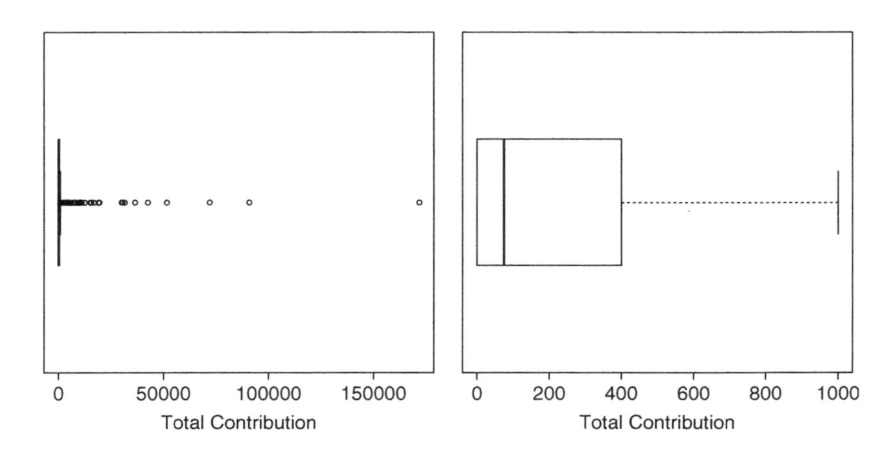

We identify below the donors who gave at least $30,000 during 2000–2004. We also list their major and their next degree. The top donor has a mathematics–physics double major with no advanced degree. Four of the top donors have law degrees.

```
ddd=don[don$TGiving>=30000,]
ddd
ddd1=ddd[,c(1:5,12)]
ddd1
ddd1[order(ddd1$TGiving,decreasing=TRUE),]
```

```
    Gender Class.Year Marital.Status                  Major Next.Degree   TGiving
99       M       1957              M  Mathematics-Physics        NONE 171870.06
123      M       1957              W   Economics-Business         MBA  90825.88
132      M       1967              M Speech (Drama, etc.)          JD  72045.31
105      M       1957              M              History         PHD  51505.84
135      M       1967              M              History          JD  42500.00
486      M       1977              M            Economics         MBA  36360.90
471      F       1977              D            Economics          JD  31500.00
1        M       1957              M              History         LLB  30460.00
2        M       1957              M              Physics          MS  30000.00
3        F       1957              M                Music        NONE  30000.00
```

For a university foundation, it is important to know who is contributing, as such information allows the foundation to target their fund-raising resources to those alumni who are most likely to donate. We show below box plots of total 5-year donation for the categories of class year, gender, marital status, and attendance at a foundation event. We have omitted in these graphs the outlying observations (those donors who contribute generously). Targeting one's effort to high contributors involves many personal characteristics that are not included in this database (such as special information about personal income and allegiance to the college). It may be a safer bet to look at the median amount of donation that can be achieved from the various groups. Class year certainly matters greatly; older alumni have access to higher life earnings, while more recent graduates may not have the resources to contribute generously. Attendance at a foundation-sponsored event certainly helps; this shows that it is important to get alumni to attend such events. This finding reminds the author about findings in his consulting work with credit card companies: if one wants someone to sign up for a credit card, one must first get that person to open up the envelope and read the advertising message. Single and divorced alumni give less; perhaps they worry about the sky-rocketing expenses of sending their own kids to college. We also provide box plots of total giving against the alumni's major and second degree. In these, we only consider those categories with frequencies exceeding a certain threshold (10); otherwise, we would have to look at the information from too many groups with low frequencies of occurrence. Alumni with an economics/business major contribute most. Among alumni with a second degree, MBAs and lawyers give the most.

```
boxplot(TGiving~Class.Year,data=don,outline=FALSE)
boxplot(TGiving~Gender,data=don,outline=FALSE)
boxplot(TGiving~Marital.Status,data=don,outline=FALSE)
boxplot(TGiving~AttendenceEvent,data=don,outline=FALSE)
```

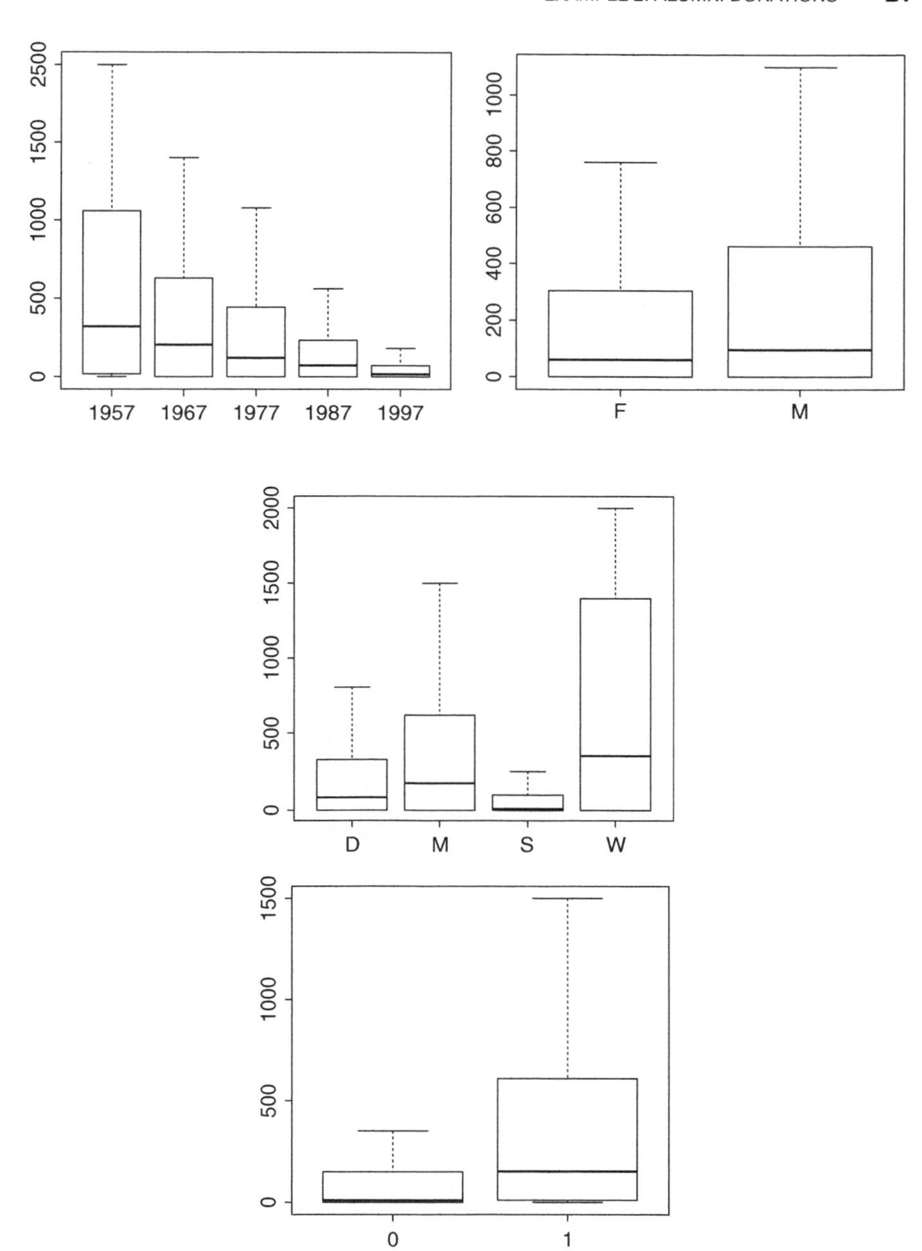


```
t4=tapply(don$TGiving,don$Major,mean,na.rm=TRUE)
t4
t5=table(don$Major)
t5
```

```
t6=cbind(t4,t5)
t7=t6[t6[,2]>10,]
t7[order(t7[,1],decreasing=TRUE),]
barchart(t7[,1],col="black")

t4=tapply(don$TGiving,don$Next.Degree,mean,na.rm=TRUE)
t4
t5=table(don$Next.Degree)
t5
t6=cbind(t4,t5)
t7=t6[t6[,2]>10,]
t7[order(t7[,1],decreasing=TRUE),]
barchart(t7[,1],col="black")
```

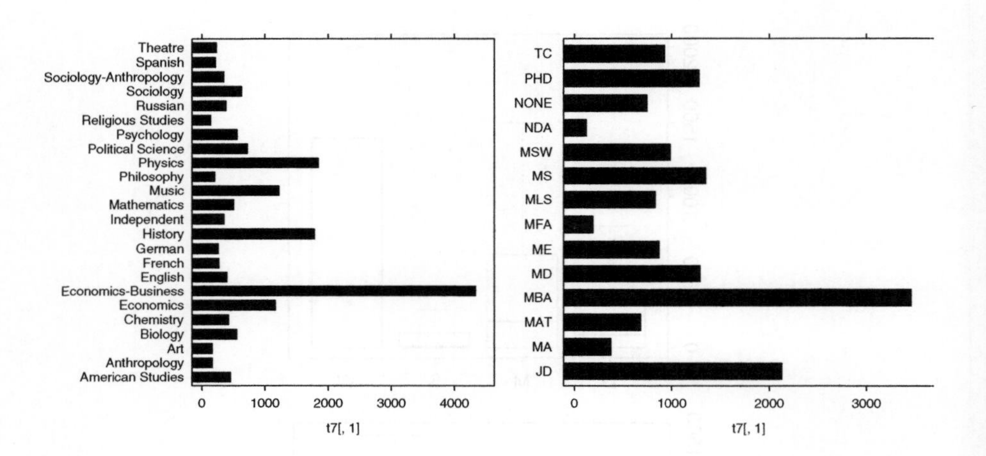

A plot of histogram densities, stratified according to year of graduation, shows the distributions of 5-year giving among alumni who gave \$1–\$1000. It gives a more detailed description of the distribution than the earlier histogram of all contributions.

```
densityplot(~TGiving|factor(Class.Year),
+     data=don[don$TGiving<=1000,][don[don$TGiving<=1000,]
+     $TGiving>0,],plot.points=FALSE,col="black")
```

We now calculate the total of the 5-year donations for the five graduation cohorts. We do this by using the tapply function (applying the summation function to the total contributions of each of the graduation classes). The result shows that the 1957 cohort has contributed \$560,000, compared to \$35,000 of the 1997 cohort.

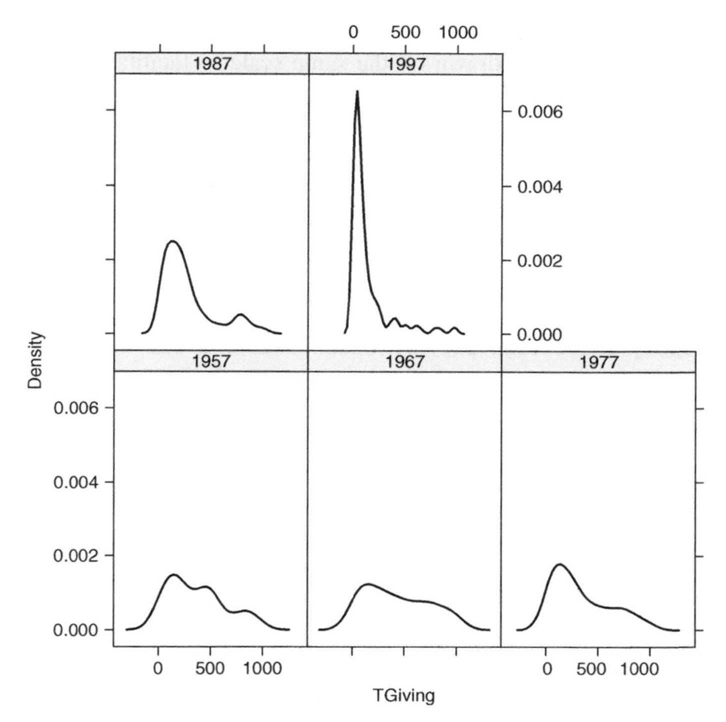


```
t11=tapply(don$TGiving,don$Class.Year,FUN=sum,na.rm=TRUE)
t11
```

```
     1957      1967      1977      1987      1997
560506.76 293750.74 210768.81 105288.37  35138.92
```

```
barplot(t11,ylab="Average Donation")
```

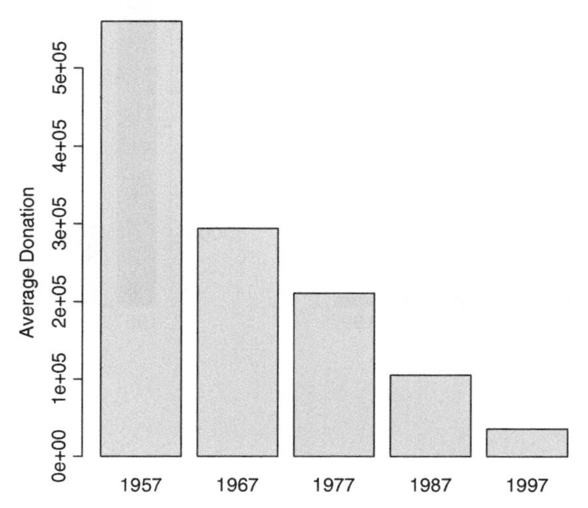

Below we calculate the annual contributions (2000–2004) of the five graduation classes. The 5 bar charts are drawn on the same scale to facilitate ready comparisons. The year 2001 was the best because of some very large contributions from the 1957 cohort.

```
barchart(tapply(don$FY04Giving,don$Class.Year,FUN=sum,
+     na.rm=TRUE),horizontal=FALSE,ylim=c(0,225000),col="black")
barchart(tapply(don$FY03Giving,don$Class.Year,FUN=sum,
+     na.rm=TRUE),horizontal=FALSE,ylim=c(0,225000),col="black")
barchart(tapply(don$FY02Giving,don$Class.Year,FUN=sum,
+     na.rm=TRUE),horizontal=FALSE,ylim=c(0,225000),col="black")
barchart(tapply(don$FY01Giving,don$Class.Year,FUN=sum,
+     na.rm=TRUE),horizontal=FALSE,ylim=c(0,225000),col="black")
barchart(tapply(don$FY00Giving,don$Class.Year,FUN=sum,
+     na.rm=TRUE),horizontal=FALSE,ylim=c(0,225000),col="black")
```

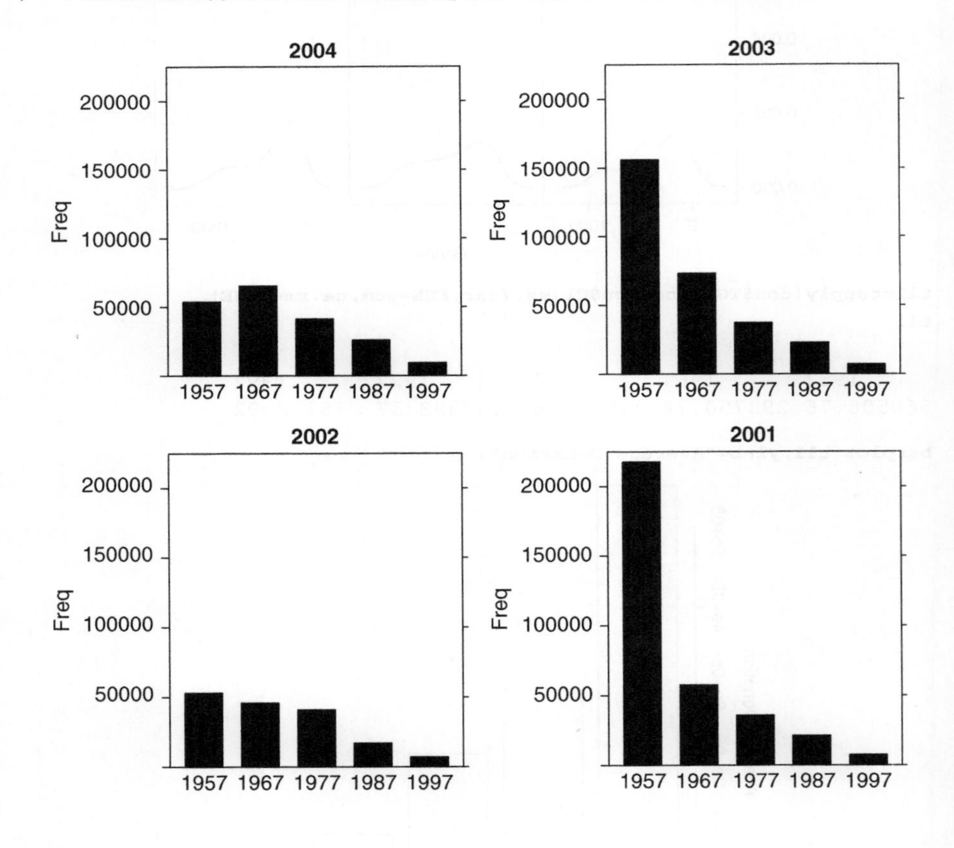

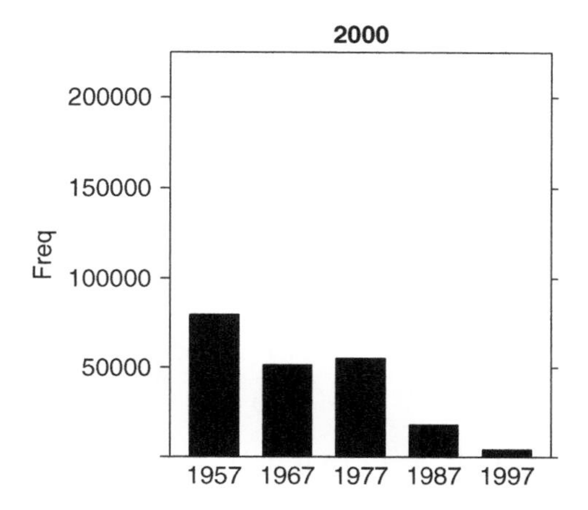

Finally, we compute the numbers and proportions of individuals who contributed. We do this by first creating an indicator variable for total giving, and displaying the numbers of the alumni who did and did not contribute. About 66% of all alumni contribute. The mosaic plot shows that the 1957 cohort has the largest proportion of contributors; the 1997 cohort has the smallest proportion of contributors, but includes the largest number of individuals (the area of the bar in a mosaic plot expresses the size of the group). The proportions of contributors shown below indicate that 75% of the 1957 cohort contributes, while only 61% of the 1997 graduating class does so. We can do the same analysis for each of the 5 years (2000–2004). The results for the most recent year 2004 are also shown.

```
don$TGivingIND=cut(don$TGiving,c(-1,0.5,10000000),
+     labels=FALSE)-1
mean(don$TGivingIND)

[1] 0.6569106

t5=table(don$TGivingIND,don$Class.Year)
t5

    1957 1967 1977 1987 1997
  0   31   71   75  105  140
  1   96  151  168  172  221

barplot(t5,beside=TRUE)
```

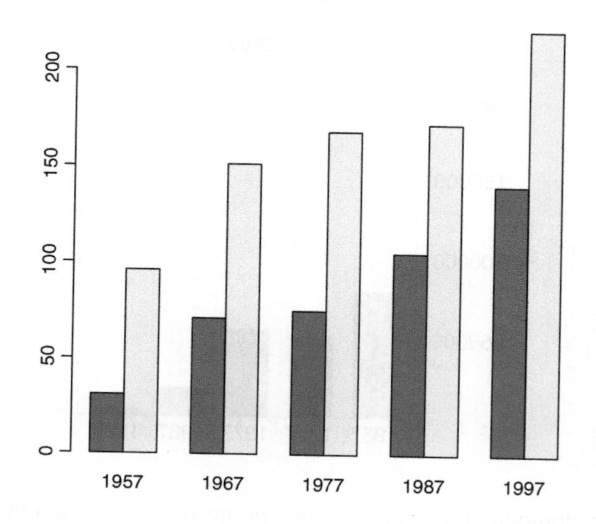


```
mosaicplot(factor(don$Class.Year)~factor(don$TGivingIND))
```

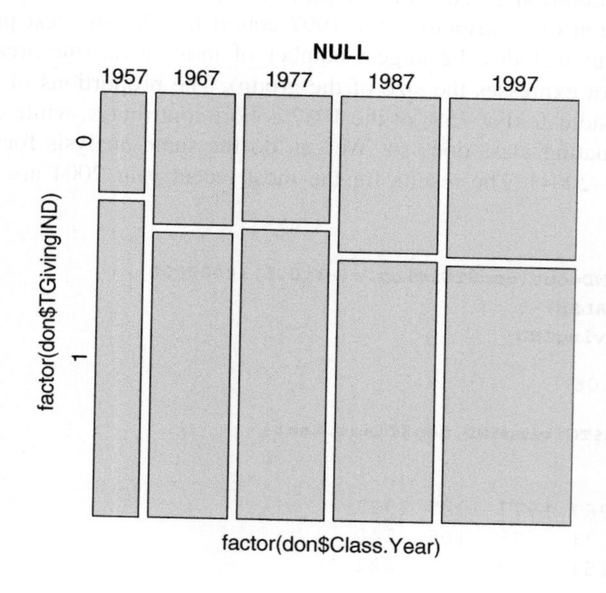


```
t50=tapply(don$TGivingIND,don$Class.Year,FUN=mean,na.rm=TRUE)
t50

     1957      1967      1977      1987      1997
0.7559055 0.6801802 0.6913580 0.6209386 0.6121884

barchart(t50,horizontal=FALSE)
```

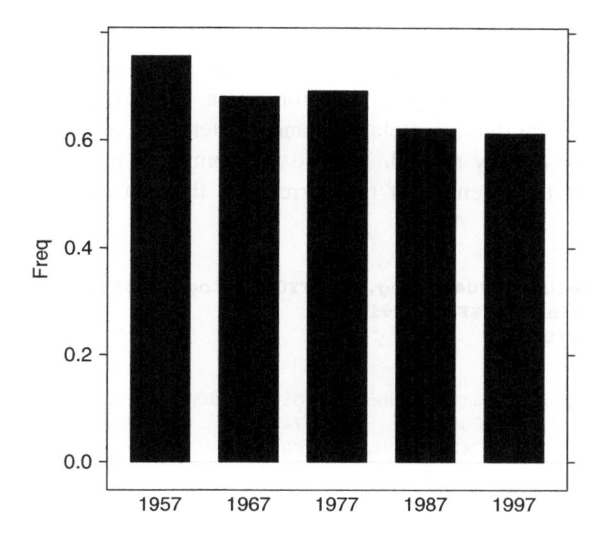


```
don$FY04GivingIND=cut(don$FY04Giving,c(-1,0.5,10000000),
+     labels=FALSE)-1
t51=tapply(don$FY04GivingIND,don$Class.Year,FUN=mean,
+     na.rm=TRUE)
t51
```

```
     1957      1967      1977      1987      1997
0.5196850 0.5000000 0.4238683 0.3610108 0.3518006
```

```
barchart(t51,horizontal=FALSE)
```

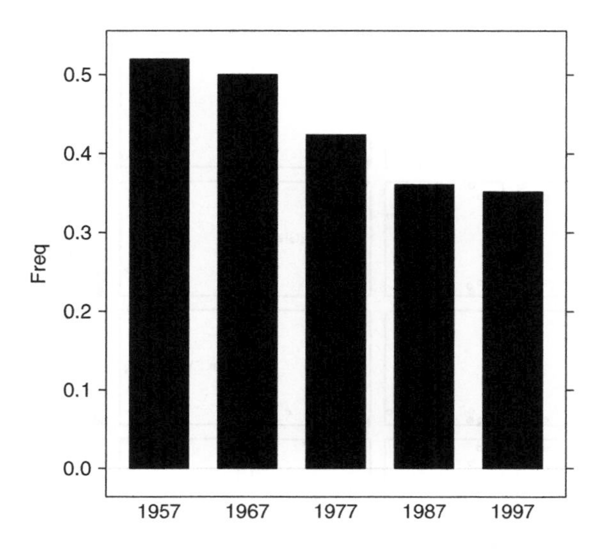

Below we explore the relationship between the alumni contributions among the 5 years. For example, if we know the amount an alumnus gives in one year (say in year 2000) does this give us information about how much that person will give in 2001? Pairwise correlations and scatter plots show that donations in different years are closely related. We use the command plotcorr in the package **ellipse** to express the strength of the correlation through ellipse-like confidence regions.

```
Data=data.frame(don$FY04Giving,don$FY03Giving,don$FY02Giving,
+     don$FY01Giving,don$FY00Giving)
correlation=cor(Data)
correlation

               don.FY04Giving don.FY03Giving don.FY02Giving don.FY01Giving
don.FY04Giving      1.0000000      0.5742938      0.8163331      0.1034995
don.FY03Giving      0.5742938      1.0000000      0.5867497      0.1385288
don.FY02Giving      0.8163331      0.5867497      1.0000000      0.2105597
don.FY01Giving      0.1034995      0.1385288      0.2105597      1.0000000
don.FY00Giving      0.6831861      0.3783280      0.8753492      0.2528295
               don.FY00Giving
don.FY04Giving      0.6831861
don.FY03Giving      0.3783280
don.FY02Giving      0.8753492
don.FY01Giving      0.2528295
don.FY00Giving      1.0000000

plot(Data)
```

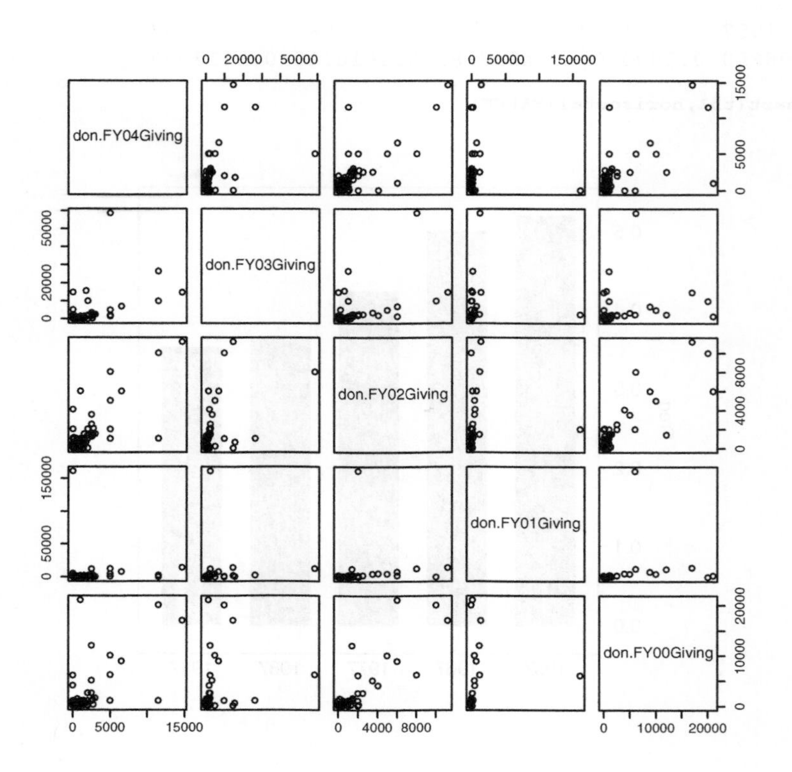

```
library(ellipse)
plotcorr(correlation)
```

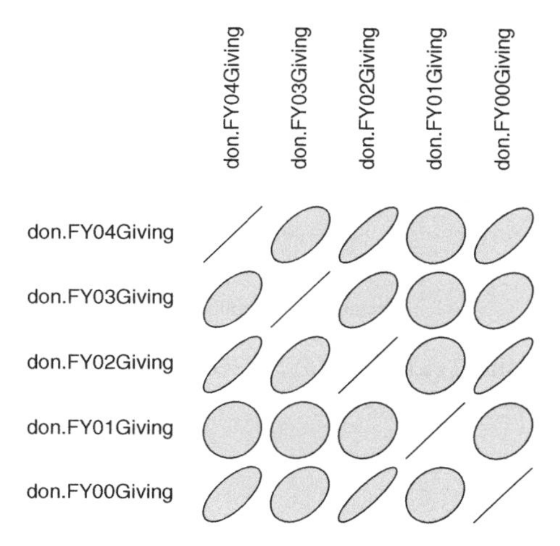

We conclude our analysis of the contribution data set with several mosaic plots that illustrate the relationships among categorical variables. The proportion of alumni making a contribution is the same for men and women. Married alumni are most likely to contribute, and the area of the bars in the mosaic plot indicates that married alumni constitute the largest group. Alumni who have attended an informational meeting are more likely to contribute, and more than half of all alumni have attended such a meeting. Separating the alumni into groups who have and have not attended an informational meeting, we create mosaic plots for giving and marital status. The likelihood of giving increases with attendance, but the relative proportions of giving across the marital status groups are fairly similar. This tells us that there is a main effect of attendance, but that there is not much of an interaction effect.

```
mosaicplot(factor(don$Gender)~factor(don$TGivingIND))
mosaicplot(factor(don$Marital.Status)~factor(don$TGivingIND))
t2=table(factor(don$Marital.Status),factor(don$TGivingIND))
mosaicplot(t2)
mosaicplot(factor(don$AttendenceEvent)~factor(don$TGivingIND))
```

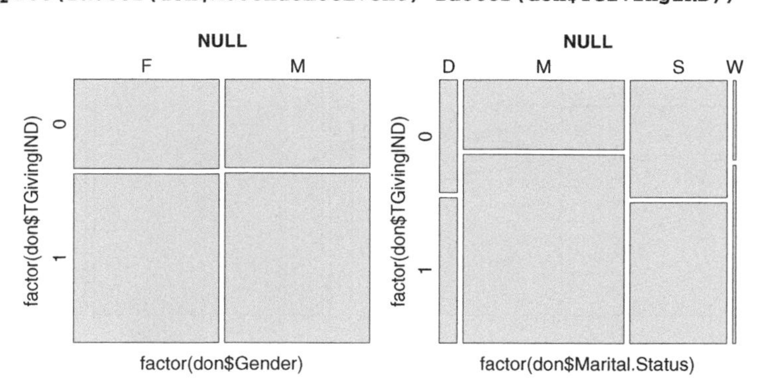

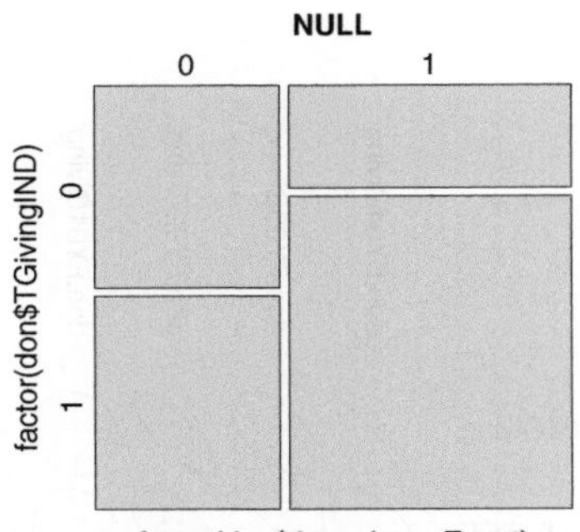


```
t2=table(factor(don$Marital.Status),factor(don$TGivingIND),
+     factor(don$AttendenceEvent))
t2

, ,  = 0
      0   1
  D  16  18
  M 106 157
  S 114  72
  W   2   2
, ,  = 1
      0   1
  D  18  26
  M  84 364
  S  80 162
  W   2   7
mosaicplot(t2[,,1])
mosaicplot(t2[,,2])
```

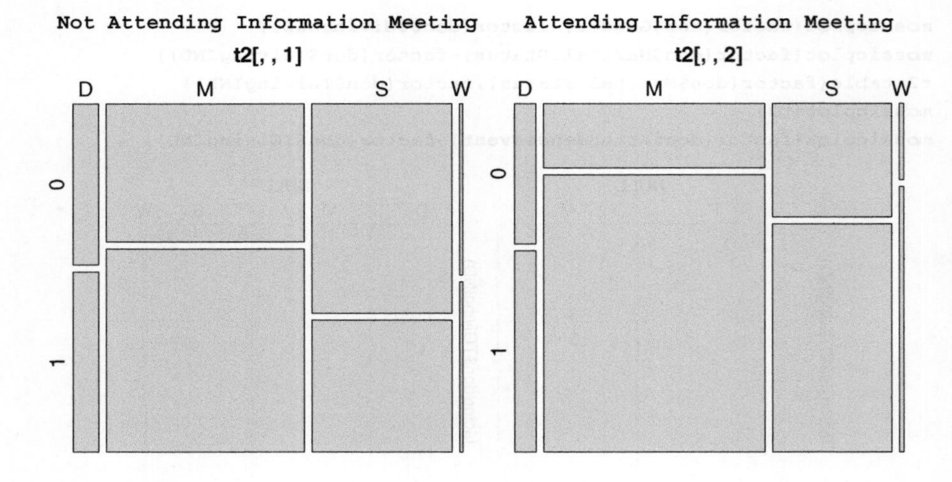

2.2.1 Modeling Issues to be Investigated in Subsequent Chapters

This discussion, with the many summaries and graphs, has told us much about the information in the data. What questions would we want to have answered with this data? It may be of interest to predict the likelihood of 2004 giving on the basis of the previous giving history (2000–2003), donor characteristics, and whether a graduate had attended an informational meeting. Logistic regression models or classification trees will be the prime models. Unfortunately, the variable "attendance at an informational meeting" does not indicate the year or years the meeting was attended, so its influence on the 2004 donation may already be incorporated in the donations of earlier years.

2.3 EXAMPLE 3: ORANGE JUICE

This section analyzes the weekly sales data of refrigerated 64-ounce orange juice containers from 83 stores in the Chicago area. There are many stores throughout the city, many time periods, and also three different brands (Dominicks, MinuteMaid, and Tropicana). The data are arranged in rows, with each row giving the recorded store sales (in logarithms; logmove), as well as brand, price, presence/absence of feature advertisement, and the demographic characteristics of the store. There are 28,947 rows in this data set. The data is taken from P. Rossi's **bayesm** package for R, and it has been used earlier in Montgomery (1987).

Time sequence plots of weekly sales, averaged over all 83 stores, are shown for the three brands. We create these plots by first obtaining the average sales for a given week and brand (averaged over the 83 stores). For this, we use the very versatile R function tapply. Time sequence plots of the averages are then graphed for each brand, and the plots are arranged on the same scale for easy comparison. An equivalent display, as three panels on the same plotting page, is produced through the xyplot function of the **lattice** package. Box plots, histograms, and smoothed density plots for sales, stratified for the three brands, are also shown. These displays average the information across the 83 stores and the 121 weeks.

```
## Install packages from CRAN; use any USA mirror
library(lattice)

oj <- read.csv("C:/DataMining/Data/oj.csv")
oj$store <- factor(oj$store)
oj[1:2,]
```

```
  store     brand week  logmove feat price     AGE60      EDUC    ETHNIC
1     2 tropicana   40 9.018695    0  3.87 0.2328647 0.2489349 0.1142799
2     2 tropicana   46 8.723231    0  3.87 0.2328647 0.2489349 0.1142799
    INCOME   HHLARGE   WORKWOM   HVAL150 SSTRDIST  SSTRVOL CPDIST5   CPWVOL5
1 10.55321 0.1039534 0.3035853 0.4638871 2.110122 1.142857 1.92728 0.3769266
2 10.55321 0.1039534 0.3035853 0.4638871 2.110122 1.142857 1.92728 0.3769266
```

```
t1=tapply(oj$logmove,oj$brand,FUN=mean,na.rm=TRUE)
t1
```

```
dominicks minute.maid   tropicana
   9.174831    9.217278    9.111483

t2=tapply(oj$logmove,INDEX=list(oj$brand,oj$week),FUN=mean,
+     na.rm=TRUE)
t2

                   40        41       42       43        44       45       46
dominicks    8.707053  7.721438 7.684779 8.220681  7.529664 7.485447 8.374706
minute.maid  8.316846 10.599174 8.350451 8.464384 10.272432 8.302100 8.975714
tropicana    8.772400  8.506540 8.859382 8.603009  8.422304 8.633549 8.579669
                   47       48        49       50        51       52        53
dominicks    8.737358 8.031447  7.790064 7.515055 10.308041 9.305908  9.136502
minute.maid  9.907359 8.238033 10.641114 8.195133  8.460606 8.340930 10.131160
tropicana    8.571572 8.739818  8.465478 8.633266  8.577919 8.827387  8.760043
. . .

plot(t2[1,],type= "l",xlab="week",ylab="dominicks",ylim=c(7,12))
plot(t2[2,],type= "l",xlab="week",ylab="minute.maid",ylim=c(7,12))
plot(t2[3,],type= "l",xlab="week",ylab="tropicana",ylim=c(7,12))
```

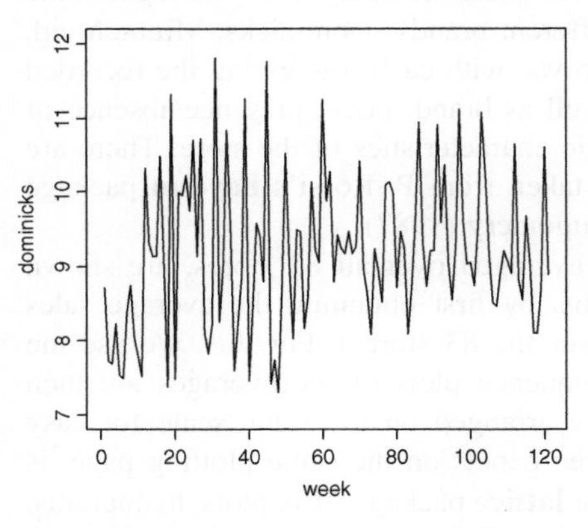

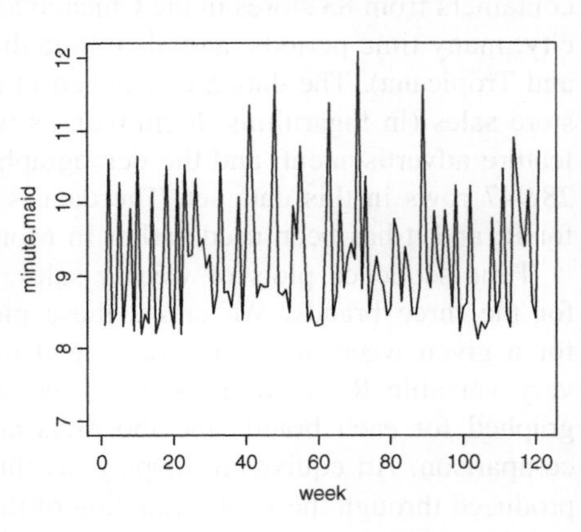

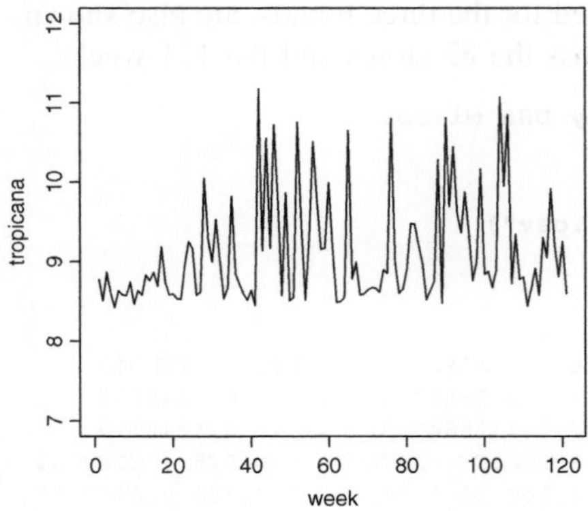


```
logmove=c(t2[1,],t2[2,],t2[3,])
week1=c(40:160)
week=c(week1,week1,week1)
```

```
brand1=rep(1,121)
brand2=rep(2,121)
brand3=rep(3,121)
brand=c(brand1,brand2,brand3)
xyplot(logmove~week|factor(brand),type= "l",layout=c(1,3),
+     col="black")
```

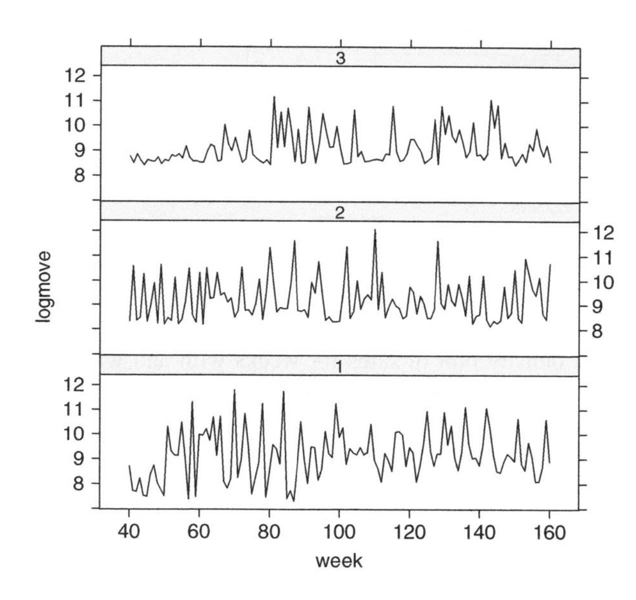


```
boxplot(logmove~brand,data=oj)
histogram(~logmove|brand,data=oj,layout=c(1,3))
densityplot(~logmove|brand,data=oj,layout=c(1,3),
+     plot.points=FALSE)
densityplot(~logmove,groups=brand,data=oj,plot.points=FALSE)
```

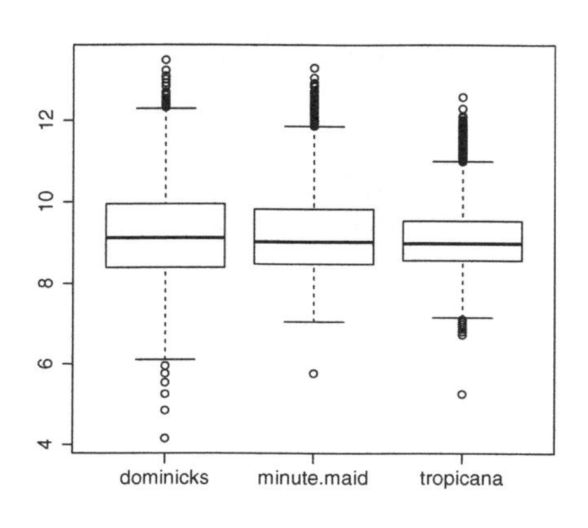

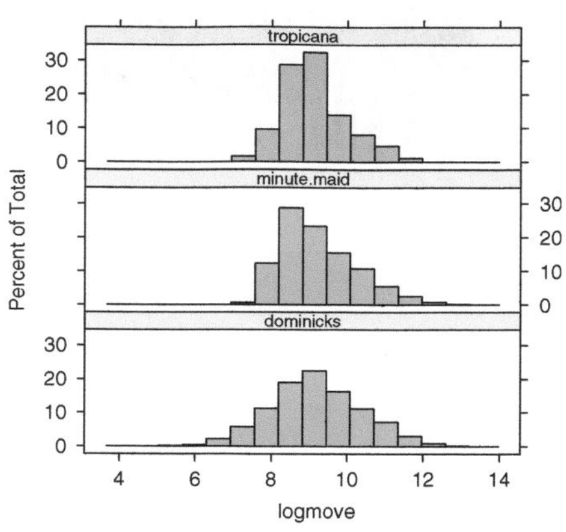

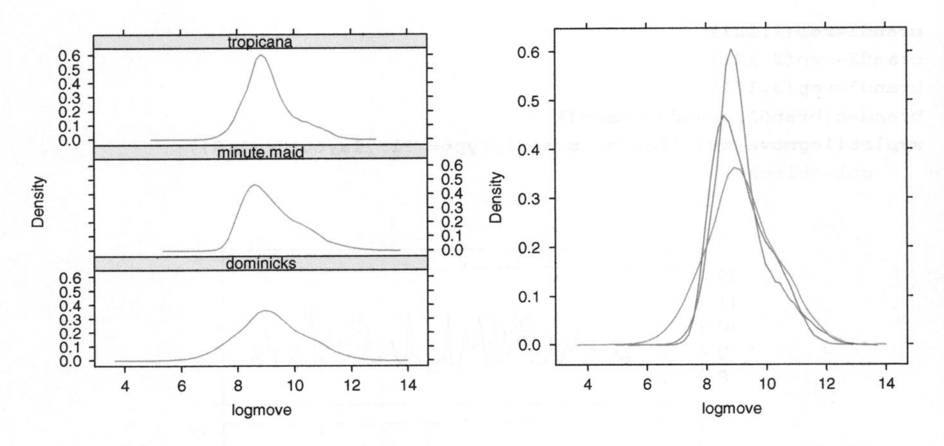

The previous displays ignore price and the presence of feature advertisement. Below we graph sales against price, and we do this for each brand separately but aggregating over weeks and stores. The graph shows that sales decrease with increasing price. A density plot of sales for weeks with and without feature advertisement, and a scatter plot of sales against price with the presence of feature advertisement indicated by the color of the plotting symbol both indicate the very positive effect of feature advertisement.

```
xyplot(logmove~week,data=oj,col="black")
xyplot(logmove~week|brand,data=oj,layout=c(1,3),col="black")
xyplot(logmove~price,data=oj,col="black")
xyplot(logmove~price|brand,data=oj,layout=c(1,3),col="black")
```

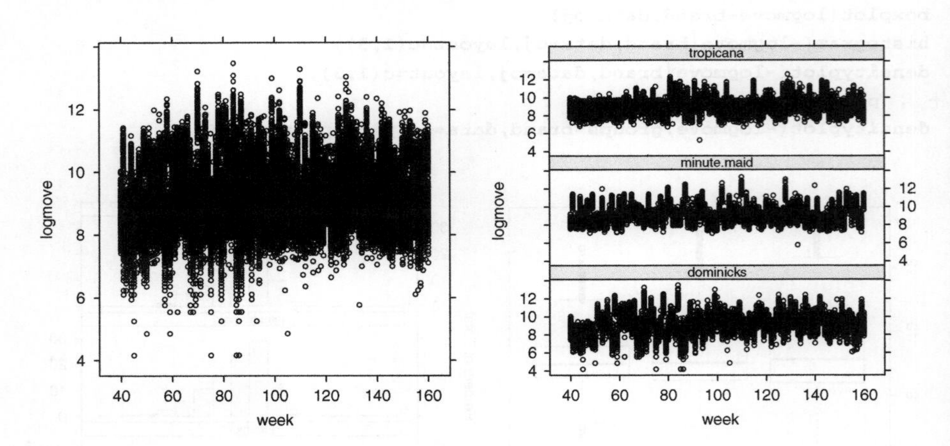

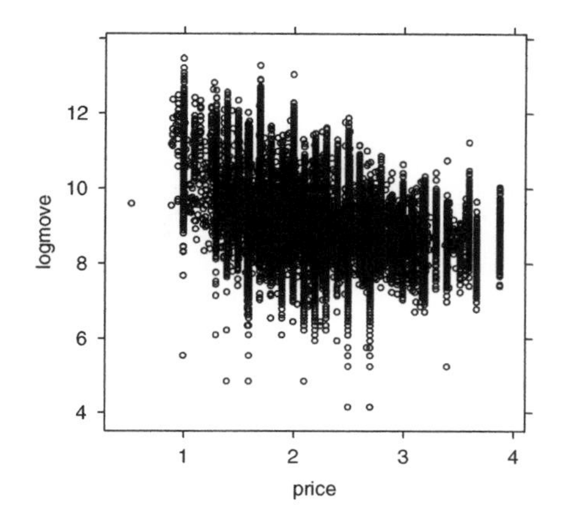

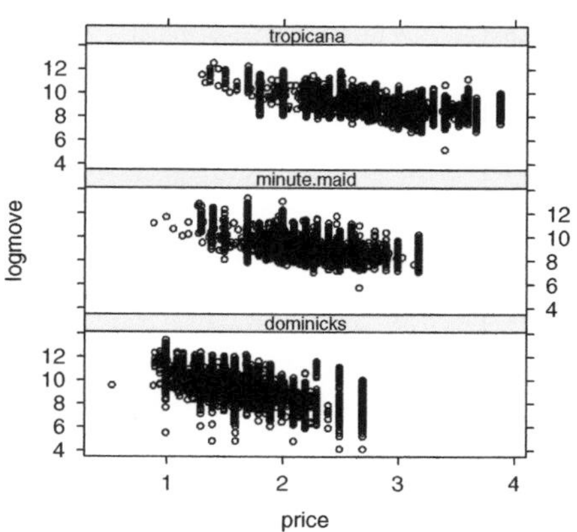


```
smoothScatter(oj$price,oj$logmove)
```

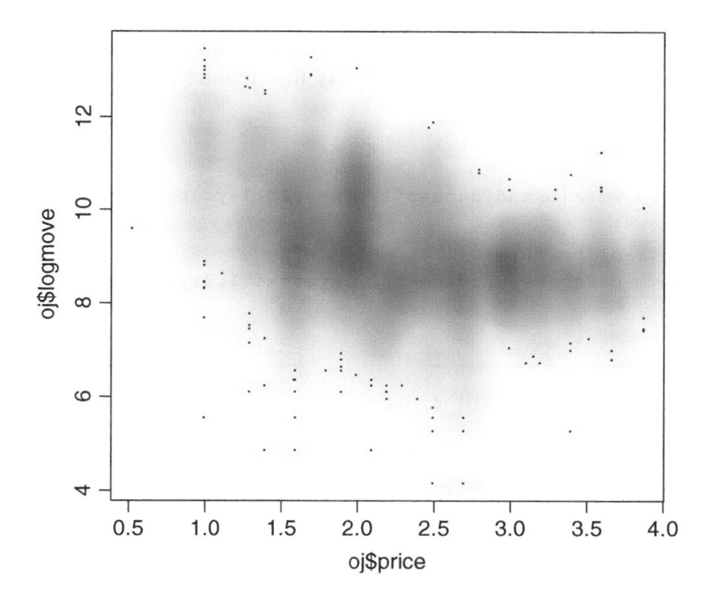


```
densityplot(~logmove,groups=feat, data=oj, plot.points=FALSE)
xyplot(logmove~price,groups=feat, data=oj)
```

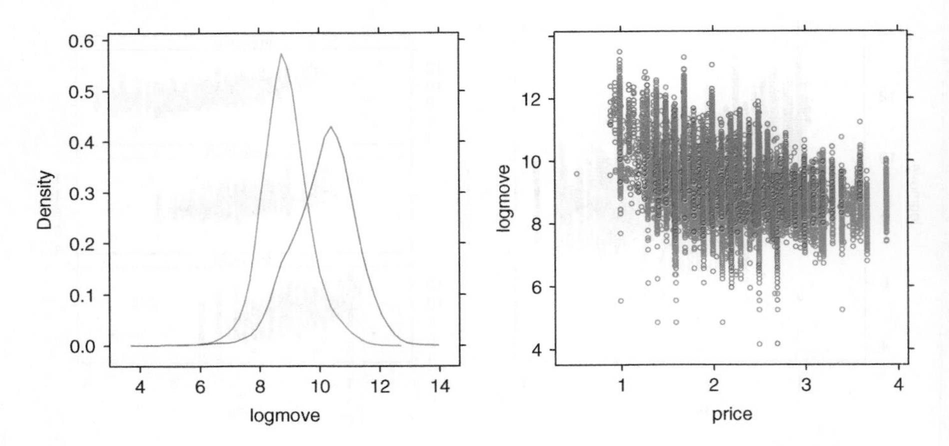

Next we consider one particular store. Time sequence plots of the sales of store 5 are shown for the three brands. Scatter plots of sales against price, separately for the three brands, are also shown; sales decrease with increasing price. Density histograms of sales and scatter plots of sales against price, with weeks with and without feature advertisement coded in color, are shown for each of the three brands. Again, these graphs show very clearly that feature advertisement increases the sales.

```
oj1=oj[oj$store == 5,]
xyplot(logmove~week|brand,data=oj1,type="l",layout=c(1,3),
+     col="black")
xyplot(logmove~price,data=oj1,col="black")
xyplot(logmove~price|brand,data=oj1,layout=c(1,3),col="black")
```

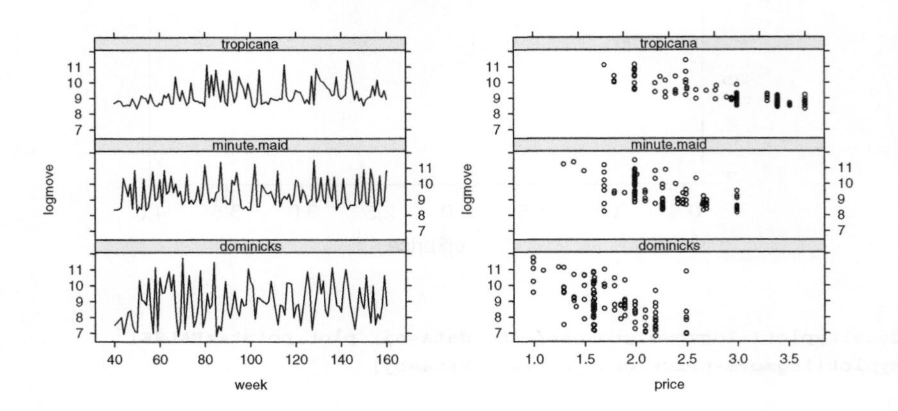


```
densityplot(~logmove|brand,groups=feat,data=oj1,
+     plot.points=FALSE)
xyplot(logmove~price|brand,groups=feat,data=oj1)
```

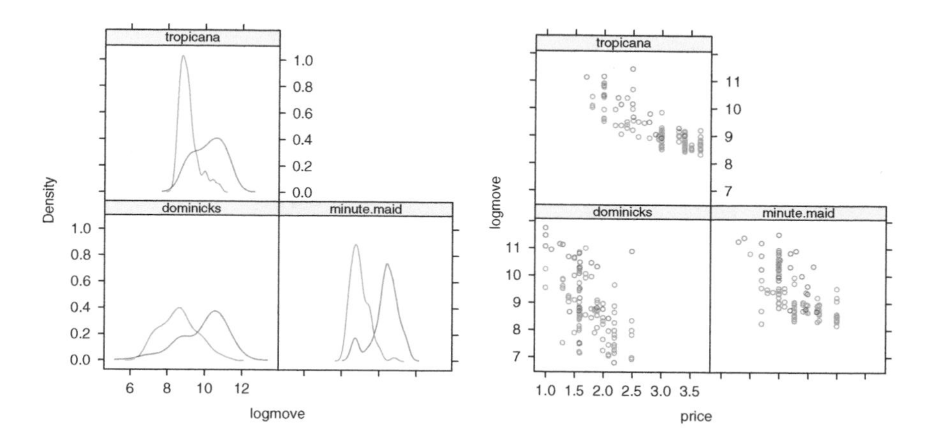

The volume of the sales of a given store certainly depends on the price that is being charged and on the feature advertisement that is being run. In addition, sales of a store may depend on the characteristics of the store such as the income, age, and educational composition of its neighborhood. We may be interested in assessing whether the sensitivity (elasticity) of the sales to changes in the price depends on the income of the customers who live in the store's neighborhood. We may expect that the price elasticity is largest in poorer neighborhoods as poorer customers have to watch their spending budgets more closely. To follow up on this hypothesis, we look for the stores in the wealthiest and the poorest neighborhoods. We find that store 62 is in the wealthiest area, while store 75 is in the poorest one. Lattice scatter plots of sales versus price, on separate panels for these two stores, with and without the presence of feature advertisments, are shown below. In order to get a better idea about the effect of price on sales, we repeat the first scatter plot and add the best fitting (least squares) line to the graph; more discussion on how to determine that best fitting line is given in Chapter 3. The slope of the fitted line is more negative for the poorest store, indicating that its customers are more sensitive to changes in the price.

```
t21=tapply(oj$INCOME,oj$store,FUN=mean,na.rm=TRUE)
t21
t21[t21==max(t21)]
t21[t21==min(t21)]
oj1=oj[oj$store == 62,]
oj2=oj[oj$store == 75,]
oj3=rbind(oj1,oj2)
xyplot(logmove~price|store,data=oj3)
xyplot(logmove~price|store,groups=feat,data=oj3)

## store in the wealthiest neighborhood
mhigh=lm(logmove~price,data=oj1)
summary(mhigh)
plot(logmove~price,data=oj1,xlim=c(0,4),ylim=c(0,13))
```

```
abline(mhigh)
## store in the poorest neighborhood
mlow=lm(logmove~price,data=oj2)
summary(mlow)
plot(logmove~price,data=oj2,xlim=c(0,4),ylim=c(0,13))
abline(mlow)
```

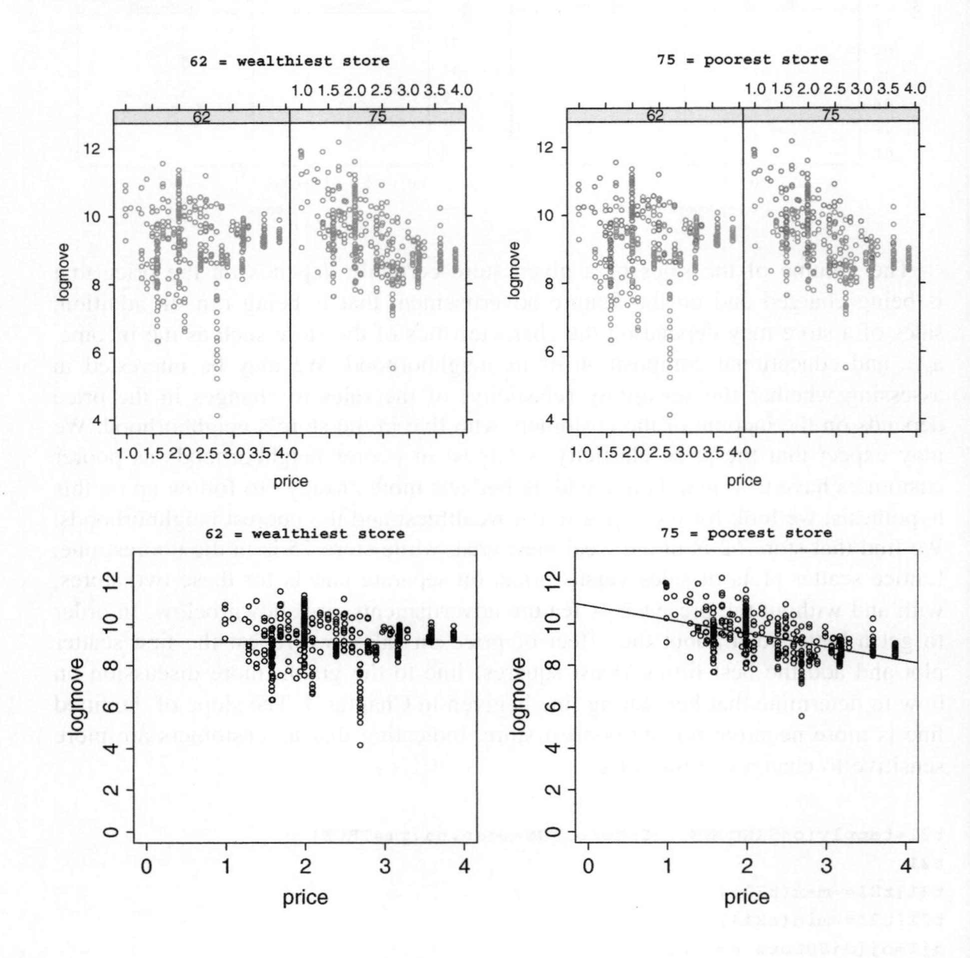

2.3.1 Modeling Issues to be Investigated in Subsequent Chapters

We can use this data set to investigate clustering. We may want to learn whether it is possible to reduce the 83 stores to a smaller number of homogeneous clusters. Furthermore, we may want to explain sales as a function of explanatory variables such as price, feature advertisements, and the characteristics of the store neighborhood. In particular, we may want to study whether the effects of price changes and feature advertisements depend on demographic characteristics of the store neighborhood. We will revisit this data set when we discuss regression (Chapter 3) and LASSO estimation (Chapter 6).

REFERENCES

Adler, J.: *R In a Nutshell: A Desktop Quick Reference*. Sebastopol, CA: O'Reilly Media, Inc., 2009.

Ledolter, J. and Swersey, A.: *Testing 1-2-3: Experimental Design with Applications in Marketing and Service Operations*. Stanford, CA: Stanford University Press, 2007.

Montgomery, A.L.: Creating micro-marketing pricing strategies using supermarket scanner data. *Marketing Science*, Vol. 16 (1987), 315–337.

CHAPTER 3

Standard Linear Regression

In the *standard linear regression model*, the response y is a continuous measurement variable such as sales or profit. We consider *linear* regression models of the form

$$y = f(x_1, x_2, \ldots, x_k) + \varepsilon = \alpha + \beta_1 x_1 + \beta_2 x_2 + \cdots + \beta_k x_k + \varepsilon,$$

where the function $f(\cdot)$ is linear in the k regressor (predictor) variables. The data on the regressor variables is collected into the design matrix $X = [x_1, x_2, \ldots, x_k]$. The error ε follows a normal distribution with mean zero and variance σ^2, implying that the conditional mean of the response is a linear function of the regressor variables,

$$E(y|X) = f(X) = \alpha + \beta_1 x_1 + \beta_2 x_2 + \cdots + \beta_k x_k.$$

If all other variables are fixed, a one-unit change in the regressor variable $x_j (j = 1, 2, \ldots, k)$ changes the expected mean response by β_j units.

The estimation of the parameters is usually achieved through least squares (which, for independent normal errors, is identical to maximum likelihood estimation). The least squares estimates $(\widehat{\alpha}, \widehat{\beta}_1, \ldots, \widehat{\beta}_k)$ minimize the sum of the squared differences between the observations and the values that are implied by the model,

$$D(\alpha, \beta_1, \ldots, \beta_k) = \sum_{i=1}^{n} \left[y_i - (\alpha + \beta_1 x_{i1} + \beta_2 x_{i2} + \cdots + \beta_k x_{ik})\right]^2.$$

This sum of squares is also referred to as the *regression deviance*. Explicit expressions for the least squares estimates can be written out in matrix form (see, e.g., Appendix 3.A). The expression $\widehat{y}_i = \widehat{\alpha} + \widehat{\beta}_1 x_{i1} + \widehat{\beta}_2 x_{i2} + \cdots + \widehat{\beta}_k x_{ik}$ is called the *fitted value* of the response y_i, and the difference $y_i - \widehat{y}_i$ is called the *residual*. The minimizing value $\widehat{D} = D(\widehat{\alpha}, \widehat{\beta}_1, \ldots, \widehat{\beta}_k) = \sum_{i=1}^{n} (y_i - \widehat{y}_i)^2$ determines the estimate of $\text{Var}(\varepsilon_i) = \sigma^2$, the R-square, and the F-statistic for testing the overall significance of the regression. The unbiased estimate of σ^2 is given by

$$\widehat{\sigma}^2 = \frac{\widehat{D}}{n - k - 1}.$$

Data Mining and Business Analytics with R, First Edition. Johannes Ledolter.

The R-square,

$$R^2 = 1 - \left[\frac{\widehat{D}}{\sum (y_i - \bar{y})^2} \right],$$

expresses the proportion of variation that is explained by the regression model.

The F-statistic,

$$F = \frac{\left[\sum_{i=1}^{n} (y_i - \bar{y})^2 - \widehat{D} \right] / k}{\widehat{D}/(n-k-1)},$$

is used test the overall significance of the regression. A large value of the F-statistic leads us to reject the null hypothesis $\beta_1 = \beta_2 = \cdots = \beta_k = 0$; the null hypothesis expresses that none of the predictor variables have an influence.

Details on regression models and their inferences, and on useful strategies for model construction and model checking are given in the text by Abraham and Ledolter (2006). Virtually all statistical packages include easy-to-use routines for the estimation of regression models. Their output provides estimates of the regression coefficients, standard errors of the estimated coefficients, summary statistics about the model fit, and predicted values and prediction intervals for new cases.

Strategies for simplifying regression models (i.e., achieving a simplified model structure without giving up too much on the fit) are also described in texts on regression. One strategy runs all possible regressions (with k predictor variables, such a strategy requires running $2^k - 1$ different models), and then evaluates and compares the explanatory power (i.e., the fit) of these models by looking at their R-squares, adjusted R-squares, and C_p-statistics.

By adding variables to a model we are bound to decrease the error sum of squares $\widehat{D}$ (in the worst case, it can stay the same) and consequently increase the R-square, $R^2 = 1 - \left[\widehat{D} / \sum (y_i - \bar{y})^2 \right]$. However, the relevant question is whether the increase in R-square is substantial or just minor.

The adjusted R-square for a model with k regressors and $k + 1$ estimated coefficients,

$$R^2_{\text{adj}} = 1 - \frac{\widehat{D}/(n-k-1)}{\sum (y_i - \bar{y})^2/(n-1)},$$

introduces a penalty for the number of estimated coefficients. While the R-square can never decrease as more variables are added to the model, the adjusted R-square of models with too many unneeded variables can actually decrease.

Mallows' C_p-statistic,

$$C_p = \frac{\widehat{D}_p}{(n-k-1)\widehat{D}_{\text{Full}}} - [n - 2(p+1)],$$

where $\widehat{D}_p$ is the error sum of squares of the regression model with p regressors (and $p + 1$ coefficients) and $\widehat{D}_{\text{Full}}$ is the error sum of squares of the full regression model with all k regressors included. If a model with $p < k$ regressors is already adequate, its value of the C_p-statistic should be about $p + 1$. It is larger (usually quite a bit larger) than $p + 1$ if a model with p regressors cannot explain the relationship. This result suggests the following strategy: Calculate the C_p-statistic for each candidate model with p regressors. This gives us k values for C_1; $k(k-1)/2$ values for C_2; ... ; and one value for C_k. The Mallows' statistic C_p measures bias. Among all models with p variables, we prefer the models with low values of C_p. We graph C_p against $p + 1$, the number of parameters, and add a line through the points (0,0) and $(k+1, k+1)$. For the largest model with k regressors, $C_k = k + 1$. We search for the simplest model (with smallest p) that gives us an acceptable model; that is, we search for a model with a C_p value close to $p + 1$ (and close to the line that has been added to the plot). Good candidate models are those with few variables and $C_p \approx p + 1$. Once we have found such a model, then there is little need to employ a more complicated model that involves more than p variables.

Automatic stepwise regression techniques (backward elimination, forward selection, and true stepwise regression) are related methods that simplify models automatically. One needs to be careful with such automatic methods as they can end up with quite different models (i.e., models including different regressor variables), but quite similar fits. In situations where regressor variables are closely related (one speaks of this as *multicollinearity*), it just happens that several models with different sets of explanatory variables can explain the data equally well.

Most data mining applications deal with the prediction of the response. We hope that the regression relationship that we have established on a training set can help us with the prediction of new cases. The key to establishing whether a regression model helps with prediction is to evaluate the predictions on a new test data set that has not been used for the estimation of the parameters. The quality of the predictions is typically assessed through statistics such as the mean (forecast) error, the root mean square error, and the mean absolute percent error.

Denote the prediction (the out-of-sample prediction, not the residual from the in-sample fit) for the response of a new case i with $\widehat{y}_i$ and its true response with y_i. Given a set of predictions for m new cases, we can evaluate the predictions according to their

Mean error: $$\text{ME} = \left(\frac{1}{m}\right)\sum_{i=1}^{m}(y_i - \widehat{y}_i),$$

Root mean square error: $$\text{RMSE} = \sqrt{\left(\frac{1}{m}\right)\sum_{i=1}^{m}(y_i - \widehat{y}_i)^2}, \quad \text{and}$$

Mean absolute percent error: $$\text{MAPE} = \frac{100}{m}\sum_{i=1}^{m}\frac{|y_i - \widehat{y}_i|}{y_i}.$$

The mean error should be close to zero; mean errors different from zero indicate a bias in the forecasts. The root mean square error expresses the magnitude of the forecast error in the units of the response variable. The mean absolute percent forecast error expresses the forecast error in percentage terms. The mean absolute percent error should not be used for a response that is close to zero as the division by a small number becomes unstable. We want unbiased forecasts with low values of RMSE and MAPE.

3.1 ESTIMATION IN R

The R function **lm** is used to fit linear (regression) models. The syntax for this command is given below:

```
lm(formula, data, subset, weights, na.action,
method = "qr", model = TRUE, x = FALSE, y = FALSE, qr = TRUE,
singular.ok = TRUE, contrasts = NULL, offset, …)
```

3.2 EXAMPLE 1: FUEL EFFICIENCY OF AUTOMOBILES

A data set on the fuel efficiencies of 38 cars (taken from Abraham and Ledolter, 2006) is used as an illustration. We try to model the fuel efficiency, measured in GPM (gallons per 100 miles), as a function of the weight of the car (in 1000 lb), cubic displacement (in cubic inches), number of cylinders, horsepower, acceleration (in seconds from 0 to 60 mph), and engine type (V-type and straight (coded as 1)). We analyze GPM instead of the usual EPA fuel efficiency measure MPG (miles per gallon). This is because the reciprocal transformation GPM = 100/MPG leads to approximate linear relationships between the response and the predictors. Scatter plots of GPM against two predictors (weight and displacement) are shown below.

```
## first we read in the data
FuelEff <- read.csv("C:/DataMining/Data/FuelEfficiency.csv")
FuelEff
```

```
   MPG   GPM    WT DIS NC  HP  ACC ET
1  16.9 5.917 4.360 350  8 155 14.9  1
2  15.5 6.452 4.054 351  8 142 14.3  1
3  19.2 5.208 3.605 267  8 125 15.0  1
4  18.5 5.405 3.940 360  8 150 13.0  1
5  30.0 3.333 2.155  98  4  68 16.5  0
. . .
```

```
plot(GPM~WT,data=FuelEff)
plot(GPM~DIS,data=FuelEff)
```

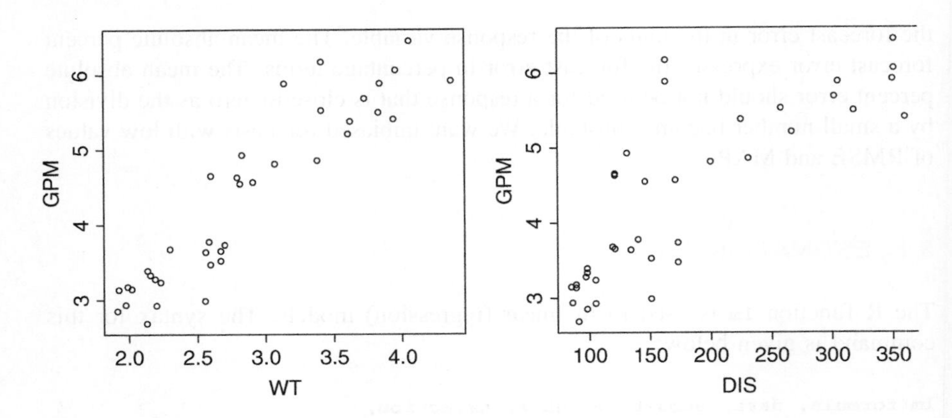

The *R* regression output shown in the following indicates that the regression model with all six explanatory variables explains 93.9 of the variation (*R*-square = 0.9386).

```
FuelEff=FuelEff[-1]
## regression on all data
m1=lm(GPM~.,data=FuelEff)
summary(m1)

Call:
lm(formula = GPM ~ ., data = FuelEff)

Residuals:
    Min      1Q  Median      3Q     Max
-0.4996 -0.2547  0.0402  0.1956  0.6455

Coefficients:
             Estimate Std. Error t value Pr(>|t|)
(Intercept) -2.599357   0.663403  -3.918 0.000458 ***
WT           0.787768   0.451925   1.743 0.091222 .
DIS         -0.004890   0.002696  -1.814 0.079408 .
NC           0.444157   0.122683   3.620 0.001036 **
HP           0.023599   0.006742   3.500 0.001431 **
ACC          0.068814   0.044213   1.556 0.129757
ET          -0.959634   0.266785  -3.597 0.001104 **
---
Signif. codes:  0 '***' 0.001 '**' 0.01 '*' 0.05 '.' 0.1 ' ' 1

Residual standard error: 0.313 on 31 degrees of freedom
Multiple R-squared: 0.9386,     Adjusted R-squared: 0.9267
F-statistic: 78.94 on 6 and 31 DF,  p-value: < 2.2e-16
```

The predictor variables in this model are themselves related. For example, one can expect that a car with large weight has a large engine size and large horsepower. The correlation matrix among all predictor variables (containing all pairwise correlations) shows this quite clearly; we have highlighted in boldface the very large correlation between weight and displacement, weight and number of cylinders, and weight and horsepower. As a consequence, we can expect that a model with fewer predictors will lead to a model representation that is almost as good.

```
cor(FuelEff)
```

```
           GPM          WT        DIS         NC         HP         ACC
GPM 1.00000000  0.92626656  0.8229098  0.8411880  0.8876992  0.03307093
WT  0.92626656  1.00000000  0.9507647  0.9166777  0.9172204 -0.03357386
DIS 0.82290984  0.95076469  1.0000000  0.9402812  0.8717993 -0.14341745
NC  0.84118805  0.91667774  0.9402812  1.0000000  0.8638473 -0.12924363
HP  0.88769915  0.91722045  0.8717993  0.8638473  1.0000000 -0.25262113
ACC 0.03307093 -0.03357386 -0.1434174 -0.1292436 -0.2526211  1.00000000
ET  0.52061208  0.66736606  0.7746636  0.8311721  0.7202350 -0.31023357
            ET
GPM  0.5206121
WT   0.6673661
DIS  0.7746636
NC   0.8311721
HP   0.7202350
ACC -0.3102336
ET   1.0000000
```

Next, we calculate all possible regressions and their R-squares, adjusted R-squares, and C_p-values. These calculations can be carried out using the function regsubsets in the R-package **leaps**.

```
## best subset regression in R
library(leaps)
X=FuelEff[,2:7]
y=FuelEff[,1]
out=summary(regsubsets(X,y,nbest=2,nvmax=ncol(X)))
tab=cbind(out$which,out$rsq,out$adjr2,out$cp)
tab
```

```
  (Intercept) WT DIS NC HP ACC ET R-Sq      R-Sq(adj) Cp
1           1  1   0  0  0   0  0 0.8579697 0.8540244 37.674750
1           1  0   0  0  1   0  0 0.7880098 0.7821212 72.979632
2           1  1   1  0  0   0  0 0.8926952 0.8865635 22.150747
2           1  1   0  0  0   0  1 0.8751262 0.8679906 31.016828
3           1  0   0  1  1   0  1 0.9145736 0.9070360 13.109930
3           1  1   1  1  0   0  0 0.9028083 0.8942326 19.047230
4           1  0   0  1  1   1  1 0.9313442 0.9230223  6.646728
4           1  1   0  1  1   0  1 0.9204005 0.9107520 12.169443
5           1  1   1  1  1   0  1 0.9337702 0.9234218  7.422476
5           1  0   1  1  1   1  1 0.9325494 0.9220103  8.038535
6           1  1   1  1  1   1  1 0.9385706 0.9266810  7.000000
```

The resulting table shows the trade-off between model size and model fit. The model with just one regressor, weight of the automobile, leads to an R-square of 85.8. Just this one variable explains most of the variability in fuel efficiency. This model is certainly easy to explain as the fuel needed to travel a certain distance must be related to the weight of the object that is being pushed forward. Adding displacement and number of cylinders to this model increases the R-square to 90.3.

Below we summarize the fitting results for the model that relates GPM to just the weight of the automobile.

```
m2=lm(GPM~WT,data=FuelEff)
summary(m2)

Call:
lm(formula = GPM ~ WT, data = FuelEff)

Residuals:
     Min       1Q   Median       3Q      Max
-0.88072 -0.29041  0.00659  0.19021  1.13164

Coefficients:
             Estimate Std. Error t value Pr(>|t|)
(Intercept) -0.006101   0.302681   -0.02    0.984
WT           1.514798   0.102721   14.75   <2e-16 ***
---
Signif. codes:  0 '***' 0.001 '**' 0.01 '*' 0.05 '.' 0.1 ' ' 1

Residual standard error: 0.4417 on 36 degrees of freedom
Multiple R-squared: 0.858,      Adjusted R-squared: 0.854
F-statistic: 217.5 on 1 and 36 DF,  p-value: < 2.2e-16
```

We use these two models, the model with all six predictor variables and the model with just weight of the automobile as explanatory variable, for cross-validation. *Cross-validation* removes one case from the data set of n cases, fits the model to the reduced data set, and predicts the response of that one case that has been removed from the estimation. This is repeated for each of the n cases. The summary statistics of the n genuine out-of-sample prediction errors (mean error, root mean square error, mean absolute percent error) help us assess the out-of-sample prediction performance. Cross-validation is very informative as it evaluates the model on new data. We find that the model with all six regressors performs better. It leads to a mean absolute percent error of about 6.75% (as compared to 8.23% for the model with weight as the only regressor).

Cross-validation for the regression on all six regressors

```
me  # mean error
[1] -0.003981948
rmse # root mean square error
[1] 0.3491357
mape # mean absolute percent error
[1] 6.75226
```

Cross-validation for the regression on weight only

```
me  # mean error
[1] -0.002960953
rmse # root mean square error
[1] 0.4517422
mape # mean absolute percent error
[1] 8.232507
```

The R program for this example, as well as R programs for all other examples in this book, is listed on the webpage that accompanies this book. The readers are encouraged to copy and paste these instructions into their own R session and check the results.

3.3 EXAMPLE 2: TOYOTA USED-CAR PRICES

The data are taken from Shmueli et al. (2010).

The data set includes sale prices and vehicle characteristics of 1436 used Toyota Corollas. The objective here is to predict the sale price of a used automobile. [We have corrected one value in the CC (cylinder volume) column, changing the obvious misprint "16,000" to "1600."]

Variable	Description
Id	Record_ID
Model	Model description
Price	Offer price in EUROs
Age_08_04	Age in months as in August 2004
Mfg_Month	Manufacturing month (1–12)
Mfg_Year	Manufacturing year
KM	Accumulated kilometers on odometer
Fuel_Type	Fuel type (petrol, diesel, CNG)
HP	Horsepower
Met_Color	Metallic color (Yes=1, No=0)
Color	Color (blue, red, gray, silver, black, and so on)
Automatic	Automatic (Yes=1, No=0)
CC	Cylinder volume in cubic centimeters
Doors	Number of doors
Cylinders	Number of cylinders
Gears	Number of gear positions

Quarterly_Tax	Quarterly road tax in EUROs
Weight	Weight in kilograms
Mfr_Guarantee	Within manufacturer's guarantee period (Yes=1, No=0)
BOVAG_Guarantee	BOVAG (Dutch dealer network) guarantee (Yes=1, No=0)
Guarantee_Period	Guarantee period in months
ABS	Anti-lock brake system (Yes=1, No=0)
Airbag_1	Driver airbag (Yes=1, No=0)
Airbag_2	Passenger airbag (Yes=1, No=0)
Airco	Airconditioning (Yes=1, No=0)
Automatic_airco	Automatic Airconditioning (Yes=1, No=0)
Boardcomputer	Board computer (Yes=1, No=0)
CD_Player	CD player (Yes=1, No=0)
Central_Lock	Central lock (Yes=1, No=0)
Powered_Windows	Powered windows (Yes=1, No=0)
Power_Steering	Power steering (Yes=1, No=0)
Radio	Radio (Yes=1, No=0)
Mistlamps	Mist lamps (Yes=1, No=0)
Sport_Model	Sport model (Yes=1, No=0)
Backseat_Divider	Backseat divider (Yes=1, No=0)
Metallic_Rim	Metallic rim (Yes=1, No=0)
Radio_cassette	Radio cassette (Yes=1, No=0)
Parking_Assistant	Parking assistance system (Yes=1, No=0)
Tow_Bar	Tow bar (Yes=1, No=0)

For this particular illustration, we do not use all variables. We use price as the response, and age (in months), accumulated kilometers on the odometer (in kilometer), fuel type (there are three: petrol, diesel, and compressed natural gas CNG), horsepower, color (whether metallic = 1, or not), transmission (whether automatic = 1, or not), cylinder volume (in cubic centimeters), doors (number of), and weight (in kilograms) as the explanatory variables.

```
toyota <- read.csv("C:/DataMining/Data/ToyotaCorolla.csv")
toyota[1:3,]

  Price Age    KM FuelType HP MetColor Automatic   CC Doors Weight
1 13500  23 46986   Diesel 90        1         0 2000     3   1165
2 13750  23 72937   Diesel 90        1         0 2000     3   1165
3 13950  24 41711   Diesel 90        1         0 2000     3   1165

## next we create indicator variables for the categorical variable
## FuelType with its three nominal outcomes: CNG, Diesel, and Petrol
v1=rep(1,length(toyota$FuelType))
v2=rep(0,length(toyota$FuelType))
toyota$FuelType1=ifelse(toyota$FuelType=="CNG",v1,v2)
toyota$FuelType2=ifelse(toyota$FuelType=="Diesel",v1,v2)
auto=toyota[-4]
auto[1:3,]
```

```
  Price Age    KM HP MetColor Automatic   CC Doors Weight FuelType1 FuelType2
1 13500  23 46986 90        1         0 2000     3   1165         0         1
2 13750  23 72937 90        1         0 2000     3   1165         0         1
3 13950  24 41711 90        1         0 2000     3   1165         0         1
```

The objective here is to predict the sale price. Scatter plots of price against age, mileage, and horsepower are shown as follows:

```
plot(Price~Age,data=auto)
plot(Price~KM,data=auto)
plot(Price~HP,data=auto)
```

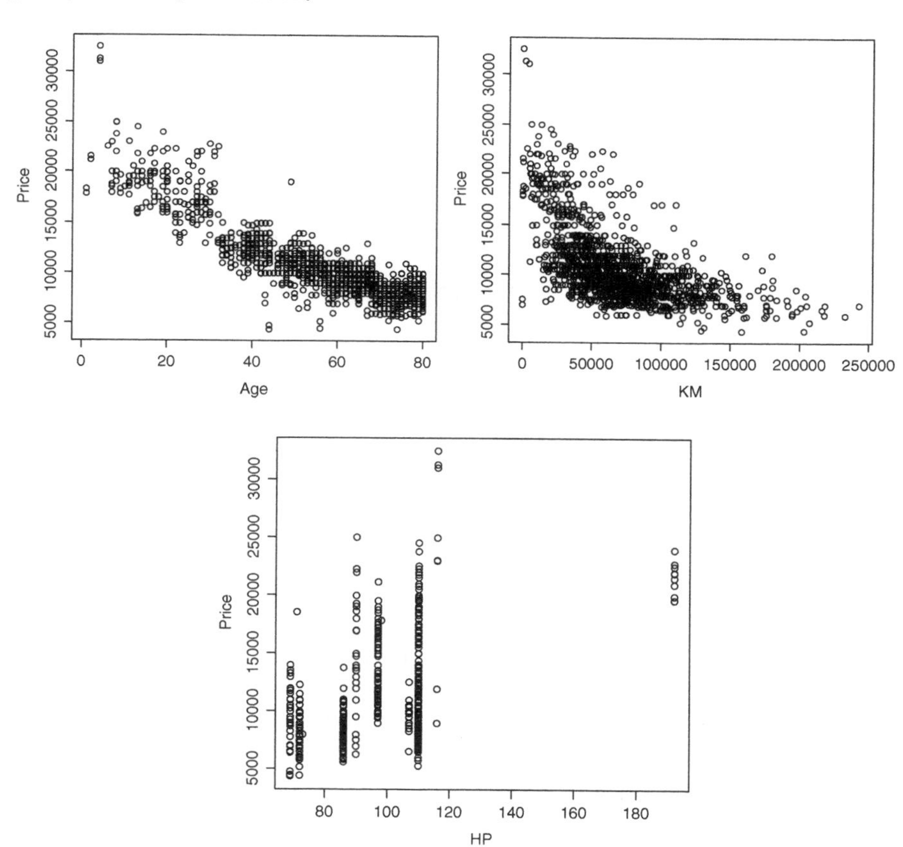

The output when fitting the regression model (relating price to the selected predictor variables) on all available data is shown on the following page. Note that the categorical variable fuel type with its three possible outcomes was converted into three indicator variables, and two of the three were included in the model (the third one is a linear combination of the two and can be omitted). The coefficients of the included indicators express the effects relative to the category that has been omitted. The model explains 86.9% of the variation in price. While this is quite good, it is probably possible to simplify the model structure; not all coefficients are statistically significant (e.g., the color of the car and the number of doors).

```
## regression on all data
m1=lm(Price~.,data=auto)
summary(m1)

Call:
lm(formula = Price ~ ., data = auto)

Residuals:
     Min       1Q   Median       3Q      Max
-10642.3   -737.7      3.1    731.3   6451.5

Coefficients:
               Estimate Std. Error t value Pr(>|t|)
(Intercept)  -2.681e+03  1.219e+03  -2.199 0.028036 *
Age          -1.220e+02  2.602e+00 -46.889  < 2e-16 ***
KM           -1.621e-02  1.313e-03 -12.347  < 2e-16 ***
HP            6.081e+01  5.756e+00  10.565  < 2e-16 ***
MetColor      5.716e+01  7.494e+01   0.763 0.445738
Automatic     3.303e+02  1.571e+02   2.102 0.035708 *
CC           -4.174e+00  5.453e-01  -7.656 3.53e-14 ***
Doors        -7.776e+00  4.006e+01  -0.194 0.846129
Weight        2.001e+01  1.203e+00  16.629  < 2e-16 ***
FuelInd…1.   -1.121e+03  3.324e+02  -3.372 0.000767 ***
FuelInd…2.    2.269e+03  4.394e+02   5.164 2.75e-07 ***
---
Signif. codes:  0 '***' 0.001 '**' 0.01 '*' 0.05 '.' 0.1 ' ' 1

Residual standard error: 1316 on 1425 degrees of freedom
Multiple R-squared: 0.8693,     Adjusted R-squared: 0.8684
F-statistic:   948 on 10 and 1425 DF,  p-value: < 2.2e-16
```

How does this model perform in out-of-sample prediction? Denote the prediction (we assume out-of-sample prediction, not the residual from the in-sample fit) for the price of new car i , y_i, with $\widehat{y}_i$. Given a set of predictions for m new cars, we again evaluate the predictions according to their

Mean error: $$\mathrm{ME} = \left(\frac{1}{m}\right)\sum_{i=1}^{m}(y_i - \widehat{y}_i),$$

Root mean square error: $$\mathrm{RMSE} = \sqrt{\frac{1}{m}\sum_{i=1}^{m}(y_i - \widehat{y}_i)^2},$$

Mean absolute percent error: $$\mathrm{MAPE} = \frac{100}{m}\sum_{i=1}^{m}\frac{|y_i - \widehat{y}_i|}{y_i}.$$

The mean error should be close to zero; a mean error different from zero indicates a bias in the forecasts. The root mean square error expresses the forecast error in the units of the response variable. The mean absolute percent forecast error expresses the forecast error in percentage terms.

Estimating the model on a randomly selected training set of 1000 cars and using the results for predicting the price of the 436 remaining cars of the evaluation data set leads to the following summary statistics:

```
me   # mean error
[1] -48.70784
rmse # root mean square error
[1] 1283.097
mape # mean absolute percent error
[1] 9.208957
```

The results for cross-validation (leaving out a single observation and predicting the response for the case that has been left out, for a total of 1436 prediction errors) are quite similar. The mean absolute percentage error is about 9.5%.

```
me   # mean error
[1] -2.726251
rmse # root mean square error
[1] 1354.509
mape # mean absolute percent error
[1] 9.530529
```

How does the last result compare with the cross-validation results for the regression model that includes just age as explanatory variable? The results show that the mean absolute percentage error for the simpler model is considerably worse (MAPE = 12.13).

3.3.1 Additional Comments

There is curvature in the scatter plot of Price against Age. This certainly makes sense as one knows that a new car loses much of its "new car" value in the first few months. A similar observation can be made for the relationship between price and driven kilometers. Including the squares of Age and KM improves the fit. Including Age2 (the square of Age) in the model improves the R-square from 79.0 (for the model with just the linear terms, Age and KM) to 83.2.

```
## Adding the squares of Age and KM to the model
auto$Age2=auto$Age^2
auto$KM2=auto$KM^2
m11=lm(Price~Age+KM,data=auto)
summary(m11)
m12=lm(Price~Age+Age2+KM+KM2,data=auto)
summary(m12)
```

```
m13=lm(Price~Age+Age2+KM,data=auto)
summary(m13)
```

The adequacy of a regression model should always be investigated, and for this, residual plots are very useful. Regression texts discuss numerous diagnostic residual plots, but in this brief regression section, we recommend just plotting the residuals against the fitted values and constructing a histogram (or a normal probability plot) of the residuals. Model inadequacies will show up through patterns in the scatter plot, and unusual cases will appear as outliers in the histogram of the residuals. Patterns in the scatter plot of residuals against fitted values indicate that not all information is extracted from the model and that some variables are missing. If the scatter plot shows no association, the model is most likely adequate. The scatter plot of residuals against fitted values for the linear model of Price on Age and KM shows patterns (some curvature). The scatter plot of the revised model with Age2 included, on the other hand, shows little association.

```
plot(m11$res~m11$fitted)
hist(m11$res)
plot(m12$res~m12$fitted)
```

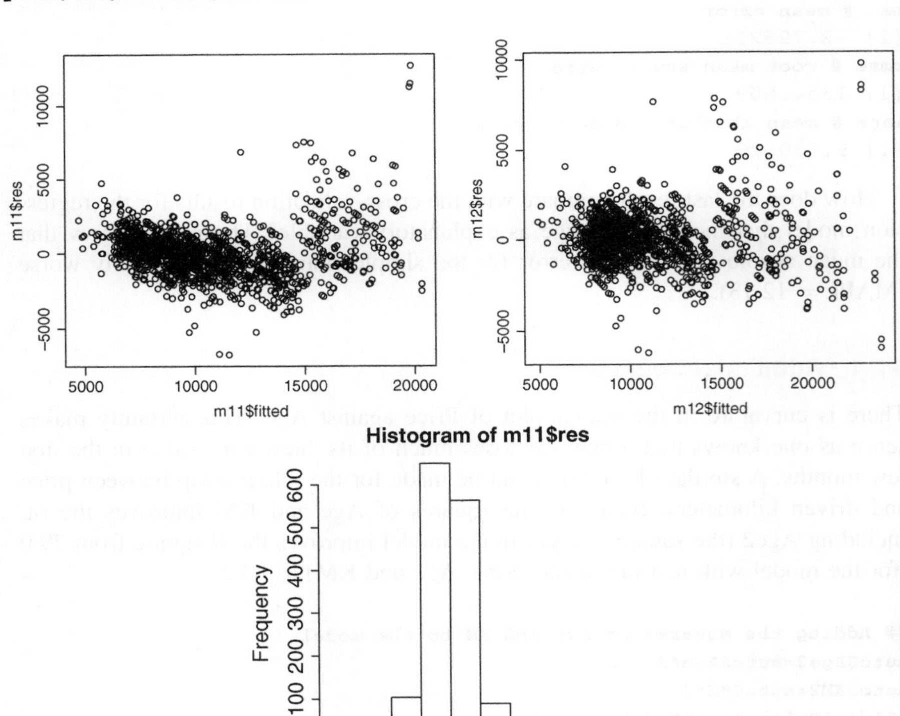

APPENDIX 3.A THE EFFECTS OF MODEL OVERFITTING ON THE AVERAGE MEAN SQUARE ERROR OF THE REGRESSION PREDICTION

Overfitting a regression model (i.e., including more covariates and estimating more regression coefficients than are actually needed) increases the variance of the prediction error. Below we provide a simple explanation of this fact. For a more thorough discussion, we refer the reader to Hawkins (2004).

From general least squares regression, we know that the least squares estimate of the $k \times 1$ vector β in the linear model $y = x'\beta + \varepsilon$ is given by $\widehat{\beta} = (X'X)^{-1}X'y$. Here

$$X = \begin{bmatrix} x_1' \\ x_2' \\ \vdots \\ x_n' \end{bmatrix}$$

represents the $n \times k$ design matrix, with each row giving the values of the regressors for one case. Note that there is no intercept in the model; but a model without intercept can always be achieved by centering the response and the predictor variables. We know that the prediction error for a new (independent) case with covariate vector x_i has mean 0 (the prediction is unbiased) and variance

$$\text{Var}(y_i - x_i'\widehat{\beta}) = \sigma^2[1 + x_i'(X'X)^{-1}x_i].$$

Assume that none of the covariates are needed. As there is no intercept in the model, the best prediction for a model with no covariate is 0 and the variance of the prediction error is σ^2. However, suppose that we incorporate all k unneeded covariates into the prediction model. We notice that the estimation of the unneeded regression parameters introduces the extra term $\sigma^2 x_i'(X'X)^{-1}x_i > 0$ into the variance. So we are worse off using the model with the unneeded covariates.

Below, we average the variance of the prediction errors over all values of the covariates in the data set, $x_1, x_2, \ldots, x_n$. In doing so, we assume that the future cases that we need to predict reflect the cases we had available for estimation. This leads to the average prediction error variance

$$\begin{aligned} \text{AvePEV} &= \left(\frac{1}{n}\right)\sum_{i=1}^{n} \text{Var}\left(y_i - x_i'\widehat{\beta}\right) \\ &= \left(\frac{1}{n}\right)\sum_{i=1}^{n} \sigma^2\left[1 + \text{tr}\left(x_i'(X'X)^{-1}x_i\right)\right] \\ &= \left(\frac{1}{n}\right)\sum_{i=1}^{n} \sigma^2\left[1 + \text{tr}\left((X'X)^{-1}x_i x_i'\right)\right] \\ &= \sigma^2\left[1 + \left(\frac{1}{n}\right)\text{tr}\left((X'X)^{-1}\left(\sum_{i=1}^{n} x_i x_i'\right)\right)\right] \end{aligned}$$

$$= \sigma^2 \left[1 + \left(\frac{1}{n}\right) \operatorname{tr}\left((X'X)^{-1} (X'X) \right) \right]$$

$$= \sigma^2 \left[1 + \left(\frac{1}{n}\right) \operatorname{tr}\left(I_{k \times k} \right) \right] = \sigma^2 \left(1 + \frac{k}{n} \right).$$

The trace of a square matrix A, denoted by $\operatorname{tr}(A)$, sums up the diagonal elements. The result implies that each unneeded covariate increases the average variance by a (multiplicative) factor of $1/n$. Assume that none of the covariates are needed, but that all k unneeded covariates are incorporated into the prediction model. As the estimate of the unneeded coefficients β is unbiased (with mean 0), the resulting mean of the prediction error from the overparametrized model is still zero (i.e., we have an unbiased prediction). However the average variance of its prediction error (which is the average mean square error as the prediction is unbiased) is increased by a factor of k/n.

A similar argument is made in Ledolter and Abraham (1981) for forecasts from ARIMA (autoregressive integrated moving average) time series models.

REFERENCES

Abraham, B. and Ledolter, J.: *Introduction to Regression Modeling*. Belmont, CA: Duxbury Press, 2006.

Hawkins, D.M.: The problem of overfitting. *Journal of Chemical Information and Computer Science*, Vol. 44 (2004), 1–12.

Ledolter, J. and Abraham, B.: Parsimony and its importance in time series forecasting. *Technometrics*, Vol. 23 (1981), 411–414.

Shmueli, G., Patel, N.R., and Bruce, P.C.: *Data Mining for Business Intelligence*. Second edition. Hoboken, NJ: John Wiley & Sons, Inc., 2010.

CHAPTER 4

Local Polynomial Regression: a Nonparametric Regression Approach

Regression models with a single explanatory variable x can be written as

$$y = \beta x + \varepsilon, \tag{4.1}$$

where $\mu(x) = \beta x$ represents the deterministic component of the model. In standard regression, the method of least squares provides a single coefficient estimate $\widehat{\beta}$ that does not change with the value of x, and the fitted value at x is given by $\widehat{\beta}x$. In *local polynomial regression*, the regression coefficient is allowed to change with the value of the explanatory variable, and it is estimated from data that lie within a certain window around x. For each selected fitting point x, we define a *bandwidth* $h(x)$ and a smoothing window $[x - h(x), x + h(x)]$. Within the smoothing window centered at the fitting point x, the regression component is approximated by the polynomial $\mu(u) = \beta_0 + \beta_1(u - x) + \cdots + \beta_p(u - x)^p$. Only observations within the smoothing window are used to estimate the coefficients of this polynomial. The estimate $\widehat{\beta}_0$ is then taken as the fitted value at x, $\widehat{\mu}(x) = \widehat{\beta}_0$; the fitted value changes with x in a nonparametric manner.

For a given fitting point x, the estimates of the regression coefficients $(\beta_0, \beta_1, \ldots, \beta_p)$ in local polynomial regression are obtained by minimizing the locally weighted least squares criterion,

$$\sum_{i=1}^{n} w_i(x)[y_i - (\beta_0 + \beta_1(x_i - x) + \cdots + \beta_p(x_i - x)^p)]^2, \tag{4.2}$$

where $w_i(x) = W[(x_i - x)/h(x)] > 0$ are weights, with $W(u)$ a weight function that assigns the largest weights to observations closest to x. Weight functions need

Data Mining and Business Analytics with R, First Edition. Johannes Ledolter.

to be nonnegative, and areas under the weight functions need to be equal to 1. The Epanechnikov weight function

$$\begin{aligned} W(u) &= \left(\frac{3}{4}\right)(1-u^2), \quad \text{for } |u| \leq 1 \\ &= 0, \quad \text{for } |u| > 1 \end{aligned} \tag{4.3}$$

is a popular choice (Epanechnikov, 1969), but other weight functions such as the rectangular one are commonly used. The bandwidth $h(x)$ controls the smoothness of the fit. The simplest choice is to take $h(x) = h$ constant, but often it is desirable to vary h with the fitting point x. The nearest neighbor bandwidth chooses $h(x)$ such that the local neighborhood contains a constant number of points. For a given nearest neighbor smoothing constant α, the nearest neighbor bandwidth is obtained by computing the distances $|x - x_i|$ between the fitting point x and the data points x_i, ordering the distances, and selecting $h(x)$ to be the kth smallest distance for which $k = n\alpha$. The bandwidth includes the $100\alpha\%$ observations that are closest to the fitting point x. A small value of the smoothing constant α implies a small bandwidth and little smoothing. A large value of the smoothing constant α implies a large bandwidth and considerable smoothing.

The weights in the least squares criterion in Equation 4.2 change with the value of x. For each selected value of x, a separate least squares estimation needs to be carried out. But the local least squares criterion involves an easy minimization and results in estimates $\widehat{\beta}_0(x), \widehat{\beta}_1(x), \ldots, \widehat{\beta}_p(x)$ that are linear functions of the response. This estimation is repeated for successive values of x, and the plot of the estimates $\widehat{\beta}_0(x)$ against x provides a nonparametric, fitted regression function.

Constant, linear, and quadratic polynomials ($p \leq 2$) are typically used. An important advantage of local linear regression [$p = 1$, with $\mu(u) = \beta_0 + \beta_1(u - x)$] over local constant regression [$p = 0$; with $\mu(u) = \beta_0$] is that local linear regression provides a better fit, especially at the data boundaries. A local quadratic regression estimate reduces the bias even further, but may increase the variance, especially at the data boundaries. Properties of local polynomial regression estimates are reviewed in Fan and Gijbels (1996).

4.1 MODEL SELECTION

Bandwidth parameters have a critical influence on the fitted curve $\widehat{\mu}(x) = \widehat{\beta}_0(x)$. A large bandwidth leads to an oversmoothed curve that may miss important features, while a small bandwidth may undersmooth the curve, resulting in a fit that is too noisy. Several tools are available to help assess the performance of local polynomial regression. Global criteria such as the generalized cross-validation (GCV) statistic (Craven and Wahba, 1979) and the C_p-statistic (Cleveland and Devlin, 1988) use the average squared prediction error as a measure of model adequacy.

At a given fitting point x, the estimates $(\widehat{\beta}_0(x), \widehat{\beta}_1(x), \ldots, \widehat{\beta}_p(x))$ and $\widehat{\mu}(x) = \widehat{\beta}_0(x)$ are linear functions of the response vector $y = (y_1, y_2, \ldots, y_n)'$, with the $n \times n$ matrix L relating the responses to the fitted values,

$$\begin{bmatrix} \widehat{\mu}(x_1) \\ \widehat{\mu}(x_2) \\ \vdots \\ \widehat{\mu}(x_n) \end{bmatrix} = Ly. \tag{4.4}$$

In ordinary regression, $L = H = X(X'X)^{-1}X'$ is the usual "hat" matrix. Its trace, $\text{tr}(H) = p + 1$, measures the "degrees of freedom" of the linear parametric fit. The degrees of freedom of a local fit, $\nu_1 = \text{tr}(L)$, provide a generalization of the number of parameters in the parametric model. A local fit increases the flexibility of the model, and $\nu_1 = \text{tr}(L)$ increases with decreasing bandwidth and decreasing nearest neighbor smoothing parameter α. In local regression with small bandwidth, the degrees of freedom increase as the procedure gives considerable flexibility to the function. A second, closely related and quite similar measure of degrees of freedom of local fit is $\nu_2 = \text{tr}(L'L)$.

The GCV statistic

$$\text{GCV}(\widehat{\mu}) = n \frac{\sum_{i=1}^{n} [y_i - \widehat{\mu}(x_i)]^2}{(n - \nu_1)^2}, \tag{4.5}$$

where $\widehat{\mu}(x_i)$ is the fitted value given in Equation 4.4, changes with the bandwidth and the smoothing parameter. In a cross-validation plot, GCV is graphed against the degrees of freedom of the local fit, and the bandwidth that minimizes GCV is adopted for the estimation.

Cleveland and Devlin extend the Mallows C_p criterion to local regression. Their C_p-statistic

$$C_p(\widehat{\mu}) = \left[\frac{1}{\widehat{\sigma}^2} \sum_{i=1}^{n} (y_i - \widehat{\mu}(x_i))^2\right] - n + 2\nu_1 \tag{4.6}$$

is used to compare several different fits with different bandwidths. Implementation of the C_p method requires an estimate of σ^2. Loader (1999, p. 30) shows that an estimate of σ^2 is given by $\widehat{\sigma}^2 = 1/(n - 2\nu_1 + \nu_2)\sum_{i=1}^{n} (y_i - \widehat{\mu}(x_i))^2$, where $\nu_1 = \text{tr}(L)$, $\nu_2 = \text{tr}(L'L)$, with the matrix L in Equation 4.4 evaluated at the smallest smoothing bandwidth for which the bias can be assumed negligible. The C_p criterion in Equation 4.6 is plotted against $\nu_2 = \text{tr}(L'L)$, the degrees of freedom of the local fit. The largest bandwidth for which $C_p(\widehat{\mu}) \approx \text{tr}(L'L)$ is taken as an acceptable local specification.

4.2 APPLICATION TO DENSITY ESTIMATION AND THE SMOOTHING OF HISTOGRAMS

Local polynomial regression can be used to estimate the density of a distribution. In this case, the response at x_i is given by the empirical frequency $y_i = 1/n$. The smoothing constant of the nearest neighbor bandwidth, as well as the order of the approximating polynomial and the weights in the local polynomial regression, determine the amount of smoothing that is being applied to the density histogram.

4.3 EXTENSION TO THE MULTIPLE REGRESSION MODEL

The local polynomial regression can be used when there are two or more explanatory variables in the model, such as $y_i = \beta x_i + \gamma z_i + \varepsilon_i$. When the explanatory variables (x_i, z_i) are measured in noncomparable units, it is important that they are standardized by their marginal standard deviations. The regression coefficients are estimated over a sliding data window. For each selected fitting point (x, z), we define a nearest neighbor bandwidth $h(x, z)$ such that the resulting two-dimensional smoothing window with corner points $[x \pm h(x, z), z \pm h(x, z)]$ covers a specified proportion of all observations. Within the two-dimensional smoothing window centered at the fitting point (x, z), the regression component is approximated by the polynomial $\mu(u, v) = \beta_0 + \sum_{j+k \le p} \beta_{jk}(u - x)^j (v - z)^k$, and observations within the smoothing window are used to estimate the regression coefficients. For a given fitting point (x, z), the estimates of the regression coefficients in local polynomial regression are obtained by minimizing the locally weighted least squares criterion,

$$\sum_{i=1}^{n} w_i(x, z) \left\{ y_i - \left[\beta_0 + \sum_{j+k \le p} \beta_{ij} (x_i - x)^j (z_i - z)^k \right] \right\}^2,$$

where $w_i(x, z) = w_i(x) w_i(z) > 0$ are weights that assign largest weights to observations closest to (x, z). The estimate $\widehat{\beta}_0$ represents the fitted value at (x, z); that is, $\widehat{\mu}(x, z) = \widehat{\beta}_0$. Contour plots connecting the values of the covariates that lead to the same fitted value provide useful displays of the fitted two-dimensional regression surface.

4.4 EXAMPLES AND SOFTWARE

The R library **locfit** can be used to carry out the computations. For illustration we use two examples: the eruption and waiting times to the next eruption of 272 eruptions of the Old Faithful geyser in the Yellowstone National Park, and the NO_x exhaust emissions when using pure ethanol as the spark-ignition fuel in a single-cylinder engine.

4.4.1 Example 1: Old Faithful

The data file *OldFaithful.csv* contains the eruption times (in minutes) and the waiting times to the next eruption (in minutes) of 272 eruptions of the Old Faithful geyser in the Yellowstone National Park. The density histograms of eruption times and of waiting times to the next eruption are shown. Next to the histograms we show the smoothed density histograms that have been obtained through local polynomial regression. We use the default parameters in the R library **locfit**, but also illustrate how one can determine the optimal smoothing (bandwidth) constants through cross-validation. The histograms for both variables are bimodal.

```
library(locfit)

## first we read in the data
OldFaithful <- read.csv("C:/DataMining/Data/OldFaithful.csv")
OldFaithful[1:3,]

  TimeEruption TimeWaiting
1        3.600          79
2        1.800          54
3        3.333          74

## density histograms and smoothed density histograms
## time of eruption
hist(OldFaithful$TimeEruption,freq=FALSE)
fit1 <- locfit(~lp(TimeEruption),data=OldFaithful)
plot(fit1)
```

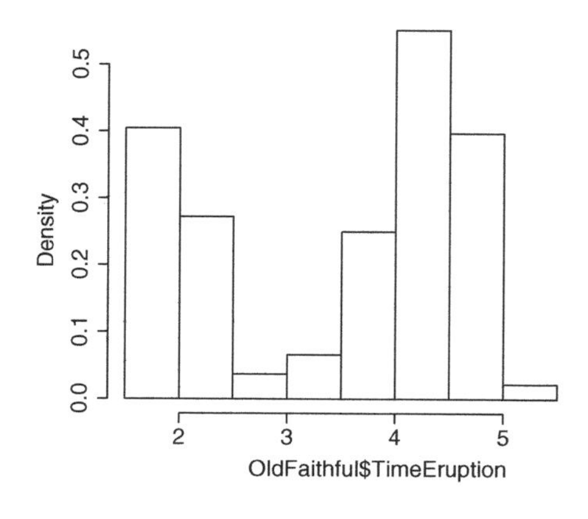

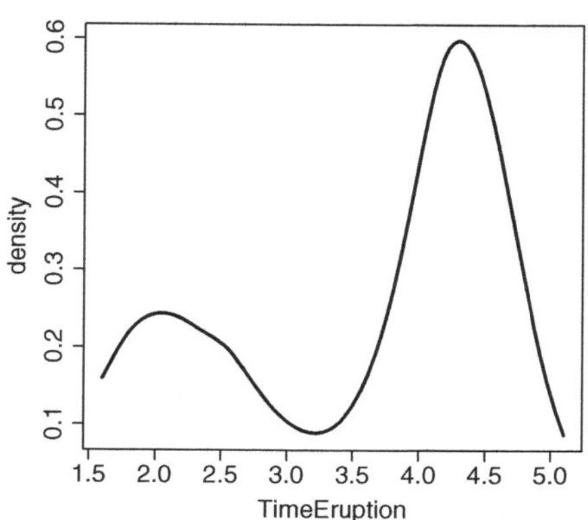


```
## waiting time to next eruption
hist(OldFaithful$TimeWaiting,freq=FALSE)
fit2 <- locfit(~lp(TimeWaiting),data=OldFaithful)
plot(fit2)
```

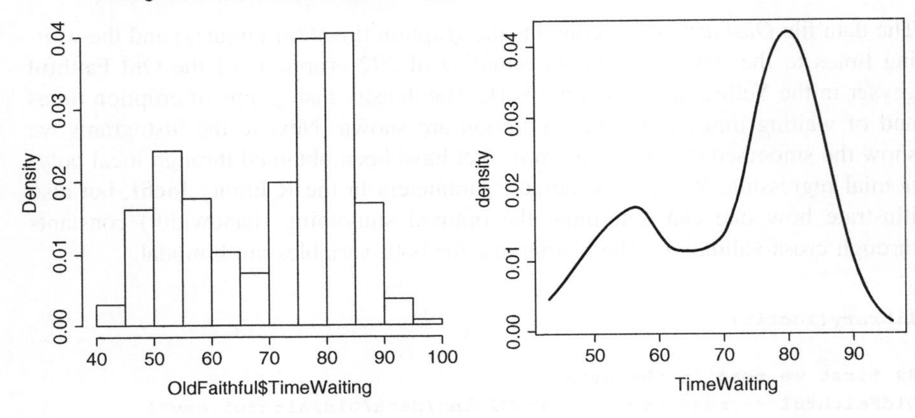


```
## cross-validation of smoothing constant
## for waiting time to next eruption
alpha<-seq(0.20,1,by=0.01)
n1=length(alpha)
g=matrix(nrow=n1,ncol=4)
for (k in 1:length(alpha)) {
g[k,]<-gcv(~lp(TimeWaiting,nn=alpha[k]),data=OldFaithful)
}
g

           [,1]       [,2]      [,3]     [,4]
 [1,] -1028.531 17.980905 16.185681 8.671283
 [2,] -1029.354 17.114850 15.407725 8.619349
 [3,] -1030.749 16.275543 14.650517 8.574473
 . . .
[42,] -1042.699  5.818206  5.365573 8.005730
[43,] -1042.497  5.758991  5.315595 8.000627
[44,] -1043.495  5.573373  5.147524 7.997128
[45,] -1043.483  5.465936  5.079868 7.990592
[46,] -1043.528  5.429989  5.045821 7.988783
[47,] -1043.557  5.385270  5.004441 7.986325          nn=0.66
[48,] -1046.023  5.323446  4.941556 8.001483
[49,] -1046.525  5.260504  4.884752 8.001542
 . . .

plot(g[,4]~g[,3],ylab="GCV",xlab="degrees of freedom")
## minimum at nn = 0.66
fit2 <- locfit(~lp(TimeWaiting,nn=0.66,deg=2),
+   data=OldFaithful)
plot(fit2)
```

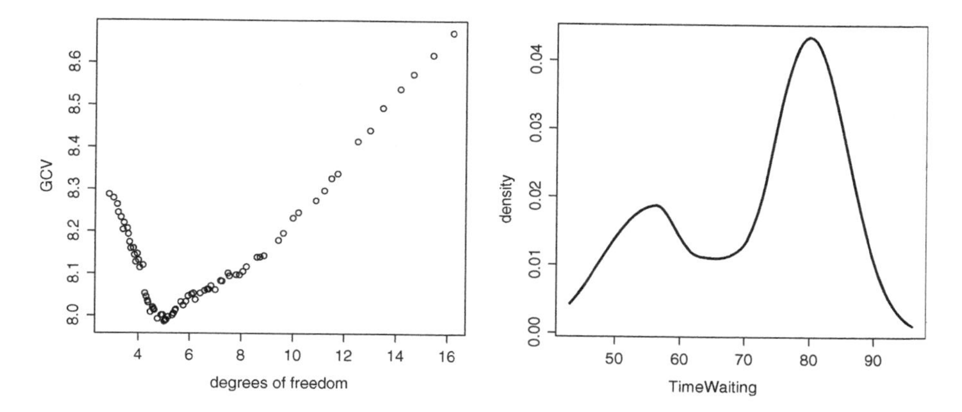

It has been suggested that the waiting time to the next eruption depends on the magnitude (i.e., the eruption time) of the current eruption. The scatter plot of waiting time against eruption time and the least squares line that passes through the data are shown below. One notices two clusters: short waiting times after small eruptions, and long waiting times after large eruptions. While the relationship is roughly linear, a closer inspection of the fit from local polynomial regression shows that the fitted waiting times to the next eruption flatten out for very short and very long eruption times.

```
## local polynomial regression of TimeEruption on TimeWaiting
plot(TimeWaiting~TimeEruption,data=OldFaithful)
# standard regression fit
fitreg=lm(TimeWaiting~TimeEruption,data=OldFaithful)
plot(TimeWaiting~TimeEruption,data=OldFaithful)
abline(fitreg)
# fit with nearest neighbor bandwidth
fit3 <- locfit(TimeWaiting~lp(TimeEruption),data=OldFaithful)
plot(fit3)
```

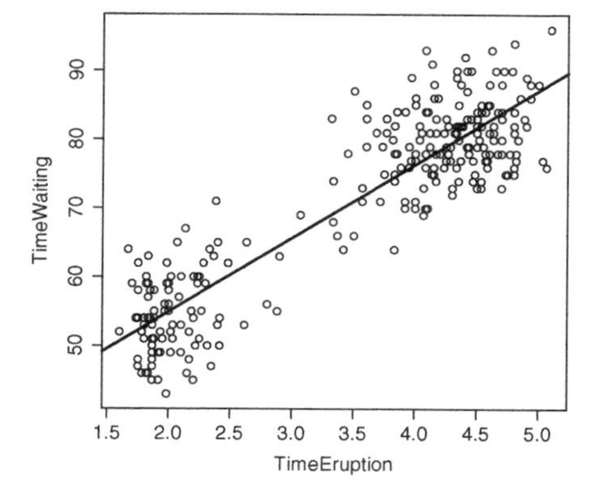

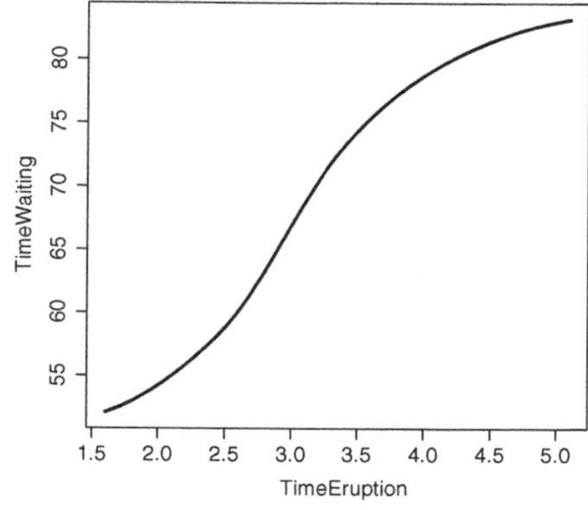

4.4.2 Example 2: NO_x Exhaust Emissions

We use the NO_x exhaust emissions from an investigation that studies pure ethanol as the spark-ignition fuel in a single-cylinder engine (Brinkman, 1981). The NO_x exhaust emissions depend on two predictor variables, the fuel–air equivalence ratio (E) and the compression ratio (R) of the engine.

The density histogram of the NO_x emissions and its smoothed version using local polynomial regression are shown as follows:

```
library(locfit)

## first we read in the data
ethanol <- read.csv("C:/DataMining/Data/ethanol.csv")
ethanol[1:3,]

    NOx CompRatio EquivRatio
1 3.741        12      0.907
2 2.295        12      0.761
3 1.498        12      1.108

## density histogram
hist(ethanol$NOx,freq=FALSE)
## smoothed density histogram
fit <- locfit(~lp(NOx),data=ethanol)
plot(fit)
```

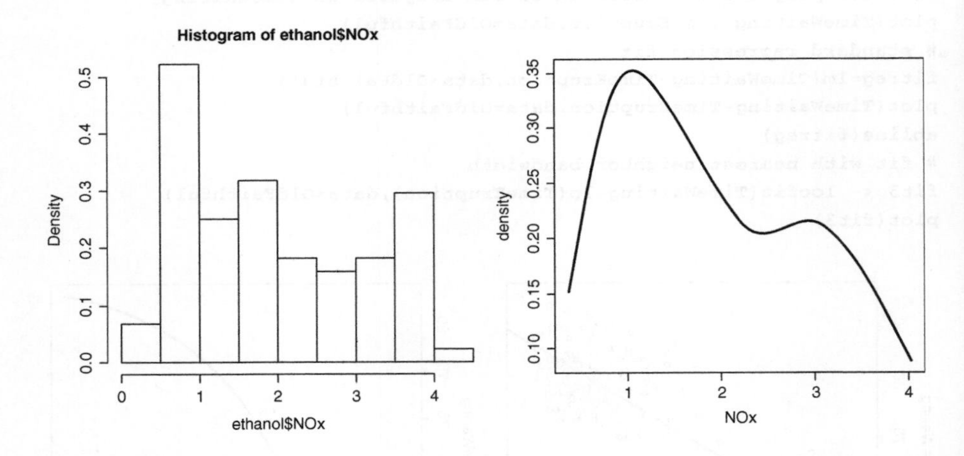

Experiment with the locfit parameters deg (the default is 2) and nn (the nearest neighbor smoothing constant α; the default is 0.7). These parameters control the order of the approximating polynomial and the nearest neighbor smoothing constant α. Larger values of nn result in a wider window and a smoother density. The R program for obtaining the smoothed density can be found on the website that accompanies this book.

Next, we relate the NO_x emissions to the engine's equivalence ratio (E). The scatter plot of NO_x emissions against the equivalence ratio is shown below. A standard regression of NO_x against E leads to an unacceptable fit. The local polynomial regression, on the other hand, provides an excellent approximation.

```
## standard regression of NOx on the equivalence ratio
fitreg=lm(NOx~EquivRatio,data=ethanol)
plot(NOx~EquivRatio,data=ethanol)
abline(fitreg)

## local polynomial regression of NOx on the equivalence ratio
## fit with a 50% nearest neighbor bandwidth.
fit <- locfit(NOx~lp(EquivRatio,nn=0.5),data=ethanol)
plot(fit)
```

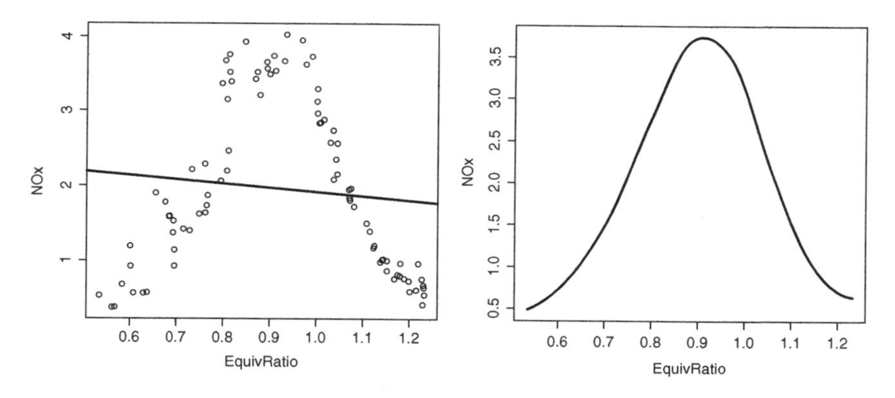

Experiment with the locfit parameters deg (default 2) and nn (default 0.7) and check their impact on the fitting results.

Cross-validation can be used to learn about the appropriate smoothing constant. The graph of the GCV statistic GCV($\widehat{\mu}$) against the degrees of freedom $\nu_1 = \text{tr}(L)$ points to a smoothing constant of around 0.30. The resulting fitted function highlights local patterns that disappear with larger nearest neighbor smoothing constants.

```
## cross-validation
alpha<-seq(0.20,1,by=0.01)
n1=length(alpha)
g=matrix(nrow=n1,ncol=4)
for (k in 1:length(alpha)) {
g[k,]<-gcv(NOx~lp(EquivRatio,nn=alpha[k]),data=ethanol)
}
g
```

```
              [,1]        [,2]        [,3]        [,4]
 [1,]   -3.220084 18.812657 16.426487 0.1183932
```

```
[2,]  -3.249601 17.616143 15.436227 0.1154507
[3,]  -3.319650 16.770041 14.752039 0.1151542
[4,]  -3.336464 15.444040 13.889209 0.1115457
[5,]  -3.373011 14.523910 13.115430 0.1099609
[6,]  -3.408908 13.967891 12.634934 0.1094681
[7,]  -3.408908 13.967891 12.634934 0.1094681
[8,]  -3.469254 12.993165 11.830996 0.1085293
[9,]  -3.504310 12.388077 11.283837 0.1078784
[10,] -3.529167 11.938379 10.928859 0.1073628
[11,] -3.546728 11.469598 10.516520 0.1065792 nn=0.30
[12,] -3.552238 11.263716 10.322329 0.1061728 nn=0.31
[13,] -3.576083 11.035752 10.135243 0.1062533 nn=0.32
[14,] -3.679128 10.540964  9.662613 0.1079229
[15,] -3.679128 10.540964  9.662613 0.1079229
[16,] -3.699044 10.465337  9.578396 0.1082955
. . .

plot(g[,4]~g[,3],ylab="GCV",xlab="degrees of freedom")
f1=locfit(NOx~lp(EquivRatio,nn=0.30),data=ethanol)
f1
plot(f1)
```

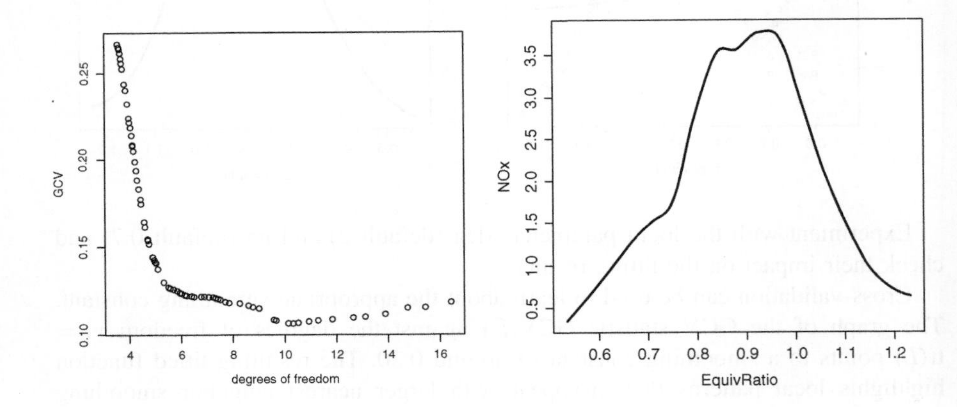

Finally, we relate the NO_x emissions to both the engine's equivalence ratio (E) and compression ratio (R). Scatter plots of NO_x emissions against the equivalence ratio and of NO_x emissions against the compression ratio are shown. The compression ratio has little marginal influence. A local polynomial regression model with both explanatory variables is fitted and the contours of the implied fitted values are also shown. The contours provide useful information, indicating that NO_x emissions are largest when E is about 0.9, but diminish when E moves away from 0.9 (in either direction).

```
## local polynomial regression on both E and C
plot(NOx~EquivRatio,data=ethanol)
plot(NOx~CompRatio,data=ethanol)
```

```
fit <- locfit(NOx~lp(EquivRatio,CompRatio,scale=TRUE),
+    data=ethanol)
plot(fit)

## experiment with the parameters of locfit
fit <- locfit(NOx~lp(EquivRatio,CompRatio,nn=0.5,scale=TRUE),
+    data=ethanol)
plot(fit)
```

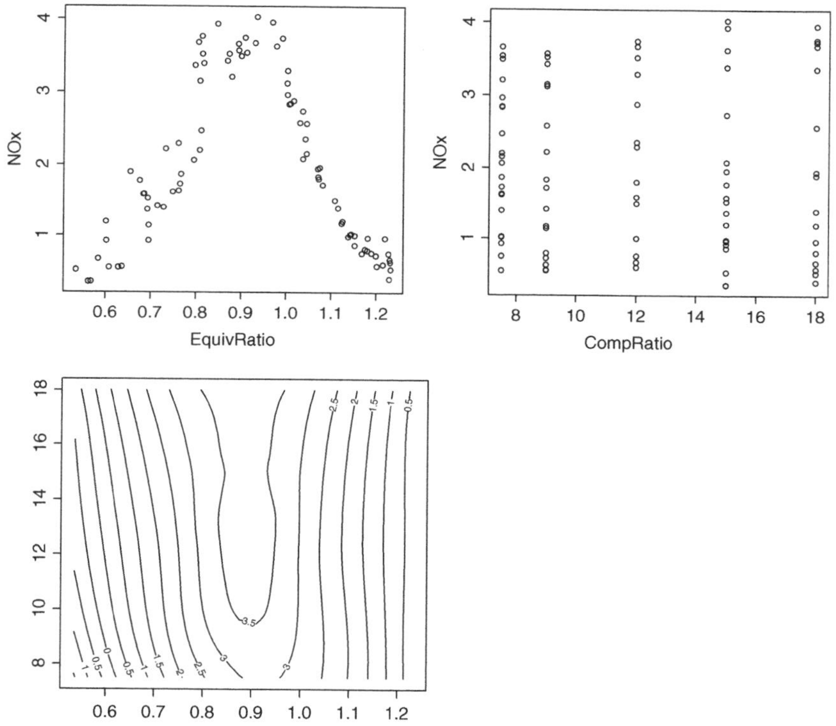

This last example with two explanatory variables illustrates that local polynomial regression can effectively summarize information that could not be gained from simple unsmoothed scatter plots or from ordinary regression. The smoothing procedures discussed in this chapter prove especially useful if one deals with large data sets, which allow for the estimation of functional representations that are more general than the standard parametric ones.

REFERENCES

Brinkman, N.D.: Ethanol fuel—a single-cylinder engine study of efficiency and exhaust emissions. *SAE Transactions*, Vol. 90 (1981).

Cleveland, W.S. and Devlin, S.J.: Locally weighted regression: an approach to regression analysis by local fitting. *Journal of the American Statistical Association*, Vol. 83 (1988), 596–610.

Craven, P. and Wahba, G.: Smoothing noisy data with spline functions. *Numerische Mathematik*, Vol. 31 (1979), 377–403.

Epanechnikov, V.A.: Nonparametric estimates of a multivariate probability density. *Theory of Probability and its Applications*, Vol. 14 (1969), 153–158.

Fan, J. and Gijbels, I.: *Local Polynomial Modelling and Its Applications*. London: Chapman and Hall, 1996.

Loader, C.: *Local Regression and Likelihood*. New York: Springer, 1999.

CHAPTER 5

Importance of Parsimony in Statistical Modeling

Let us suppose that we are fitting a huge regression model with 100 covariates. We assume that only 5 of 100 (regression) coefficients are influential and that we are able to identify all five influential ones. We test the remaining useless coefficients at the 5% significance level ($\alpha = 0.05$). As we reject the null hypothesis of no influence for 5% of the useless 95 variables, $100(4.75/9.75) \approx 50\%$ of the significant regression coefficients are false positives. We refer to this as the *false discovery rate* (FDR). Many large data mining applications involving business data probably have fewer than 5–10% influential factors. Genetics and web analytics are generally far worse. A sequence of multiple tests of hypotheses at the 5% level is bound to run into problems as it will end up with many false positives. We refer to this as the *multiplicity problem*.

5.1 HOW DO WE GUARD AGAINST FALSE DISCOVERY

Controlling the familywise error rate is one approach to addressing the multiplicity problem. The Bonferroni method, for example, adjusts the significance (alpha) level for individual comparisons. If it is desired that the significance level for the whole family of m tests be (at most) α, then the Bonferroni correction tests each individual hypothesis at significance level α/m. Several other adjustment methods are available, but are not discussed here.

An alternative approach controls the *FDR*, the expected proportion of falsely rejected hypotheses. This is described below.

Assume that the general testing setup involves m tests of hypotheses and m specified random test statistics. Assume that m_0 is the number of true null hypotheses (where a variable has no effect) and $m - m_0$ is the number of true alternative hypotheses (where a variable does have an effect). Assume that V is the number of false positives (type I error decisions) and T is the number of false negatives

Data Mining and Business Analytics with R, First Edition. Johannes Ledolter.

(type II error decisions). Let U be the number of true negatives. It is easiest to summarize this in the following table.

	Null Hypothesis Is True (H_0)	Alternative Hypothesis Is True (H_1)	Total
Declared significant	V	S	R
Declared nonsignificant	U	T	$m - R$
Total	m_0	$m - m_0$	m

In m hypothesis tests of which m_0 are true null hypotheses, R (the number of significant tests) is an observable random variable, and S, T, U, and V are unobservable random variables. The FDR (the expected proportion of falsely rejected null hypotheses) is given by $\text{FDR} = E[V/(V+S)] = E[V/R]$. One wants to keep this value below a certain specified threshold α.

The following procedure is valid when the m tests are *independent*. Let $H_1, H_2, \ldots, H_m$ be the m null hypotheses and let $p_1, p_2, \ldots, p_m$ be their corresponding probability values. Order these values in increasing order and denote the ordered probability values by $p_{(1)} \leq p_{(2)} \leq \cdots \leq p_{(m)}$. For a given desired false discovery rate $\text{FDR} = \alpha$, find the largest k such that $p_{(k)} \leq (k/m)\alpha$. Then reject (i.e., declare positive) all $H_{(i)}$ for $i = 1, \ldots, k$. In other words, the p-value rejection cutoff is not α (as it would be for a single hypothesis test), but $\alpha^* = \max_j \{p_{(j)} \leq \alpha(j/m)\}$. Accept all other null hypotheses (i.e., "zero out" everything else) with probability value greater than α^*. Benjamini and Hochberg (1995) show that this approach will guarantee that $\text{FDR} \leq \alpha$.

A simple way to carry out this approach graphically is by overlaying the plot of the ordered $p_{(j)}$ against j with a line plot of $\alpha(j/m)$ against j, and determine the largest probability value that lies below the added line. It is easier to plot quantities on a log–log scale as then $\log[\alpha(j/m)] = \log(\alpha/m) + \log(j)$ is linear in $\log(j)$.

EXAMPLE 5.1 We generate 100 $N(0,1)$ columns of length 25. We add 1 to the first five columns to force the first 5 hypotheses tests $H_i : \mu_i = 0$ $(i = 1, 2, \ldots, 5)$ to come out significant. We conduct significance tests on all 100 columns and calculate 100 probability values according to the one-sample t-test on 25 observations. For our simulation, 10 of the 100 tests turn out to be significant at the individual 0.05 level. There are five false positives, for a false positive rate of 5/10, or 50%.

Next, we select $\alpha = 0.10$ as the FDR to be controlled. A plot of the ordered $p_{(j)}$ against j, overlaid with a plot of $\alpha(j/m) = (0.10)(j/100)$ against j with both plots on a log–log scale, is shown on the following page. We calculate the p-value rejection cutoff $\alpha^* = \max_j \{p_{(j)} \leq 0.10(j/100)\}$, and find this to be 0.0015; we get this by taking the largest p-value that lies below the added straight line. It turns out that we declare six tests (the ones from the first five columns and one other) significant. This is much better!

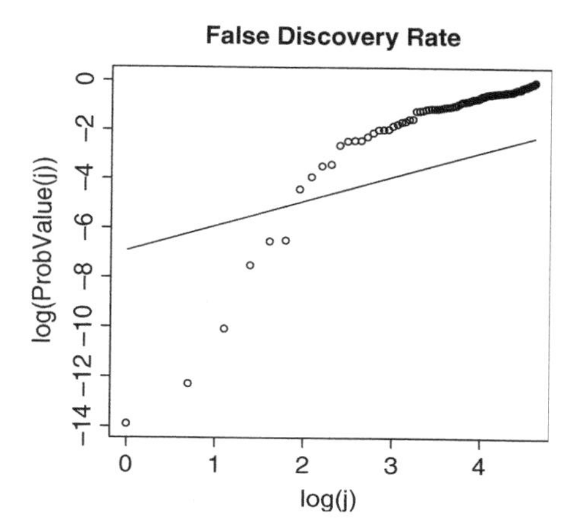

EXAMPLE 5.2 Generate 500 $N(0,1)$ columns of length 25. Again add 1 to make the first five tests of hypotheses come out significant. Conduct significance tests on all 500 columns and calculate the resulting 500 probability values. Thirty two turn out to be significant at the individual 0.05 level, for a false positive rate of 27/32, or 84%. This is not good at all.

Now select $\alpha = 0.20$ as the FDR to be controlled. Calculate the p-value rejection cutoff $\alpha^* = \max_j\{p_{(j)} \leq 0.20(j/500)\}$. The plot of the ordered $p_{(j)}$ against j and the plot of $\alpha(j/m) = (0.20)(j/500)$ against j (both plots on a log–log scale) are shown below. The line is crossed after $j = 7$ resulting in $\alpha^* \approx 0.002$. Using $\alpha^* \approx 0.002$ as significance level, only seven tests are significant; the ones from the first five columns and two others. This is much better!

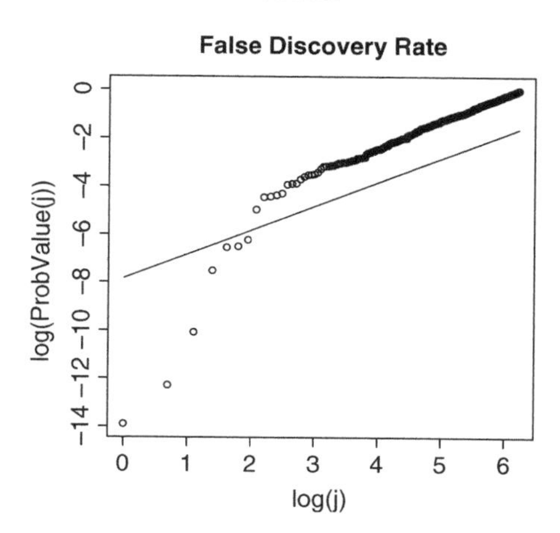

The previous discussion applies for independent tests. But what about dependent tests? The *Benjamini–Hochberg–Yekutieli procedure* controls the FDR under dependence assumptions; see Benjamini and Yekutieli (2001). Their refinement modifies the threshold, and the p-value rejection cutoff becomes $\alpha^{**} = \max_j \{p_{(j)} \leq \alpha[j/(m \times c(m))]\}$, where $c(m) = 1$ if the tests are independent or positively correlated, and $c(m) = \sum_{i=1}^{m} 1/i \approx \ln(m) + \gamma$ and $\gamma \approx 0.57721$ (the Euler–Mascheroni constant) if the tests are negatively correlated.

The R programs that generated the analysis in Examples 5.1 and 5.2 can be found on the website that accompanies this book.

REFERENCES

Benjamini, Y. and Hochberg, Y.: Controlling the false discovery rate: a practical and powerful approach to multiple testing. *Journal of the Royal Statistical Society: Series B*, Vol. 57 (1995), No 1, 289–300.

Benjamini, Y. and Yekutieli, D.: The control of the false discovery rate in multiple testing under dependency. *Annals of Statistics*, Vol. 29 (2001), No 4, 1165–1188.

CHAPTER 6

Penalty-Based Variable Selection in Regression Models with Many Parameters (LASSO)

Many regression models include a very large number of covariates, especially when some of the covariates are categorical variables that have to be converted into indicator variables. Recall that in Section 3.3, we included in the regression model two indicator variables for the three possible outcomes of the categorical variable fuel type. If one had a categorical variable with say 10 possible outcomes, one would need nine indicator variables. We also mentioned earlier, in our discussion of regression models (Appendix 3.A) and in the context of false discovery rates (Chapter 5), that one needs to simplify models as the estimation of unneeded coefficients degrades the forecasting performance.

One can apply multiple-comparison adjustments (such as the Bonferroni adjustment) or methods that control the false discovery rate when testing the significance of a very large number of predictors. Alternatively, one can adopt a regularized version of the least squares solution that constrains the size of the estimates of the coefficients $\beta = (\beta_1, \beta_2, \ldots, \beta_k)$ in the regression model

$$y = \beta_1 x_1 + \beta_2 x_2 + \ldots + \beta_k x_k + \varepsilon.$$

Note that this regression model does not contain an intercept. But a model without intercept can always be accomplished by centering both the response and the explanatory variables.

The LASSO (least absolute shrinkage and selection operator) algorithm finds a least squares solution under the constraint that $\sum_{j=1}^{k} |\widehat{\beta}_j|$, the L1 norm of the estimated parameter vector, is no greater than a certain given value; that is, the LASSO estimate of $\beta = (\beta_1, \beta_2, \ldots, \beta_k)$ is such that

$$\widehat{\beta}^{\text{LASSO}} = \arg\min_{\beta} \sum_{i=1}^{n} \left(y_i - \sum_{j=1}^{k} \beta_j x_{ij} \right)^2 \quad \text{subject to} \quad \sum_{j=1}^{k} |\beta_j| \leq t.$$

Data Mining and Business Analytics with R, First Edition. Johannes Ledolter.

Here $t \geq 0$ is a tuning parameter that controls the shrinkage that is being applied to the estimates. Let the vector $\widehat{\beta}$ be the standard least squares estimate and let $t_0 = \sum_{j=1}^{k} |\widehat{\beta}_j|$ be its L1 norm. Values of $t \geq t_0$ do not affect the least squares minimization, and the least squares and the LASSO estimates are the same. Values of $t < t_0$, on the other hand, lead to a shrinkage of the least squares solution toward 0; and some coefficients will be 0 exactly, leading to variable selection and a simplification of the model. If $t = 0$, all estimated coefficients are shrunk to zero (one may want to say, are "zeroed out"), and none of the variables are selected into the model.

The computation of the LASSO estimate involves the solution of a quadratic programming problem with linear inequality constraints. Solutions to such problems have been studied quite extensively in the optimization literature. The LASSO optimization problem can be rewritten in its equivalent Lagrangian form

$$\widehat{\beta}^{\text{LASSO}} = \arg\min_{\beta} \left[\sum_{i=1}^{n} \left(y_i - \sum_{j=1}^{k} \beta_j x_{ij} \right)^2 + \lambda \sum_{j=1}^{k} |\beta_j| \right],$$

where the Lagrange multiplier λ serves the role of a penalty coefficient. For a fixed value of the penalty λ, we obtain the LASSO estimate, $\widehat{\beta}^{\text{LASSO}}(\lambda)$, and its L1 norm, $\sum_{j=1}^{k} |\widehat{\beta}_j^{\text{LASSO}}(\lambda)|$. Large values of λ penalize large coefficients, and the resulting LASSO estimates will be shrunk to 0. Small values of the penalty λ, on the other hand, result in little shrinkage; for $\lambda = 0$, there is no shrinkage at all. Note that the shrinkage applies only to the slope coefficients of the regression model, but not the intercept. The estimate of the intercept can always be calculated from the mean-corrected version of the regression model, and it is given by $\overline{y} - \sum_{j=1}^{k} \widehat{\beta}_j^{\text{LASSO}}(\lambda)\overline{x}_j$.

The term $\lambda \sum_{j=1}^{k} |\beta_j|$ in the earlier minimization has two major consequences: (i) it makes the LASSO solutions nonlinear in the response observations y_i and (ii) there is no closed-form expression any more for the estimates. This differs from the closed-form and linear solutions in ordinary least squares estimation as well as in *ridge regression*, another penalty-based estimation approach. But, instead of constraining the sum of the absolute values of the coefficients as done in LASSO, ridge regression constrains the sum of the squared coefficients, $\sum_{j=1}^{k} (\beta_j)^2$.

However, there exist efficient algorithms for computing the entire path of solutions as the LASSO penalty λ is varied. The *least angle regression* (LARS) algorithm (Efron et al., 2004) computes the entire LASSO path, and it does so very efficiently, requiring essentially the same order of computations as that of a single least squares fit on the k predictors. Details are given in Efron et al. (2004), as well as in the book by Hastie et al. (2009).

The L1-regularized estimation formulation of the LASSO has a tendency to prefer solutions with fewer nonzero parameter values, thus effectively reducing the number of variables upon which the given solution is dependent. Thus LASSO can be thought of as a penalty-based variable selection approach that selects variables to be included into the model. Such an approach is certainly advantageous in regression situations where one works with extremely large models that contain many variables and many coefficients, but where one knows a priori that many of these variables are not needed. LASSO shrinks the estimates and sets some of the coefficients equal to 0.

When using the LASSO approach, one needs to select λ (or, equivalently, the coefficient t in the L1 norm constraint, $\sum_{j=1}^{k} |\beta_j| \leq t$). Using the available very efficient algorithms, LASSO estimates are readily obtained for any value of λ. Forecasts can be calculated with the resulting constrained regression estimates, and the out-of-sample prediction performance can be evaluated for any value of λ. This makes cross-validation a very practical approach for selecting the penalty parameter λ.

The **lars** package in R provides LASSO estimates of linear regression coefficients for a range of λ's. The mode = "fraction" argument of this package, with the fraction s representing a number between 0 and 1, provides regression estimates of various degrees of shrinkage. The number s expresses the ratio of the L1 norm of the LASSO estimate of the coefficient vector, relative to the L1 norm of the least squares solution. The fraction $s = 0$, for example, implies that all coefficient estimates are 0 (complete shrinkage to 0); $s = 1$ leads to the least squares solution. Values in between reflect various degrees of shrinkage. Using the R function *plot*, one can trace the behavior of the standardized estimates (estimates divided by their standard errors) for changing values of s.

The *predict* function of the lars library can be used to predict the response with LASSO estimates that have been shrunk to a certain fraction of the least squares estimates. This results in fitted values for in-sample evaluation and in genuine out-of-sample predictions for new cases.

The optimal value for s can be obtained through cross-validation and the R command *cv.lars*. V-fold cross-validation with $K = 10$ folds, for example, divides the cases of the data set into $K = 10$ nonoverlapping parts; it uses 9 of the 10 parts for estimation and the tenth part for forecast evaluation. This is repeated 10 times, for the 10 different segments of the holdout sample. Cross-validation mean square errors, plotted for changing values of s, tell us about the shrinkage (the value of s) that should be used. The R commands for carrying out these computations are shown as follows:

```
lasso <- lars(x=x, y=y, trace=TRUE)                  # fit lasso
coef(lasso, s=c(1/4,1/2,3/4), mode="fraction")  # coefficients
predict(lasso, x, s=1/2, mode="fraction")        # predict
cv.lars(x=x,y=y,K=10)    ## cross-validation using 10 folds
```

The LASSO estimates and the implied penalty-based variable selection have several advantages over an approach that controls the false discovery rate. LASSO is almost automatic and works very well for models that contain many covariates. The estimation algorithms are very fast and efficient, and good software is available to carry out the calculations. LASSO also calculates standard errors of the estimates, and estimation and testing can be carried out at the same time. Multicollinearity does not cause problems as in the case of closely related variables, LASSO chooses one variable and "zeros out" the rest.

However, there are also disadvantages. LASSO is scale dependent, and because of this one must be careful with the units of the explanatory variables. If one unit is in the tens and one in the millions and if their associated coefficients differ by several orders of magnitude, the penalty on the sum of the absolute values of the coefficients may not have much meaning. One should standardize the explanatory variables if their units are very different. In addition, and this is another disadvantage, LASSO only works for regression models, while the approach of controlling the false discovery rate applies to any testing situation.

6.1 EXAMPLE 1: PROSTATE CANCER

Let us consider an example on prostate cancer. The data, taken from Stamey et al. (1989), contains the results of biopsies on 97 men of various ages. This data set has been analyzed by Tibshirani (1996) in his introduction of the LASSO and by many texts on data mining. The biopsy information includes:

Gleason score (gleason): scores are assigned to the two most common tumor patterns ranging from 2 to 10; in this data set, the range is from 6 to 9.
Prostate-specific antigen (psa): laboratory results on protein production.
Capsular penetration (cp): reach of cancer into the gland lining.
Benign prostatic hyperplasia amount (bph): size of the prostate.

The goal is to predict the tumor log volume (which measures the tumor's size or spread). We try to predict this variable from five covariates (age; logarithms of bph, cp, and psa; and the Gleason score). The predicted size of the tumor has important implications for the subsequent treatment options, which include chemotherapy, radiation treatment, and surgical removal of the prostate.

```
prostate <- read.csv("C:/DataMining/Data/prostate.csv")
prostate[1:3,]

       lcavol age      lbph       lcp gleason       lpsa
1 -0.5798185  50 -1.386294 -1.386294       6 -0.4307829
2 -0.9942523  58 -1.386294 -1.386294       6 -0.1625189
3 -0.5108256  74 -1.386294 -1.386294       7 -0.1625189
```

The output of both the standard regression analysis and the LASSO estimation are shown as follows:

```
m1=lm(lcavol~.,data=prostate)
summary(m1)

Call:
lm(formula = lcavol ~ ., data = prostate)
```

```
Residuals:
     Min       1Q   Median       3Q      Max
-1.88964 -0.52719 -0.07263  0.57834  1.98728

Coefficients:
            Estimate Std. Error t value Pr(>|t|)
(Intercept) -1.49371    0.94261  -1.585   0.1165
age          0.01902    0.01063   1.789   0.0769 .
lbph        -0.08918    0.05376  -1.659   0.1006
lcp          0.29727    0.06762   4.396 2.98e-05 ***
gleason      0.05240    0.11965   0.438   0.6625
lpsa         0.53955    0.07648   7.054 3.30e-10 ***
---
Signif. codes:  0 '***' 0.001 '**' 0.01 '*' 0.05 '.' 0.1 ' ' 1

Residual standard error: 0.7015 on 91 degrees of freedom
Multiple R-squared: 0.6642,     Adjusted R-squared: 0.6457
F-statistic:    36 on 5 and 91 DF,  p-value: < 2.2e-16
```

```
## the model.matrix statement defines the model to be fitted
x <- model.matrix(lcavol~age+lbph+lcp+gleason+lpsa,
+    data=prostate)
x=x[,-1]
## stripping off the column of 1s as LASSO includes the
## intercept automatically
library(lars)
## lasso on all data
lasso <- lars(x=x,y=prostate$lcavol,trace=TRUE)
## trace of lasso (standardized) coefficients for varying
## penalty
plot(lasso)
```

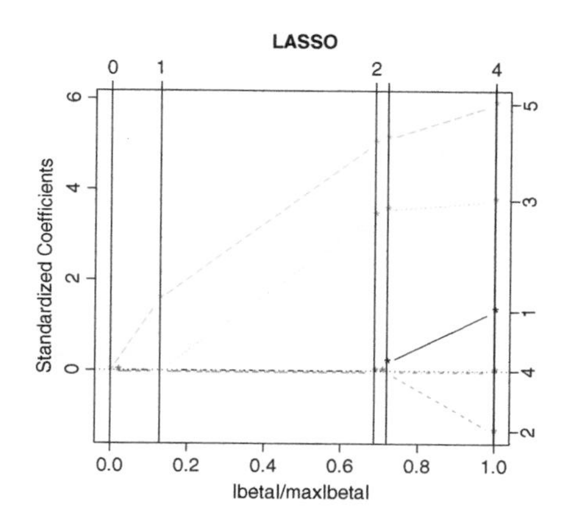

```
lasso

Call:
lars(x = x, y = prostate$lcavol, trace = TRUE)
R-squared: 0.664
Sequence of LASSO moves:
     lpsa lcp age gleason lbph
Var     5   3   1       4    2
Step    1   2   3       4    5
```

The graph of the LASSO estimates as a function of the shrinkage illustrates the order in which variables enter the model as one relaxes the constraint on the L1 norm of their estimates. Initially there is nothing in the model (look to the left of the graph, where $s = 0$). Moving to the right on this graph, one finds that the first variable to enter is variable 5 (lpsa); then variable 3 (lcp) enters; then variable 1 (age). Or, scanning the graph from the right-hand side to the left, we notice that variables 2 (lbph), 4 (gleason), and 1 (age) are "zeroed out" in that order. Below, we list the parameter estimates for selected values of shrinkage; $s = 0.25, 0.50, 0.75,$ and 1.00. The LASSO estimates for $s = 1$ are the ordinary least squares estimates (see the prior regression output).

```
coef(lasso,s=c(.25,.50,0.75,1.0),mode="fraction")

     (Intercept)         age          lbph        lcp    gleason      lpsa
[1,]           0 0.000000000  0.000000000 0.06519506 0.00000000 0.2128290 # s=0.25
[2,]           0 0.000000000  0.000000000 0.18564339 0.00000000 0.3587292 # s=0.50
[3,]           0 0.005369985 -0.001402051 0.28821232 0.01136331 0.4827810 # s=0.75
[4,]           0 0.019023772 -0.089182565 0.29727207 0.05239529 0.5395488 # s=1.00

## cross-validation using 10 folds
cv.lars(x=x,y=prostate$lcavol,K=10)
```

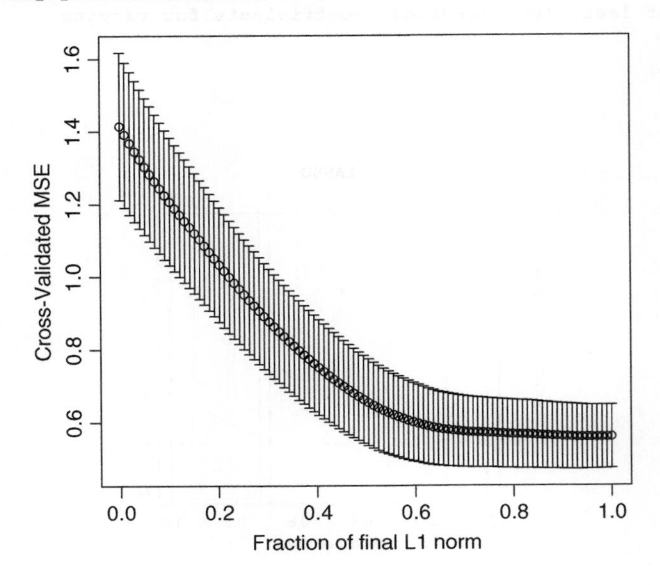

The output of cross-validation (average mean square errors and their associated standard error bounds) shows that the mean square error increases quite rapidly if we shrink the coefficients too aggressively. The mean square curve is smallest for $s = 1$ (the least squares solution), but is actually quite flat for all values of s larger than 0.6. The present example does not call for much shrinkage. This is not too surprising as the number of estimated coefficients (five coefficients, excluding the intercept) is rather small relative to the size of the sample ($n = 97$).

A similar conclusion about the required amount of shrinkage is obtained when selecting at random 80 of the 97 men for the training data set, and applying the LASSO estimates to predict the log volume of the remaining 17 men. Repeating the random sampling 10 times (each time leaving out 17 different randomly selected men) leads to the mean square prediction errors that are shown in the box plot. Again, one is best off staying with the least squares estimates.

```
## another way to evaluate lasso's out-of-sample
## prediction performance
MSElasso25=dim(10)
MSElasso50=dim(10)
MSElasso75=dim(10)
MSElasso100=dim(10)
set.seed(1)
for(i in 1:10){
      train <- sample(1:nrow(prostate),80)
      lasso <- lars(x=x[train,],y=prostate$lcavol[train])
      MSElasso25[i]=
+           mean((predict(lasso,x[-train,],s=.25,
+           mode="fraction")$fit-
+           prostate$lcavol[-train])^2)
      MSElasso50[i]=
+           mean((predict(lasso,x[-train,],s=.50,
+           mode="fraction")$fit-
+           prostate$lcavol[-train])^2)
      MSElasso75[i]=
+           mean((predict(lasso,x[-train,],s=.75,
+           mode="fraction")$fit-
+           prostate$lcavol[-train])^2)
      MSElasso100[i]=
+           mean((predict(lasso,x[-train,],s=1.00,
+           mode="fraction")$fit-
+           prostate$lcavol[-train])^2)
      }
mean(MSElasso25)
[1] 1.021938
mean(MSElasso50)
[1] 0.6723226
mean(MSElasso75)
[1] 0.5410033
```

```
mean(MSElasso100)
[1] 0.5352386
boxplot(MSElasso25,MSElasso50,MSElasso75,MSElasso100,
+     ylab="MSE", sub="LASSO model",
+     xlab="s=0.25            s=0.50            s=0.75            s=1.0(LS)")
```

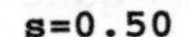

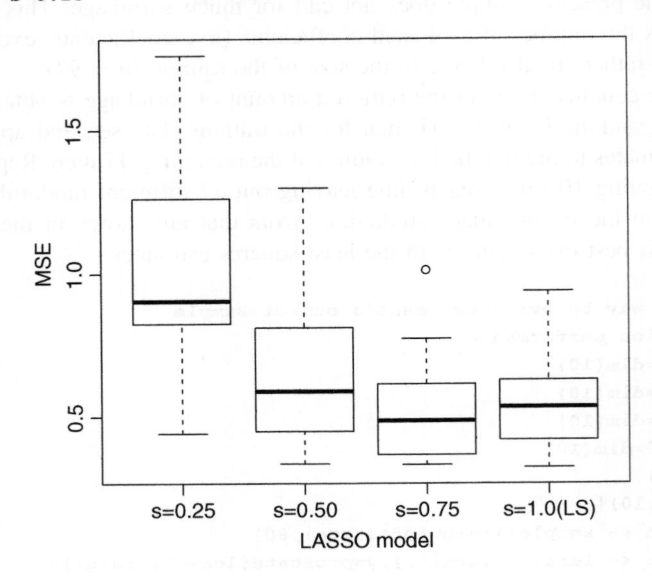

6.2 EXAMPLE 2: ORANGE JUICE

This section analyzes the weekly sales data of refrigerated 64-ounce orange juice containers from 83 stores in the Chicago area. There are many stores throughout the city, many time periods, and three brands (Dominicks, MinuteMaid, and Tropicana). The data are arranged in rows with each row giving the recorded sales (in logarithms; logmove), as well as brand, price, presence/absence of feature advertisement, and the demographic characteristics of the stores. In total, there are 28,947 rows in this data set. The data is taken from P. Rossi's **bayesm** package for R, and it has been used earlier in Montgomery (1987). We have looked at this data set in Chapter 2 of this text.

```
oj <- read.csv("C:/DataMining/Data/oj.csv")
```

```
STORE      store number
BRAND      brand indicator
WEEK       week number
LOGMOVE    log of the number of 64oz units sold
PRICE      price of 64oz unit
FEATURE    feature advertisement
AGE60      proportion of the population that is aged 60 or older
EDUC       proportion of the population that has a college degree
```

```
ETHNIC     proportion of the population that is black or Hispanic
INCOME     log median income
HHLARGE    proportion of households with 5 or more persons
WORKWOM    proportion of women with full-time jobs
HVAL150    proportion of households worth more than $150,000
SSTRDIST   distance to the nearest warehouse store
SSTRVOL    ratio of sales of this store to the nearest warehouse store
CPDIST5    average distance in miles to the nearest 5 supermarkets
CPWVOL5    ratio of sales of this store to the average of the nearest
           five stores
```

```
oj[1:2,]
```

```
  store     brand week  logmove feat price     AGE60      EDUC    ETHNIC
1     2 tropicana   40 9.018695    0  3.87 0.2328647 0.2489349 0.1142799
2     2 tropicana   46 8.723231    0  3.87 0.2328647 0.2489349 0.1142799
    INCOME   HHLARGE   WORKWOM   HVAL150 SSTRDIST  SSTRVOL CPDIST5   CPWVOL5
1 10.55321 0.1039534 0.3035853 0.4638871 2.110122 1.142857 1.92728 0.3769266
2 10.55321 0.1039534 0.3035853 0.4638871 2.110122 1.142857 1.92728 0.3769266
```

```
x <- model.matrix(logmove ~ log(price)*(feat + brand
+     + AGE60 + EDUC + ETHNIC + INCOME + HHLARGE + WORKWOM
+     + HVAL150 + SSTRDIST + SSTRVOL + CPDIST5 + CPWVOL5)^2, data=oj)
dim(x)
```

```
[1] 28947   210
```

The model that is specified here contains as explanatory variables the logarithm of price and its interaction with linear and quadratic components for feature, brand, and the demographic characteristics of a store's neighborhood. We mentioned in Chapter 2 that price elasticities are most likely affected by demographic characteristics such as the average income of a store's immediate neighborhood. The *model.matrix* statement in R allows us to specify the model without having to write out all its terms in detail. The model $y \sim z * (x1 + x2 + x3)^\wedge 2$, for example, includes the intercept and the following 13 terms: $\mathrm{z, x1, x2, x3, x1 * x2, x1 * x3, x2 * x3, z * x1, z * x2, z * x3, z * x1 * x2, z * x1 * x3, z * x2 * x3}$. Our model, with the three brands represented by two indicator variables, contains 210 covariates (including the intercept). This is a very large number, suggesting a shrinkage approach such as LASSO for the estimation of its parameters. We know that LASSO is sensitive to scale. Hence we normalize the covariates as they are of very different magnitudes, and we transform the covariates such that each covariate has mean 0 and standard deviation 1.

```
## First column of x consists of ones (the intercept)
## We strip the column of ones as intercept is included
## automatically
x=x[,-1]
## We normalize the covariates as they are of very different
## magnitudes
## Each normalized covariate has mean 0 and standard
## deviation 1
for (j in 1:209) {
```

```
x[,j]=(x[,j]-mean(x[,j]))/sd(x[,j])
}

## One could consider the standard regression model
reg <- lm(oj$logmove~x)
summary(reg)
p0=predict(reg)

## Or, one could consider LASSO
library(lars)
lasso <- lars(x=x, y=oj$logmove, trace=TRUE)
coef(lasso, s=c(.25,.50,0.75,1.00), mode="fraction")
## creates LASSO estimates as function of lambda
## gives you the estimates for four shrinkage coef

## Check that predictions in regression and lars (s=1) are the
## same
p1=predict(lasso,x,s=1,mode="fraction")
p1$fit
pdiff=p1$fit-p0
pdiff  ## zero differences

## out of sample prediction; estimate model on 20,000 rows
MSElasso10=dim(10)
MSElasso50=dim(10)
MSElasso90=dim(10)
MSElasso100=dim(10)
set.seed(1) ## fixes seed to make random draws reproducible
for(i in 1:10){
     train <- sample(1:nrow(oj), 20000)
     lasso <- lars(x=x[train,], y=oj$logmove[train])
     MSElasso10[i]=mean((predict(lasso,x[-train,], s=.10,
+          mode="fraction")$fit -
+          oj$logmove[-train])^2)
     MSElasso50[i]=mean((predict(lasso,x[-train,], s=.50,
+          mode="fraction")$fit -
+          oj$logmove[-train])^2)
     MSElasso90[i]=mean((predict(lasso,x[-train,], s=.90,
+          mode="fraction")$fit -
+          oj$logmove[-train])^2)
     MSElasso100[i]=mean((predict(lasso,x[-train,], s=1.0,
+          mode="fraction")$fit -
+          oj$logmove[-train])^2)
     }
mean(MSElasso10)
[1] 0.3494492
mean(MSElasso50)
[1] 0.3456142
mean(MSElasso90)
```

```
[1] 0.3455313
mean(MSElasso100)
[1] 0.3455732

boxplot(MSElasso10,MSElasso50,MSElasso90,MSElasso100,
+    ylab="MSE", xlab="LASSO model")
```

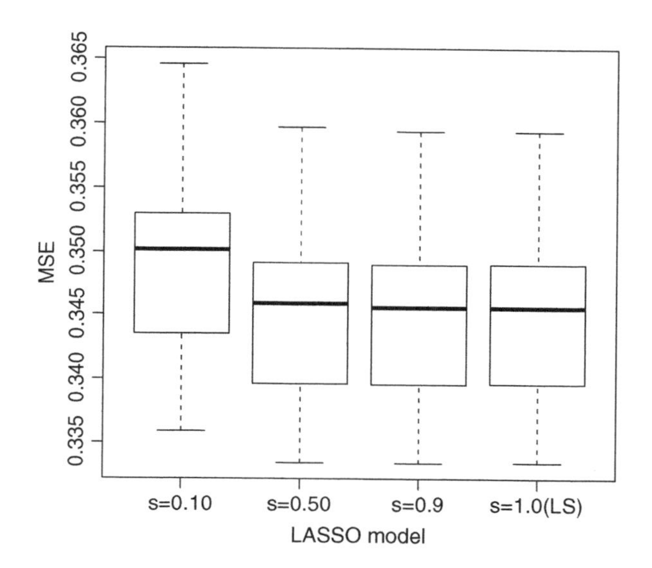

The box plot summarizes the results for LASSO estimates that use different amounts of shrinkage (fraction $s = 0.10, 0.50, 0.90$, and 1.00). Estimates are obtained for 10 random samples of size 20,000 (always taken from the 28,947 units), and for each sample, the resulting estimates are used to predict the response of the remaining 8947 units in the holdout sample. The mean of the 8947 squared forecast errors is obtained for each random sample, and box plots of the 10 mean square errors are shown for each of the four shrinkage fractions that have been considered. The finding that least squares with little or no shrinkage is appropriate is not very surprising as we use lots of data (20,000 records) to estimate just 210 coefficients. With so many observations, ordinary least squares provides very reliable information about the coefficients, and little or no shrinkage is needed.

It is a different story if we estimate the 210 coefficients on just 1000 data points. We repeat the calculations, but now sample only 1000 rows (instead of the 20,000 considered earlier) for inclusion into the estimation data set. The predictions are now evaluated on the remaining $28,947-1000 = 27,947$ rows. The results shown in the following graph indicate that in this case shrinkage clearly helps; reliable estimation of that many parameters from just 1000 records is not possible without some shrinkage.

```
mean(MSElasso10)
[1] 0.4309619      BEST
mean(MSElasso50)
[1] 0.4436273
mean(MSElasso90)
[1] 0.4475674
mean(MSElasso100)
[1] 0.448235
```

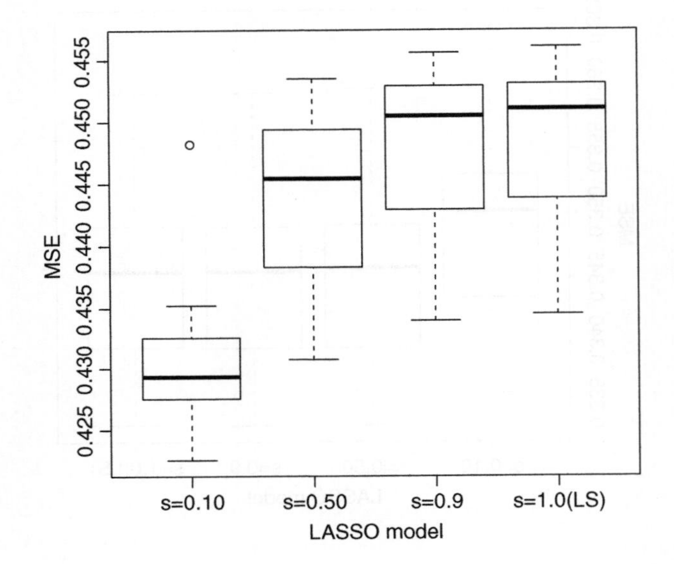

REFERENCES

Efron, B., Johnstone, I., Hastie, T., and Tibshirani, R.: Least angle regression. *Annals of Statistics*, Vol. 32 (2004), No 2, 407–499.

Hastie, T., Tibshirani, R., and Friedman, J.: *The Elements of Statistical Learning: Data Mining, Inference and Prediction.* Second edition. New York: Springer, 2009.

Montgomery, A.L.: Creating micro-marketing pricing strategies using supermarket scanner data. *Marketing Science*, Vol. 16 (1987), 315–337.

Stamey, T., Kabalin, J., McNeal, J., Johnstone, I., Freiha, F., Redwine, E., and Yang, N.: Prostate specific antigen in the diagnosis and treatment of adenocarcinoma of the prostate, ii: radical prostatectomy treated patients. *Journal of Urology*, Vol. 141 (1989), 1076–1083.

Tibshirani, R.: Regression shrinkage and selection via the LASSO. *Journal of the Royal Statistical Society: Series B*, Vol. 58 (1996), No 1, 267–288.

CHAPTER 7

Logistic Regression

In the standard regression model, the response is a continuous measurement variable such as sales or profit. There we consider linear regression models of the form

$$y = f(x_1, x_2, \ldots, x_k) + \varepsilon = \alpha + \beta_1 x_1 + \beta_2 x_2 + \cdots + \beta_k x_k + \varepsilon,$$

where the function $f(\cdot)$ is linear in the regressor (predictor) variables $X = [x_1, x_2, \ldots, x_k]$. The error ε follows a normal distribution with mean zero and variance σ^2, implying that the conditional mean of the response is a linear function of the regressor variables

$$E(y|X) = f(X) = \alpha + \beta_1 x_1 + \beta_2 x_2 + \cdots + \beta_k x_k.$$

In *logistic regression* the response variable is *binary*. The response can be either true (success) or false (failure), usually coded as 1 and 0, such as buy a product or do not buy; get the death penalty for a crime or not; and become insolvent or not.

Especially in high dimensions, it is often convenient to phrase the problem in binary form. In many data mining problems, the target is a binary response such as

- profit or loss, response greater or less than a certain value, being able to pay back a loan or default;
- thumbs up or down, buy or not buy, become a potential customer or not;
- win or lose, sick or healthy, Republican or Democrat.

7.1 BUILDING A LINEAR MODEL FOR BINARY RESPONSE DATA

For a binary response, the conditional mean in the regression model becomes

$$E(y|X) = 1 \times P(y = 1|X) + 0 \times P(y = 0|X) = P(y = 1|X).$$

Data Mining and Business Analytics with R, First Edition. Johannes Ledolter.

The expectation is now a probability that must always be between 0 and 1. Hence, we cannot use just any linear regression function $f(\cdot)$, as an acceptable function must always have values between 0 and 1. We want a response model where

$$p = P(y = 1|X) = f(\alpha + \beta_1 x_1 + \beta_2 x_2 + \cdots + \beta_k x_k)$$

is between 0 and 1. For a single regressor variable x, we want a function $f(\cdot)$ that looks like an "S-shaped" curve such as the ones shown below; for the moment, we ignore the equations that are given next to the curves. Here, we have drawn the functions starting at 0 and increasing to 1. Alternatively, we could have switched the sign of the slope coefficient and made them decreasing functions of the explanatory variable. Note that on these curves, p changes more quickly in the middle (when p is 0.5) than at either end of the x-axis.

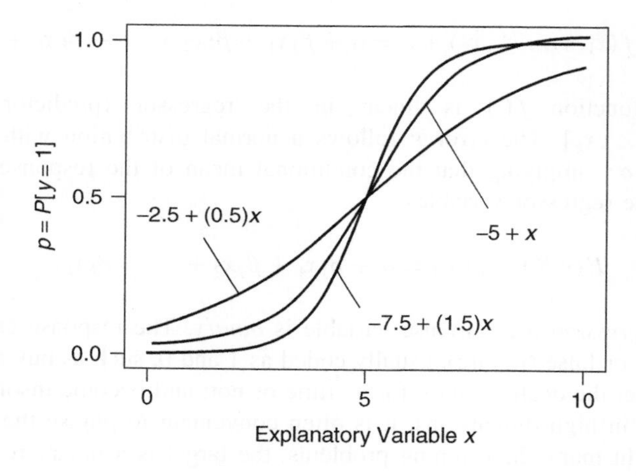

The *logistic regression model* links the predictor variables to probabilities through the equation

$$p = f(\alpha + \beta_1 x_1 + \beta_2 x_2 + \cdots + \beta_k x_k) = \frac{\exp(\alpha + \beta_1 x_1 + \beta_2 x_2 + \cdots + \beta_k x_k)}{1 + \exp(\alpha + \beta_1 x_1 + \beta_2 x_2 + \cdots + \beta_k x_k)}.$$

It can be verified that this function leads to S-shaped curves between 0 and 1 such as those shown in the earlier graph. The curves in the figure shown earlier involve a single predictor and result from the equation with ($\alpha = -2.5$, $\beta_1 = 0.5$), ($\alpha = -5.0$, $\beta_1 = 1.0$), and ($\alpha = -7.5$, $\beta_1 = 1.5$). For large values of $\alpha + \beta_1 x_1 + \beta_2 x_2 + \cdots + \beta_k x_k$, the probability approaches 1. The probability approaches 0 if $\alpha + \beta_1 x_1 + \beta_2 x_2 + \cdots + \beta_k x_k$ is small. Further, the probability becomes 0.5 if $\alpha + \beta_1 x_1 + \beta_2 x_2 + \cdots + \beta_k x_k = 0$.

Simple algebra shows that

$$\log \frac{p}{1-p} = \alpha + \beta_1 x_1 + \beta_2 x_2 + \cdots + \beta_k x_k.$$

The quantity $p/(1-p)$ relates the probability of success, p, to the probability of failure, $1-p$, and we refer to $p/(1-p)$ as the *odds of success*. The quantity $\log[p/(1-p)]$ is referred to as the *logit* of p and expresses the log odds of success. The logistic regression model specifies a linear model for the log odds of success.

Logistic regression, with its logit "link" $\log[p/(1-p)]$ modeled as a linear function of the predictor variables, is the most popular model for a binary outcome variable. Another useful model for binary outcome variables is the probit model. It models the *probit* of p, the inverse cumulative probability distribution function of a normal distribution, as a linear function of the predictor variables. The cumulative probability distribution function of the normal distribution has the desired S-shaped form between 0 and 1, which makes the model with a probit link useful. The model with the probit link leads to very similar conclusions and, for this reason, it is not considered in this text.

7.2 INTERPRETATION OF THE REGRESSION COEFFICIENTS IN A LOGISTIC REGRESSION MODEL

All other variables remaining fixed, a change of one unit in the regressor variable x_1 changes the log odds of success by β_1 units. This implies that the odds of success are changed by the multiplicative factor $\exp(\beta_1)$, which is called the *odds ratio*. Consider a regression coefficient $\beta_1 = -0.2$ and $\exp(-0.2) = 0.82$. It implies that a change from x_1 to $x_1 + 1$ changes the odds of occurrence by the factor 0.82. It reduces the odds of occurrence by $100(1-0.82) = 18\%$. A value $\beta_1 = 0$ and $\exp(0) = 1$ implies that a change in the explanatory variable has no effect on the odds of occurrence. A value $\beta_1 = 1.5$ and $\exp(1.5) = 4.48$ implies that a change from x_1 to $x_1 + 1$ changes the odds of occurrence by the multiplicative factor 4.48. It increases the odds by $100(4.48-1) = 348\%$.

7.3 STATISTICAL INFERENCE

The estimation of the parameters can be carried out through maximum likelihood estimation. We skip the details; the interested reader may refer to Abraham and Ledolter (2006) for the details. Virtually all statistical packages include routines for the estimation of logistic regression models. The output from these packages provides estimates as well as standard errors of the estimates.

For illustration, consider the model with a single regressor variable x. Assume that we have n cases; that is, there are n pairs of observations containing the value of the covariate x_i and the success indicator $y_i = 0/1$. Maximum likelihood estimation maximizes the likelihood (probability) of the realized observations. In the Bernouilli model setup, where the outcome of case i is either 1 or 0 with probabilities $p_i = [\exp(\alpha + \beta_1 x_i)]/[1 + \exp(\alpha + \beta_1 x_i)]$ and $1 - p_i =$

$1/[1+\exp(\alpha+\beta_1 x_i)]$, the likelihood function is

$$\prod_{i=1}^{n} p(y_i|x_i) = \prod_{i=1}^{n} (p_i)^{y_i}(1-p_i)^{1-y_i}$$

$$= \prod_{i=1}^{n} \left[\frac{\exp(\alpha+\beta_1 x_i)}{1+\exp(\alpha+\beta_1 x_i)}\right]^{y_i} \left[\frac{1}{1+\exp(\alpha+\beta_1 x_i)}\right]^{1-y_i}.$$

The maximization of this function with respect to the (logistic) regression parameters α and β_1 is equivalent to the minimization of the *deviance* (the negative logarithm of the likelihood)

$$D = -\left[\sum_{i=1}^{n} y_i \log(p_i) + \sum_{i=1}^{n} (1-y_i)\log(1-p_i)\right],$$

with probabilities

$$p_i = \frac{\exp(\alpha+\beta_1 x_i)}{1+\exp(\alpha+\beta_1 x_i)}.$$

The deviance in standard regression, $D = \sum_{i=1}^{n} [y_i - (\alpha+\beta_1 x_i)]^2$, was discussed in Chapter 3. The minimizing value determines the estimate of $\text{Var}(\varepsilon_i) = \sigma^2$ (and also the R-square and the F-statistic for testing the overall significance of the regression). In logistic regression there is no σ^2 to estimate. But we use the deviance in logistic regression in the same manner that we use the deviance in standard regression. We find the parameter estimates by minimizing the deviance, and we use deviances to compare the fits of different models. For example, we compare the deviance of the model that includes covariates (the fitted model) to the deviance of the null model where the estimate of p is simply the sample proportion of successes (i.e., the average of the 0/1 responses) that ignores the information of covariates.

7.4 CLASSIFICATION OF NEW CASES

Logistic regression models are quite useful for classifying new cases into one of two outcome categories ("success" or "failure"). The estimated logistic model, applied to new cases of a test (evaluation) data set, provides predictions of success probabilities. With a certain cutoff on the predicted success probabilities (in later sections we will say more about the appropriate selection of the cutoff), the logistic regression provides a rule for classifying new cases. One can use the actual realizations of the cases in the test (evaluation) data set to investigate whether the logistic regression (or any other classification method) is in fact capable of identifying the actual outcomes. The *lift curve*, introduced in Section 7.7, assesses whether cases with the largest predicted probabilities of success are actually true successes. If this is true, we say that the model gives us a nice "lift" in identifying the most likely candidates for success.

7.5 ESTIMATION IN R

We look at several examples. We use the statistical software R for the estimation, but also use Minitab, a popular spreadsheet-based software program, in the first illustrative example.

The R command **glm** is used to fit logistic regression models. glm is a very general routine as it can be used to fit *generalized linear models* (hence its name) that are specified by listing the linear predictors, defining the error distribution, and specifying a link function. For logistic regression, the family of the error distribution is the *binomial*. The default, family = binomial [or family = binomial(link = "logit")] specifies the logistic regression model. The probit model is obtained by specifying family = binomial(link = "probit"). The general glm syntax is shown below; the following examples illustrate its use.

```
glm(formula, family = gaussian, data, weights, subset,
    na.action, start = NULL, etastart, mustart, offset,
    control = list(...), model = TRUE, method = "glm.fit",
    x = FALSE, y = TRUE, contrasts = NULL, ...)

glm.fit(x, y, weights = rep(1, nobs),
    start = NULL, etastart = NULL, mustart = NULL,
    offset = rep(0, nobs), family = binomial,
    control = list(), intercept = TRUE)
```

7.6 EXAMPLE 1: DEATH PENALTY DATA

Is the death penalty more likely if the victim is white? Is the death penalty more likely if the crime is horrible with many aggravating features? The data set, taken from Abraham and Ledolter (2006) and listed here in summarized form, includes the following variables:

Aggravation index measuring the severity of the crime; from 1 through 6 (really bad);

VRace: race of victim (White/Black); VRaceC: race of victim (coded);

DY: number of people getting death (death = Yes);

DN: number of people getting life (death = No).

Aggrav	VRace	VRaceC	DY	DN	Number
1	White	1	2	60	62
1	Black	0	1	181	182
2	White	1	2	15	17
2	Black	0	1	21	22
3	White	1	6	7	13
3	Black	0	2	9	11
4	White	1	9	3	12
4	Black	0	2	4	6
5	White	1	9	0	9
5	Black	0	4	3	7
6	White	1	17	0	17
6	Black	0	4	0	4

Among the 62 cases where the victim was white (coded as 1) and where the aggravation index was 1 (the lowest possible value), 2 defendants received the death penalty, while 60 got life in prison. For the six cases with aggravation index 4 and black victims (coded as 0), two received the death penalty, while four got life in prison.

The output of *Minitab's* logistic regression program (Minitab, a popular statistical software, is used in Abraham and Ledolter, 2006) is shown as follows:

7.6.1 Binary Logistic Regression: Minitab Program Output

```
Link Function: Logit

Response Information

Variable  Value       Count
DY        Event          59
          Non-event     303
Number    Total         362

Logistic Regression Table
                                                      Odds     95% CI
Predictor      Coef    SE Coef       Z      P  Ratio  Lower  Upper
Constant   -6.67598   0.757445  -8.81  0.000
VRaceC      1.81065   0.536116   3.38  0.001   6.11   2.14  17.49
Aggrav      1.53966   0.186726   8.25  0.000   4.66   3.23   6.72

Log-Likelihood = -56.738
Test that all slopes are zero: G = 208.402, DF = 2, P-Value = 0.000

Goodness-of-Fit Tests

Method          Chi-Square  DF      P
Pearson            3.09395   9  0.960
```

```
Deviance                3.88158   9  0.919
Hosmer-Lemeshow         0.11147   2  0.946
```

7.6.2 Interpretation of Results and Analysis with R

The maximum likelihood estimate of the effect of race of victim is 1.81, leading to the odds ratio $\exp(1.81) = 6.11$. The estimate is statistically different from 0; the probability value of the standardized test statistic is 0.001 and much smaller than any reasonable significance level. Conditioning on the severity of the crime (the other variable in the model), the odds of receiving the death sentence when the victim is White (coded as 1) are 6.11 times the odds of getting the death sentence when the victim is Black (coded as 0). There is no doubt that the victim's race makes a big difference.

Furthermore, the aggravation of the crime has a significant impact. The severity of the crime increases the odds of receiving the death penalty. Each extra unit on the aggravation scale multiplies the odds of receiving the death penalty by the factor 4.66. This effect can be seen clearly from the figure shown below, which graphs the estimated probabilities of receiving the death penalty,

$$p = \frac{\exp(-6.68 + 1.81\text{VRaceC} + 1.54\text{Aggrav})}{1 + \exp(-6.68 + 1.81\text{VRaceC} + 1.54\text{Aggrav})},$$

as a function of aggravation, and does so separately for a White (VRaceC = 1) and Black (VRaceC = 0) victim.

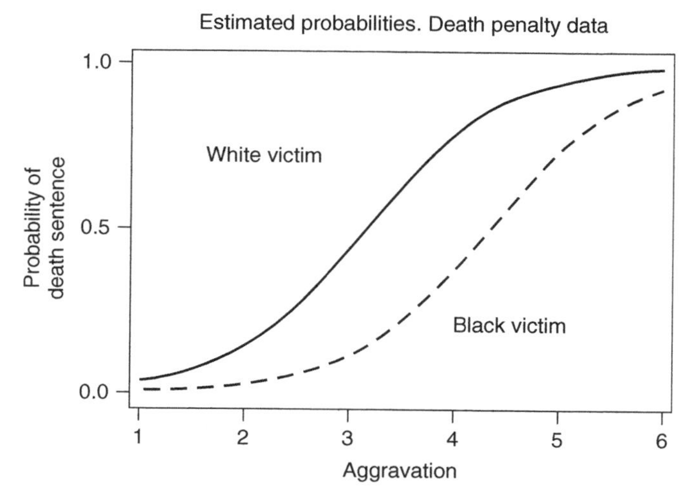

Next, we duplicate the results with the statistical software package R. We list the data again, but now after having converted the summarized data shown earlier back into the original 362 individual cases. In many applications, the data are given in terms of individual cases (such as shown in the following), and not in summarized form. The individual cases are stored in the file *DeathPenalty.csv*.

```
## analyzing individual observations
dpen <- read.csv("C:/DataMining/Data/DeathPenalty.csv")
dpen[1:4,]

  Agg VRace Death
1  1    1     1
2  1    1     1
3  1    1     0
4  1    1     0

dpen[359:362,]

359   6     0    1
360   6     0    1
361   6     0    1
362   6     0    1

m1=glm(Death~VRace+Agg,family=binomial,data=dpen)
m1

summary(m1)

Call:  glm(formula = Death ~ VRace + Agg, family = binomial, data = dpen)

Coefficients:
(Intercept)         VRace          Agg
     -6.676         1.811        1.540

Degrees of Freedom: 361 Total (i.e. Null);  359 Residual
Null Deviance:       321.9
Residual Deviance:   113.5         AIC: 119.5

summary(m1)

Call:
glm(formula = Death ~ VRace + Agg, family = binomial, data = dpen)

Deviance Residuals:
    Min        1Q    Median        3Q       Max
-1.7526  -0.2658  -0.1083   -0.1083    3.2069

Coefficients:
            Estimate Std. Error z value Pr(>|z|)
(Intercept)  -6.6760     0.7574  -8.814  < 2e-16 ***
VRace         1.8106     0.5361   3.377 0.000732 ***
Agg           1.5397     0.1867   8.246  < 2e-16 ***
---
Signif. codes:  0 '***' 0.001 '**' 0.01 '*' 0.05 '.' 0.1 ' ' 1

(Dispersion parameter for binomial family taken to be 1)

    Null deviance: 321.88  on 361  degrees of freedom
```

```
Residual deviance: 113.48  on 359  degrees of freedom
AIC: 119.48

Number of Fisher Scoring iterations: 7
```

The results are the same, with odds ratios $\exp(1.8106) = 6.11$ and $\exp(1.5397) = 4.66$. The estimated probabilities are easily plotted in R as well.

```
## calculating logits
exp(m1$coef[2])

 VRace
6.1144

exp(m1$coef[3])

     Agg
4.663011

## plotting probability of getting death penalty as a function of
## aggravation separately for black (in black) and white (in red)
## victim
fitBlack=dim(501)
fitWhite=dim(501)
ag=dim(501)
for (i in 1:501) {
ag[i]=(99+i)/100
fitBlack[i]=exp(m1$coef[1]+ag[i]*m1$coef[3])/(1+exp(m1$coef[1]+
+    ag[i]*m1$coef[3]))
fitWhite[i]=exp(m1$coef[1]+m1$coef[2]+ag[i]*m1$coef[3])/
+    (1+exp(m1$coef[1]+m1$coef[2]+ag[i]*m1$coef[3]))
}
plot(fitBlack~ag,type="l",col="black",ylab="Prob[Death]",
+    xlab="Aggravation",ylim=c(0,1),
+    main="red line for white victim; black line for black victim")
points(fitWhite~ag,type="l",col="red")
```

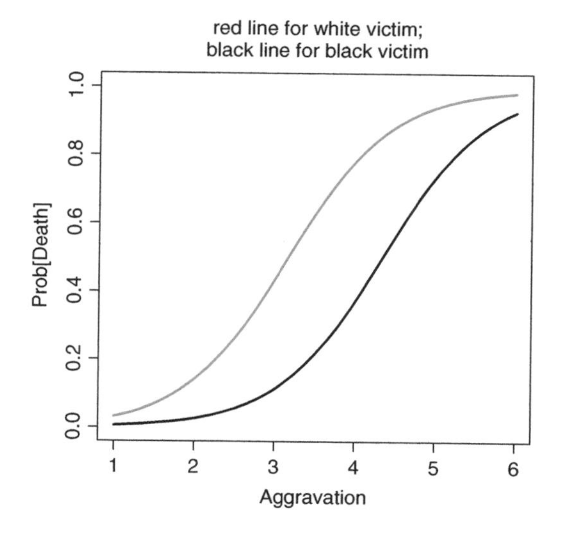

Below, we illustrate the analysis of the aggregated data using the weights option in R. The aggregated data are stored in the data file *DeathPenaltyOther.csv*.

```
## analyzing summarized data
dpenother <- read.csv("C:/DataMining/Data/DeathPenaltyOther.csv")
dpenother

   Agg VicRace VRace Death Freq
1    1   White     1     1    2
2    1   White     1     0   60
3    1   Black     0     1    1
4    1   Black     0     0  181
5    2   White     1     1    2
6    2   White     1     0   15
7    2   Black     0     1    1
8    2   Black     0     0   21
9    3   White     1     1    6
10   3   White     1     0    7
11   3   Black     0     1    2
12   3   Black     0     0    9
13   4   White     1     1    9
14   4   White     1     0    3
15   4   Black     0     1    2
16   4   Black     0     0    4
17   5   White     1     1    9
18   5   Black     0     1    4
19   5   Black     0     0    3
20   6   White     1     1   17
21   6   Black     0     1    4

m1=glm(Death~VRace+Agg,family=binomial,weights=Freq,data=dpenother)
m1
summary(m1)
exp(m1$coef[2])
exp(m1$coef[3])
```

7.7 EXAMPLE 2: DELAYED AIRPLANES

The data are taken from Shmueli et al. (2010). The data set consists of 2201 airplane flights in January 2004 from the Washington DC area into the NYC area. The characteristic of interest (the response) is whether or not a flight has been delayed by more than 15 min (coded as 0 for no delay, and 1 for delay).

The explanatory variables include

three different arrival airports (Kennedy, Newark, and LaGuardia);

three different departure airports (Reagan, Dulles, and Baltimore);

eight carriers;

a categorical variable for 16 different hours of departure (6 AM to 10 PM);

weather conditions (0 = good/1 = bad);
day of week (1 for Sunday and Monday; and 0 for all other days).

Here the objective is to identify flights that are likely to be delayed. The binary classification problem, which amounts to deciding whether a new case with given features is either a "success" (in this case, delayed) or a "failure" (in this case, on-time) is very common in data mining, and this is what makes the logistic regression approach so useful. The logistic regression model provides an estimate of the probability of success, and this probability is then used to classify cases into one of two groups, success or failure. For many classifications, it makes sense to use cutoff 0.5 on the probability of success. With this cutoff, we classify a case into the success group if its probability of success is 0.5 or larger, and we classify it as a failure otherwise. This is known as the majority rule as it classifies a case on the larger of the two probabilities.

The selection of the probability cutoff has important consequences for classification as it changes (i.e., trades off) the frequencies of the two possible misclassification errors: misclassifying a true success as failure, and misclassifying a true failure as success. In the next chapter, we will say more about this, and we will introduce a curve, the *receiver-operating characteristic (ROC) function*, to describe this trade-off. For a symmetric cost structure where either misclassification is equally costly, a cutoff 0.5 is appropriate. For asymmetric costs, where one type of misclassification is more costly than the other, it makes sense to select a probability cutoff that is different from 0.5 (see Section 7.9).

Evaluating the quality of a method or a model (here, the logistic regression) is important, and we discuss the evaluation of a classification rule in the context of this example. It is important that the classification is evaluated on future cases (out-of-sample evaluation), and not just on cases that have been used to estimate the model coefficients (in-sample evaluation). A retrospective in-sample classification on cases that have been used for model fitting tends to given an overly positive assessment.

With the large data sets that are commonly available in data mining, we can easily split the cases into two groups. One set of cases is used for model fitting, while the remaining set of cases is used for evaluation (testing). We use random sampling without replacement to select the cases for model fitting. Quite often, we split the data roughly in half (50/50 split). We saw in earlier examples that R includes very convenient methods to do this. Predictions for the cases in the evaluation set can then be compared to the actual outcomes, and the number of correct classifications and the number of misclassifications can be determined. This process can be repeated for different randomly chosen training sets, and the resulting misclassifications can be averaged to obtain stable estimates of misclassification errors that are not sensitive to just one particular selection of a training set.

Another way to evaluate the classifications on new data is to leave out just a single case from all available n cases and then obtain and evaluate the prediction of the case that has been held back. This process is repeated for a total of n times, with each of the n cases being held back once. An overall assessment of the quality of the classification is obtained by averaging the misclassification errors. This method is referred to as *cross-validation*. We have already used this approach in previous chapters.

For our illustration, we select 60% of the cases of our data set (1320 cases) for the fitting (training) data set; the remaining 40% of the cases (881 cases) become the evaluation data set. The success probability (proportion of delayed planes) in the training set is 0.198; the failure probability (proportion of on-time flights) is 0.802.

We assume a symmetric cost structure and probability cutoff 0.5. The naïve rule that does not incorporate any covariate information classifies every flight as being on-time as the estimated unconditional probability of a flight being on-time, 0.802, is larger than the cutoff 0.5. This rule never makes an error predicting a flight that is on-time, but it makes a 100% error when the flight is delayed. The naïve rule fails to identify the 167 delayed flights among the 881 flights of the evaluation data set; its misclassification error rate in the holdout sample is 167/881 = 0.189.

We now use for classification the logistic regression model that includes all explanatory variables as covariates. We estimate the logistic regression model on the training data set and classify a new case as success if its predicted probability of success is larger than 0.5. The attached R program illustrates how this can be done. The predictor variables are categorical variables or factors; the columns of factor levels need to be transformed into indicator variables, and the resulting indicator variables are then included in the logistic regression as explanatory variables. The logistic regression reduces the overall misclassification error in the holdout (evaluation/test) data set to 0.176, for a somewhat modest improvement over the naïve rule (0.189). A table of correct classifications and misclassifications is shown later. Among the 167 delayed flights, logistic regression identifies correctly 14 delayed flights (8.4%), but it misses 153/167 delayed flights (92.6%). Furthermore, the logistic regression model predicts 2 of the 714 on-time flights as being delayed.

```
library(car) ## needed to recode variables
set.seed(1)

## read and print the data
del <- read.csv("C:/DataMining/Data/FlightDelays.csv")
del[1:3,]

schedtime carrier deptime dest distance     date flightnumber origin weather
1      1455      OH    1455  JFK      184 1/1/2004         5935    BWI       0
2      1640      DH    1640  JFK      213 1/1/2004         6155    DCA       0
3      1245      DH    1245  LGA      229 1/1/2004         7208    IAD       0
  dayweek daymonth tailnu  delay
1       4        1 N940CA ontime
2       4        1 N405FJ ontime
3       4        1 N695BR ontime

## define hours of departure
del$sched=factor(floor(del$schedtime/100))
table(del$sched)
table(del$carrier)
```

```
table(del$dest)
table(del$origin)
table(del$weather)
table(del$dayweek)
table(del$daymonth)
table(del$delay)
del$delay=recode(del$delay,"'delayed'=1;else=0")
del$delay=as.numeric(levels(del$delay)[del$delay])
table(del$delay)
## Delay: 1=Monday; 2=Tuesday; 3=Wednesday; 4=Thursday;
## 5=Friday; 6=Saturday; 7=Sunday
## 7=Sunday and 1=Monday coded as 1
del$dayweek=recode(del$dayweek,"c(1,7)=1;else=0")
table(del$dayweek)
## omit unused variables
del=del[,c(-1,-3,-5,-6,-7,-11,-12)]
del[1:3,]

  carrier dest origin weather dayweek delay sched
1      OH  JFK    BWI       0       0     0    14
2      DH  JFK    DCA       0       0     0    16
3      DH  LGA    IAD       0       0     0    12

n=length(del$delay)
n
[1] 2201
n1=floor(n*(0.6))
n1
[1] 1320
n2=n-n1
n2
[1] 881
train=sample(1:n,n1)

## estimation of the logistic regression model
## explanatory variables: carrier, destination, origin, weather,
## day of week (weekday/weekend), scheduled hour of departure
## create design matrix; indicators for categorical variables
## (factors)
Xdel <- model.matrix(delay~.,data=del)[,-1]
Xdel[1:3,]

  carrierDH carrierDL carrierMQ carrierOH carrierRU carrierUA carrierUS destJFK
1         0         0         0         1         0         0         0       1
2         1         0         0         0         0         0         0       1
3         1         0         0         0         0         0         0       0
  destLGA originDCA originIAD weather dayweek sched7 sched8 sched9 sched10
1       0         0         0       0       0      0      0      0       0
2       0         1         0       0       0      0      0      0       0
3       1         0         1       0       0      0      0      0       0
  sched11 sched12 sched13 sched14 sched15 sched16 sched17 sched18 sched19
1       0       0       0       1       0       0       0       0       0
2       0       0       0       0       0       1       0       0       0
3       0       1       0       0       0       0       0       0       0
  sched20 sched21
1       0       0
2       0       0
3       0       0

xtrain <- Xdel[train,]
```

```
xnew <- Xdel[-train,]
ytrain <- del$delay[train]
ynew <- del$delay[-train]
m1=glm(delay~.,family=binomial,data=data.frame(delay=ytrain,xtrain))
summary(m1)

Call:
glm(formula = delay ~ ., family = binomial, data = data.frame(delay = ytrain,
    xtrain))

Deviance Residuals:
    Min       1Q   Median       3Q      Max
-1.3065  -0.6850  -0.5193  -0.2764   2.6671

Coefficients:
              Estimate Std. Error z value Pr(>|z|)
(Intercept)  -0.506462   0.674045  -0.751 0.452426
carrierDH    -1.109683   0.587743  -1.888 0.059020 .
carrierDL    -1.687451   0.521459  -3.236 0.001212 **
carrierMQ    -0.510547   0.496381  -1.029 0.303697
carrierOH    -2.262073   0.908804  -2.489 0.012808 *
carrierRU    -0.838344   0.424301  -1.976 0.048175 *
carrierUA    -1.602981   0.955408  -1.678 0.093387 .
carrierUS    -1.941613   0.529565  -3.666 0.000246 ***
destJFK      -0.005843   0.317887  -0.018 0.985336
destLGA       0.171454   0.327708   0.523 0.600841
originDCA    -0.800683   0.413647  -1.936 0.052908 .
originIAD    -0.319476   0.401076  -0.797 0.425714
weather      17.881818 500.451538   0.036 0.971497
dayweek       0.669711   0.161234   4.154 3.27e-05 ***
sched7       -0.168093   0.515374  -0.326 0.744305
sched8        0.338228   0.487051   0.694 0.487406
sched9       -0.450550   0.602829  -0.747 0.454826
sched10      -0.382502   0.601444  -0.636 0.524794
sched11      -0.578642   0.828878  -0.698 0.485113
sched12       0.614203   0.467384   1.314 0.188803
sched13      -0.232381   0.504607  -0.461 0.645143
sched14       0.973601   0.430169   2.263 0.023617 *
sched15       0.778466   0.454437   1.713 0.086706 .
sched16       0.528690   0.452783   1.168 0.242951
sched17       0.605440   0.422702   1.432 0.152056
sched18       0.169869   0.587675   0.289 0.772541
sched19       0.682830   0.500860   1.363 0.172783
sched20       0.922454   0.680325   1.356 0.175131
sched21       0.883470   0.441079   2.003 0.045180 *
---
Signif. codes:  0 '***' 0.001 '**' 0.01 '*' 0.05 '.' 0.1 ' ' 1

(Dispersion parameter for binomial family taken to be 1)

    Null deviance: 1312.7  on 1319  degrees of freedom
Residual deviance: 1134.9  on 1291  degrees of freedom
AIC: 1192.9

Number of Fisher Scoring iterations: 15

## prediction: predicted default probabilities for cases in test set
ptest <- predict(m1,newdata=data.frame(xnew),type="response")
data.frame(ynew,ptest)[1:10,]
## first column in list represents the case number of the test
## element
```

```
    ynew     ptest
1      0 0.1417566
5      0 0.1046429
6      0 0.1675298
9      0 0.2081648
10     0 0.2576931
12     0 0.1164139
15     0 0.1359269
17     0 0.1300306
21     0 0.1094095
22     0 0.1325476

plot(ynew~ptest)

## coding as 1 if probability 0.5 or larger
gg1=floor(ptest+0.5) ## floor function; see help command
ttt=table(ynew,gg1)
ttt

gg1
    0   1
0 712   2     # on-time flight in test data set
1 153  14     # delayed flight in test data set

error=(ttt[1,2]+ttt[2,1])/n2
error

0.1759364
```

7.7.1 The Lift

The lift chart explained and shown below gives us a quick way of identifying those flights that are most likely to be delayed. For that, we order the cases in the test (evaluation) sample according to their predicted probabilities of success. Cases with large probabilities of success are listed first. Next to them we print the actual results; we assume that we know these as we are evaluating what would have happened if we had used this rule. We see that the cases we predict as successes (cases with probabilities 0.5 or larger) are in fact actual successes (delayed flights). The *lift curve* graphs the cumulative number of successes (after having sorted the cases according to their predicted values in decreasing order) against the number of cases. The reference line on this plot graphs the expected number of delayed flights, assuming that the probability of delay is estimated by the proportion of delayed flights in the evaluation sample, against the number of cases. The reference line expresses the performance of the naïve model. With 10 flights, for example, the expected number of delayed flights is $10\,\overline{p}$, where $\overline{p}$ is the proportion of delayed flights in the evaluation sample (here it is 0.189). At the very end, the lift curve and the reference line meet. However, at the beginning, our logistic regression leads to a “lift.” Pick the 10 cases with the largest estimated success probabilities, as an example. All 10 cases turn out to be delayed. A good lift curve is one that has a very steep incline at the beginning of the curve. If the lift is close to the reference line, then there is not much point in using the estimated model for classification.

A major objective of classification rules is to identify those flights that will be delayed, the buyers who are most likely to buy, or the companies that are most likely to go bankrupt. We are not particularly interested in finding potential customers who do not end up buying, as with tight budgets, we are not going to waste precious selling efforts on people who are unlikely to buy. Defaults create big losses and we want to identify those companies that are likely to go bankrupt. But we may not be interested in identifying companies that do not go bankrupt, as those companies will not create large losses for us. Such circumstances may actually indicate that the costs of the two misclassification errors are not the same and that the probability cutoff should be different from 0.5; more on that is discussed in Section 7.9.

In our example of airline delays, the overall misclassification rate of the logistic regression is not that different from that of the naïve strategy that considers all flights as being on-time. But, as the lift curve shows, flights with the largest probabilities of being delayed are classified correctly. The logistic regression is quite successful in identifying those flights as being delayed. One can think of those cases as the "low hanging fruit" for easy harvest.

Below we list the 20 cases with the largest probabilities of success, together with their actual outcomes. Among the 16 cases with probabilities larger than 0.50, 14 flights are actually delayed. The lift curve shows that the model gives us an advantage (i.e., a nice "lift") in detecting the most obvious flights that are going to be delayed.

```
bb=cbind(ptest,ynew)
bb
bb1=bb[order(ptest,decreasing=TRUE),]
bb1
## order cases in test set according to their success prob
## actual outcome shown next to it
## overall success (delay) prob in the evaluation data set
xbar=mean(ynew)
xbar

## calculating the lift
## cumulative 1's sorted by predicted values
## cumulative 1's using the average success prob from
## evaluation set
axis=dim(n2)
ax=dim(n2)
ay=dim(n2)
axis[1]=1
ax[1]=xbar
ay[1]=bb1[1,2]
for (i in 2:n2) {
axis[i]=i
ax[i]=xbar*i
ay[i]=ay[i-1]+bb1[i,2]
}
```

```
aaa=cbind(bb1[,1],bb1[,2],ay,ax)
aaa[1:100,]
```

```
     Prob(S)  Actual
1908 1.0000000 1  1   0.1895573
1850 1.0000000 1  2   0.3791146
1829 1.0000000 1  3   0.5686720
1877 1.0000000 1  4   0.7582293
1893 0.9999999 1  5   0.9477866
1868 0.9999999 1  6   1.1373439
1834 0.9999999 1  7   1.3269012
2102 0.9999999 1  8   1.5164586
1911 0.9999999 1  9   1.7060159
1823 0.9999999 1 10   1.8955732
1916 0.9999999 1 11   2.0851305
209  0.5740773 1 12   2.2746879
820  0.5740773 1 13   2.4642452
1254 0.5740773 1 14   2.6538025
1325 0.5740773 0 14   2.8433598
1849 0.5740773 0 14   3.0329171
288  0.4920023 1 15   3.2224745
821  0.4825933 0 15   3.4120318
1785 0.4728407 0 15   3.6015891
167  0.4698229 1 16   3.7911464
695  0.4698229 0 16   3.9807037
. . .
```

```
plot(axis,ay,xlab="number of cases",ylab="number of
+       successes", main="Lift: Cum successes sorted by
+       pred val/success prob")
points(axis,ax,type="l")
```

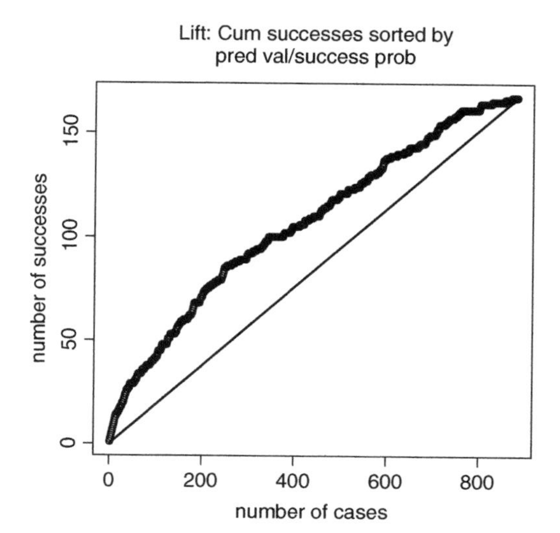

The R program that we used to carry out the calculations, the logistic regression as well as the lift curve, can be found on the webpage that accompanies this book.

7.8 EXAMPLE 3: LOAN ACCEPTANCE

The data are also taken from Shmueli et al. (2010). The data set contains information on 5000 loan applications. The response is whether or not an offered loan had been accepted on an earlier occasion. The explanatory variables include

Age of customer;
Experience: professional experience in years;
Income of customer;
Family size of customer;
CCAvg: average monthly credit card spending;
Mortgage: size of mortgage;
SecuritiesAccount: No/Yes;
CDAccount: No/Yes;
Online: No/Yes;
CreditCard: No/Yes;
Educational level: three categories (undergraduate, graduate, professional).

Sixty percent of the data (i.e., 3000 of the 5000 cases) are used for estimation (training set); the remaining data (40%, or 2000 cases) are used for evaluation (as a test data set). The data are separated into these two sets at random.

The overall success probability (with success defined as the acceptance of an earlier loan) among all 5000 observations is 0.096; the success probability in the training set is also 0.096. The logistic regression model, estimated on the training set of 3000 cases, is used to predict the probabilities of success in the evaluation data set.

The detailed logistic regression output (from R) shows that the relationships are fairly strong.

```
summary(m2)

Call:
glm(formula = response ~ ., family = binomial,
data = data.frame(response = ytrain, xtrain))

Deviance Residuals:
    Min        1Q    Median        3Q       Max
-2.9615  -0.1912  -0.0664  -0.0199    4.0850
```

```
Coefficients:
              Estimate Std. Error z value Pr(>|z|)
(Intercept) -1.255e+01  2.337e+00  -5.370 7.89e-08 ***
Age         -5.384e-04  8.602e-02  -0.006 0.995006
Exp          5.044e-03  8.535e-02   0.059 0.952877
Inc          6.140e-02  3.942e-03  15.577  < 2e-16 ***
Fam2        -8.712e-02  2.972e-01  -0.293 0.769448
Fam3         1.948e+00  3.286e-01   5.927 3.08e-09 ***
Fam4         1.551e+00  3.088e-01   5.022 5.11e-07 ***
CCAve        1.903e-01  6.079e-02   3.130 0.001747 **
Mort         3.017e-04  7.525e-04   0.401 0.688445
SecAcc      -1.064e+00  4.188e-01  -2.541 0.011067 *
CD           3.660e+00  4.640e-01   7.888 3.08e-15 ***
Online      -7.200e-01  2.157e-01  -3.338 0.000844 ***
CreditCard  -1.051e+00  2.972e-01  -3.537 0.000404 ***
Educ2        4.022e+00  3.596e-01  11.183  < 2e-16 ***
Educ3        4.005e+00  3.580e-01  11.186  < 2e-16 ***
---
Signif. codes:  0 '***' 0.001 '**' 0.01 '*' 0.05 '.' 0.1 ' ' 1

(Dispersion parameter for binomial family taken to be 1)

    Null deviance: 1910.65  on 2999  degrees of freedom
Residual deviance:  692.86  on 2985  degrees of freedom
AIC: 722.86

Number of Fisher Scoring iterations: 8
```

The estimated model is used to predict the probability of success for the cases in the evaluation data set. Cases with success probabilities larger than 0.50 are classified as "success." The logistic regression classification approach is able to reduce the classification error in the holdout set to 0.038. Of the 189 previously accepted loans in the test (evaluation) data set, we predict 128 correctly (68%). Of the 1811 nonaccepted loans, we predict 1796 correctly (99%).

```
## coding as 1 if probability 0.5 or larger
gg1=floor(ptest+0.5)
ttt=table(ynew,gg1)
```

```
        gg1
ynew     0    1
   0 1796   15     ## non accepted loans
   1   61  128     ## accepted loans
```

```
error=(ttt[1,2]+ttt[2,1])/n2
error
```

```
[1] 0.038
```

Below we show the 20 cases with the largest probabilities of success, as well as their actual outcomes. All 20 cases are successes. The lift that we get from the logistic regression is quite strong. The lift curve in the graph is quite steep in the beginning and exceeds the reference line that connects the expected number of successes using the relative frequency of success in the holdout period as an estimate of the probability of success.

```
aaa[1:20,]

     Prob(S)   Actual
1133 0.9999017 1   1 0.0945
1180 0.9997186 1   2 0.1890
758  0.9996639 1   3 0.2835
626  0.9995217 1   4 0.3780
300  0.9994912 1   5 0.4725
216  0.9993812 1   6 0.5670
1598 0.9991014 1   7 0.6615
840  0.9987773 1   8 0.7560
1725 0.9987519 1   9 0.8505
373  0.9984118 1  10 0.9450
1605 0.9983513 1  11 1.0395
1121 0.9977243 1  12 1.1340
764  0.9975621 1  13 1.2285
21   0.9972330 1  14 1.3230
888  0.9969690 1  15 1.4175
1531 0.9969300 1  16 1.5120
643  0.9968237 1  17 1.6065
714  0.9961473 1  18 1.7010
1741 0.9960146 1  19 1.7955
1837 0.9949948 1  20 1.8900
```

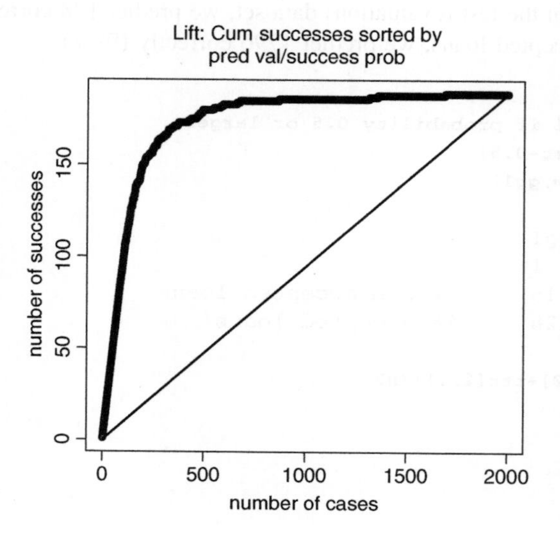

7.9 EXAMPLE 4: GERMAN CREDIT DATA

The German credit data set was obtained from the UCI (University of California at Irwin) Machine Learning Repository (Asuncion and Newman, 2007). The data set, which contains attributes and outcomes on 1000 loan applications, was provided in 1994 by Professor Dr. Hans Hofmann of the Institut fuer Statistik und Oekonometrie at the University of Hamburg. It has served as an important test data set for several credit-scoring algorithms. A description of the variables is given in the word file *germancreditDescription* that can also be found on the webpage for this book.

Assume that, on average, lending into default is five times as costly as not lending to a good debtor (assume that this latter cost is 1). Here default is defined as "success." Suppose we estimate a certain p for the probability of default. Then the expected costs are $5p$ if we make the loan, and $1(1-p)$ if we refuse the loan. Hence if $5p < 1-p$, we expect to lose less by loaning than by turning away business. This implies the following decision rule: make the loan if the probability of default $p < 1/6$. Predict default ("success") whenever $p > 1/6$. This is an example where knowledge about the relative costs of misclassification impacts the choice of the probability cutoff.

Below we analyze the German credit data; this data set will also be used in later chapters (e.g., Chapter 9 on nearest neighbor classification). The two outcomes are success (defaulting on the loan), and failure (not defaulting). We use logistic regression to estimate the probability of default, using continuous variables (duration, amount, installment, age) and categorical variables (loan history, purpose, foreign, rent) as explanatory variables.

We randomly select 900 of the 1000 cases for the training set, and put the remaining 100 cases into the test set. The performance of the logistic regression model for the test data set is as follows: the logistic regression recognizes 23 of the 28 defaults (or 82%), but predicts as defaults 42 of the 72 good loans (or 58%).

The R program and its output are listed below. The R program can also be found on the webpage that accompanies this book.

```
#### ******* German Credit Data ******* ####
#### ******* data on 1000 loans ******* ####

## read data and create some 'interesting' variables
credit <- read.csv("C:/DataMining/Data/germancredit.csv")
credit

Default: 0 (no) and 1 (yes)

Attribute 1: (qualitative)     Status of existing checking account
    A11 : ... < 0 DM
    A12 : 0 <= ... < 200 DM
    A13 : ... >= 200 DM / salary assignments for at least 1 year
    A14 : no checking account

Attribute 2: (numerical)       Duration in month
```

```
Attribute 3: (qualitative)    Credit history
    A30 : no credits taken/ all credits paid back duly
    A31 : all credits at this bank paid back duly
    A32 : existing credits paid back duly till now
    A33 : delay in paying off in the past
    A34 : critical account/ other credits existing (not at this bank)

Attribute 4: (qualitative)    Purpose
    A40 : car (new)
    A41 : car (used)
    A42 : furniture/equipment
    A43 : radio/television
    A44 : domestic appliances
    A45 : repairs
    A46 : education
    A47 : (vacation - does not exist?)
    A48 : retraining
    A49 : business
    A410 : others

Attribute 5: (numerical)      Credit amount

Attribute 6: (qualitative)    Savings account/bonds
    A61 : … < 100 DM
    A62 : 100 <= … < 500 DM
    A63 : 500 <= … < 1000 DM
    A64 : .. >= 1000 DM
    A65 : unknown/ no savings account

Attribute 7: (qualitative)    Present employment since
    A71 : unemployed
    A72 : … < 1 year
    A73 : 1 <= … < 4 years
    A74 : 4 <= … < 7 years
    A75 : .. >= 7 years

Attribute 8: (numerical)      Installment rate in percentage of
    disposable income

Attribute 9: (qualitative)    Personal status and sex
    A91 : male : divorced/separated
    A92 : female : divorced/separated/married
    A93 : male : single
    A94 : male : married/widowed
    A95 : female : single

Attribute 10: (qualitative)   Other debtors / guarantors
    A101 : none
    A102 : co-applicant
    A103 : guarantor

Attribute 11: (numerical)     Present residence since

Attribute 12: (qualitative)   Property
    A121 : real estate
    A122 : if not A121 : building society savings agreement/ life insurance
    A123 : if not A121/A122 : car or other, not in attribute 6
    A124 : unknown / no property

Attribute 13: (numerical)     Age in years
```

```
Attribute 14: (qualitative)   Other installment plans
    A141 : bank
    A142 : stores
    A143 : none

Attribute 15: (qualitative)   Housing
    A151 : rent
    A152 : own
    A153 : for free

Attribute 16: (numerical)     Number of existing credits at this bank

Attribute 17: (qualitative) Job
    A171 : unemployed/ unskilled - non-resident
    A172 : unskilled - resident
    A173 : skilled employee / official
    A174 : management/ self-employed/
highly qualified employee/ officer

Attribute 18: (numerical)     Number of people being liable to provide
    maintenance for

Attribute 19: (qualitative)   Telephone
    A191 : none
    A192 : yes, registered under the customers name

Attribute 20: (qualitative)   foreign worker
    A201 : yes
    A202 : no
```

```
credit$Default <- factor(credit$Default)


## re-level the credit history and a few other variables
credit$history = factor(credit$history, levels=c("A30","A31","A32",
+     "A33","A34"))
levels(credit$history) = c("good","good","poor","poor","terrible")
credit$foreign <- factor(credit$foreign, levels=c("A201","A202"),
+     labels=c("foreign","german"))
credit$rent <- factor(credit$housing=="A151")
credit$purpose <- factor(credit$purpose, levels=c("A40","A41","A42",
+     "A43","A44","A45","A46","A47","A48","A49","A410"))
levels(credit$purpose) <- c("newcar","usedcar",rep("goods/repair",4),
+     "edu",NA, "edu","biz","biz")


## for demonstration, cut the dataset to these variables
credit <- credit[,c("Default","duration","amount","installment",
+     "age","history","purpose","foreign","rent")]
credit[1:3,]
```

```
  Default duration amount installment age  history      purpose foreign  rent
1       0        6   1169           4  67 terrible goods/repair foreign FALSE
2       1       48   5951           2  22     poor goods/repair foreign FALSE
3       0       12   2096           2  49 terrible          edu foreign FALSE
```

```
summary(credit) # check out the data
```

```
Default    duration         amount         installment          age
 0:700   Min.   : 4.0   Min.   :  250   Min.   :1.000   Min.   :19.00
```

```
1:300   1st Qu.:12.0   1st Qu.: 1366   1st Qu.:2.000   1st Qu.:27.00
        Median :18.0   Median : 2320   Median :3.000   Median :33.00
        Mean   :20.9   Mean   : 3271   Mean   :2.973   Mean   :35.55
        3rd Qu.:24.0   3rd Qu.: 3972   3rd Qu.:4.000   3rd Qu.:42.00
        Max.   :72.0   Max.   :18424   Max.   :4.000   Max.   :75.00
     history              purpose          foreign        rent
 good    : 89   newcar       :234   foreign:963   FALSE:821
 poor    :618   usedcar      :103   german : 37   TRUE :179
 terrible:293   goods/repair:495
                edu          : 59
                biz          :109

## create a design matrix
## factor variables are turned into indicator variables
## the first column of ones is omitted
Xcred <- model.matrix(Default~.,data=credit)[,-1]
Xcred[1:3,]

  duration amount installment age historypoor historyterrible purposeusedcar
1        6   1169           4  67           0               1              0
2       48   5951           2  22           1               0              0
3       12   2096           2  49           0               1              0
  purposegoods/repair purposeedu purposebiz foreigngerman rentTRUE
1                   1          0          0             0        0
2                   1          0          0             0        0
3                   0          1          0             0        0

## creating training and prediction datasets
## select 900 rows for estimation and 100 for testing
set.seed(1)
train <- sample(1:1000,900)
xtrain <- Xcred[train,]
xnew <- Xcred[-train,]
ytrain <- credit$Default[train]
ynew <- credit$Default[-train]
credglm=glm(Default~.,family=binomial,data=data.frame(Default=ytrain,
+   xtrain))
summary(credglm)

Call:
glm(formula = Default ~ ., family = binomial, data = data.frame
    (Default = ytrain, xtrain))

Deviance Residuals:
    Min       1Q   Median       3Q      Max
-2.2912  -0.7951  -0.5553   0.9922   2.2601

Coefficients:
                     Estimate Std. Error z value Pr(>|z|)
(Intercept)         -2.705e-01  4.833e-01  -0.560 0.575693
duration             2.721e-02  8.464e-03   3.215 0.001303 **
amount               9.040e-05  3.854e-05   2.346 0.018987 *
installment          2.228e-01  8.064e-02   2.763 0.005722 **
age                 -1.327e-02  7.704e-03  -1.723 0.084961 .
historypoor         -1.102e+00  2.641e-01  -4.173 3.01e-05 ***
historyterrible     -1.860e+00  3.007e-01  -6.184 6.25e-10 ***
purposeusedcar      -1.793e+00  3.555e-01  -5.043 4.58e-07 ***
purposegoods.repair -7.447e-01  1.976e-01  -3.769 0.000164 ***
purposeedu          -6.809e-02  3.401e-01  -0.200 0.841325
purposebiz          -7.342e-01  2.916e-01  -2.518 0.011812 *
```

```
foreigngerman        -1.363e+00  6.638e-01  -2.053 0.040054 *
rentTRUE              7.011e-01  2.075e-01   3.378 0.000730 ***
---
Signif. codes:  0 '***' 0.001 '**' 0.01 '*' 0.05 '.' 0.1 ' ' 1

(Dispersion parameter for binomial family taken to be 1)

    Null deviance: 1102.92  on 899  degrees of freedom
Residual deviance:  955.21  on 887  degrees of freedom
AIC: 981.21

Number of Fisher Scoring iterations: 5

## Now to prediction: what are the underlying default probabilities
## for cases in the test set

ptest <- predict(credglm, newdata=data.frame(cnew),type="response")
data.frame(ynew,ptest)
   ynew      ptest
16     1 0.29285579
18     0 0.65821629
19     1 0.25582878
26     0 0.11059259
27     0 0.39183380
43     0 0.23061764
44     0 0.12544267
50     0 0.25283649
61     0 0.16128056
. . .

## What are our misclassification rates on that training set?
## We use probability cutoff 1/6
## coding as 1 (predicting default) if probability 1/6 or larger
gg1=floor(ptest+(5/6))
ttt=table(ynew,gg1)
ttt

      gg1
ynew  0  1
   0 30 42 ## no default in evaluation (test) data set
   1  5 23 ## default in evaluation (test) data set

error=(ttt[1,2]+ttt[2,1])/100
error
[1] 0.47
```

REFERENCES

Abraham, B. and Ledolter, J.: *Introduction to Regression Modeling*. Belmont, CA: Duxbury Press, 2006.

Asuncion, A. and Newman, D.J.: *UCI Machine Learning Repository*. Irvine, CA: University of California, School of Information and Computer Science, 2007. Available at http://www.ics.uci.edu/~mlearn/MLRepository.html. Accessed 2013 Jan 10.

Shmueli G., Patel, N.R., and Bruce, P.C.: *Data Mining for Business Intelligence*. Second edition. Hoboken, NJ: John Wiley & Sons, Inc., 2010.

CHAPTER 8

Binary Classification, Probabilities, and Evaluating Classification Performance

8.1 BINARY CLASSIFICATION

Many decision problems can be reduced to a binary classification. Let us define 1 as "success" and 0 as "failure." In order to make a decision on a new case, we need to estimate its success probability $p = P[y = 1]$. In a previous chapter (Chapter 7), we have illustrated how to estimate this probability through logistic regression. Once we know $\widehat{p}$, the estimated probability of $y = 1$, and after having decided on an appropriate cutoff for the probability, we can make a decision and assess its risk.

8.2 USING PROBABILITIES TO MAKE DECISIONS

There are two ways to go wrong in a binary problem: A false positive error occurs if we predict $\widehat{y} = 1$ when $y = 0$. A false negative error occurs if we predict $\widehat{y} = 0$ when $y = 1$.

The *false positive rate* is defined as the number of misclassified negatives (actual negatives classified as positives) divided by the number of negatives. Or, to put this in another way, the false positive rate is the proportion of absent events that yield positive test outcomes. It is the conditional probability of a positive test result given that the event we look for is absent.

The *false negative rate* is the number of misclassified positives (actual positives classified as negatives) divided by the number of positives. The false negative rate is the proportion of present events that yield negative test outcomes. It is the conditional probability of a negative test result given that the event we look for has taken place.

The cutoff on the (estimated) probability of success determines how items are classified. How do we decide on this cutoff? In Section 7.9 (German credit data),

Data Mining and Business Analytics with R, First Edition. Johannes Ledolter.

we discussed whether or not to extend a loan. Assume that, on average, lending into default is 5 times as costly as not lending to a good debtor. Here default is defined as "success." Suppose p is the probability of default. Then the expected costs are $5p$ if we make the loan, and $1(1-p)$ if we refuse to make the loan. If $5p < 1-p$, we expect to lose less by loaning than by turning away business. This implies the following decision rule: make the loan if the probability of default $p < 1/6$. Refuse the loan and predict default ("success") whenever $p > 1/6$. In this example, the cutoff depends on the two different costs of misclassifications. If the two misclassification costs are the same, we classify a new case as success if its estimated probability of success exceeds 0.5 (we refer to this as the *majority rule*).

8.3 SENSITIVITY AND SPECIFICITY

The classification rule depends on the class probability and on the probability cutoff.

We call a classification rule *sensitive* if it predicts 1 for most $y = 1$ observations, and *specific* if it predicts 0 for most $y = 0$ observations. *Sensitivity* is the proportion of observed positives classified correctly (this is the same as 1 − the false negative rate). *Specificity* is the proportion of observed negatives classified correctly (this is the same as 1 − the false positive rate). Sensitivity and specificity depend on the cutoff for the estimated probability. It is common to calculate sensitivity and specificity for changing cutoffs, and plot sensitivity (i.e., the proportion of observed positives classified correctly) against 1 − specificity (i.e., the false positive rate). The resulting function is called the *receiver operating characteristic (ROC) function* (the terminology comes from signal processing). The cutoff on the probability affects the trade-off between sensitivity and specificity, and the ROC curve illustrates this graphically. Ideally, we want the ROC curve to be attracted to the top-left corner of the graph, with a very fast rise to the upper horizontal line at 1 and a large area under the curve. We want high sensitivity (predicting positives correctly) for a low false positive rate.

We can use the ROC curve to select among several classification methods. A commonly used design criterion looks at the area under the ROC curve and selects the classification method that maximizes this area.

8.4 EXAMPLE: GERMAN CREDIT DATA

Let us go back to Section 7.9 on logistic regression where we analyzed the German credit data. For cutoff 1/6, the decision rule performed as follows in the holdout sample of 100 firms: 23 of 28 (82%) of defaulters are being recognized, while 30 of 72 (42%) of good loans are being made. This implies sensitivity 0.82 and specificity $30/72 = 0.42$. The coordinates $(\text{sensitivity} = 0.82, 1 - \text{specificity} = 0.58)$ define one point on the ROC curve. Another point on this curve is obtained with cutoff 0.50. Then, sensitivity $= 8/28 = 0.29$ and specificity $= 68/72 = 0.94$, for another point on the ROC curve (sensitivity = 0.29, 1 − specificity = 0.06). The R program given below, *roc*, calculates and graphs the ROC curve. We get this curve by varying the cutoff on the probability. The ROC curves that assess the predictive quality of the classification rule on the holdout sample of 100 observations (the left graph) and on the complete data set of all 1000 cases (in-sample evaluation; right graph) are shown below. Specified values on sensitivity and specificity imply a certain value for the probability cutoff. Of course, for certain data and models no cutoff may achieve the given desired properties on sensitivity and specificity; this implies that the desired sensitivity and specificity cannot be attained.

We have written the R macro *roc* for graphing the ROC curve. Alternatively, one can use functions in the R package **ROCR**. The syntax for **ROCR** is also attached.

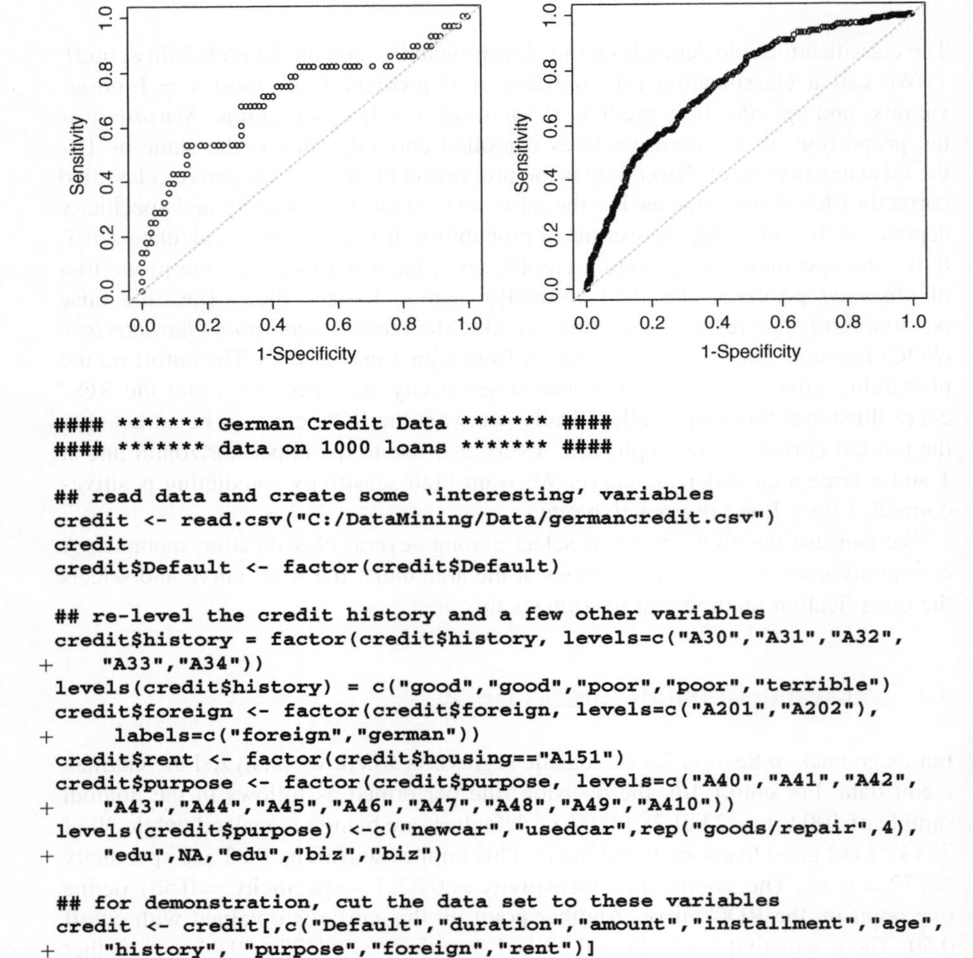


```
#### ******* German Credit Data ******* ####
#### ******* data on 1000 loans ******* ####

## read data and create some 'interesting' variables
credit <- read.csv("C:/DataMining/Data/germancredit.csv")
credit
credit$Default <- factor(credit$Default)

## re-level the credit history and a few other variables
credit$history = factor(credit$history, levels=c("A30","A31","A32",
+     "A33","A34"))
levels(credit$history) = c("good","good","poor","poor","terrible")
credit$foreign <- factor(credit$foreign, levels=c("A201","A202"),
+     labels=c("foreign","german"))
credit$rent <- factor(credit$housing=="A151")
credit$purpose <- factor(credit$purpose, levels=c("A40","A41","A42",
+     "A43","A44","A45","A46","A47","A48","A49","A410"))
levels(credit$purpose) <-c("newcar","usedcar",rep("goods/repair",4),
+   "edu",NA,"edu","biz","biz")

## for demonstration, cut the data set to these variables
credit <- credit[,c("Default","duration","amount","installment","age",
+   "history", "purpose","foreign","rent")]
```

```
credit
summary(credit) # check out the data

## create a design matrix
## factor variables are turned into indicator variables
## the first column of ones is omitted
Xcred <- model.matrix(Default~.,data=credit)[,-1]
Xcred[1:3,]

  duration amount installment age historypoor historyterrible purposeusedcar
1        6   1169           4  67           0               1              0
2       48   5951           2  22           1               0              0
3       12   2096           2  49           0               1              0
  purposegoods/repair purposeedu purposebiz foreigngerman rentTRUE
1                   1          0          0             0        0
2                   1          0          0             0        0
3                   0          1          0             0        0

## creating training and prediction datasets
## select 900 rows for estimation and 100 for testing
set.seed(1)
train <- sample(1:1000,900)
xtrain <- Xcred[train,]
xnew <- Xcred[-train,]
ytrain <- credit$Default[train]
ynew <- credit$Default[-train]
credglm=glm(Default~.,family=binomial,data=data.frame(Default=ytrain,
+   xtrain))
summary(credglm)

Call:
glm(formula = Default ~ ., family = binomial,
    data = data.frame(Default = ytrain, xtrain))

Deviance Residuals:
    Min       1Q   Median       3Q      Max
-2.2912  -0.7951  -0.5553   0.9922   2.2601

Coefficients:
                     Estimate Std. Error z value Pr(>|z|)
(Intercept)        -2.705e-01  4.833e-01  -0.560 0.575693
duration            2.721e-02  8.464e-03   3.215 0.001303 **
amount              9.040e-05  3.854e-05   2.346 0.018987 *
installment         2.228e-01  8.064e-02   2.763 0.005722 **
age                -1.327e-02  7.704e-03  -1.723 0.084961 .
historypoor        -1.102e+00  2.641e-01  -4.173 3.01e-05 ***
historyterrible    -1.860e+00  3.007e-01  -6.184 6.25e-10 ***
purposeusedcar     -1.793e+00  3.555e-01  -5.043 4.58e-07 ***
purposegoods.repair -7.447e-01 1.976e-01  -3.769 0.000164 ***
purposeedu         -6.809e-02  3.401e-01  -0.200 0.841325
purposebiz         -7.342e-01  2.916e-01  -2.518 0.011812 *
foreigngerman      -1.363e+00  6.638e-01  -2.053 0.040054 *
rentTRUE            7.011e-01  2.075e-01   3.378 0.000730 ***
---
Signif. codes:  0 '***' 0.001 '**' 0.01 '*' 0.05 '.' 0.1 ' ' 1

(Dispersion parameter for binomial family taken to be 1)
```

```
    Null deviance: 1102.92  on 899  degrees of freedom
Residual deviance:  955.21  on 887  degrees of freedom
AIC: 981.21

Number of Fisher Scoring iterations: 5
```

```
## Now to prediction: what are the underlying default probabilities
## for cases in the test set
ptest <- predict(credglm, newdata=data.frame(xnew),type="response")
data.frame(ynew,ptest)
```

```
   ynew      ptest
16    1 0.29285579
18    0 0.65821629
19    1 0.25582878
26    0 0.11059259
27    0 0.39183380
43    0 0.23061764
44    0 0.12544267
50    0 0.25283649
61    0 0.16128056
. . .
```

```
## What are our misclassification rates on that training set?
## We use probability cutoff 1/6
## coding as 1 (predicting default) if probability 1/6 or larger
gg1=floor(ptest+(5/6))
ttt=table(ynew,gg1)
ttt
```

```
     gg1
ynew  0  1
   0 30 42 ## no default in evaluation (test) data set
   1  5 23 ## default in evaluation (test) data set
```

```
truepos < - ynew==1 & ptest >=cut
trueneg < - ynew==0 & ptest < cut
# Sensitivity (predict default when it does happen)
sum(truepos)/sum(ynew==1)
```

```
[1] 0.8214286
```

```
# Specificity (predict no default when it does not happen)
sum(trueneg)/sum(ynew==0)
```

```
[1] 0.4166667
```

```
## Next, we use probability cutoff 1/2
## coding as 1 if probability 1/2 or larger
cut=1/2
gg1=floor(ptest+(1-cut))
ttt=table(ynew,gg1)
ttt
error=(ttt[1,2]+ttt[2,1])/100
error
```

```
     gg1
ynew  0  1
```

```
   0 68  4
   1 20  8

truepos < - ynew==1 & ptest > =cut
trueneg < - ynew==0 & ptest < cut
# Sensitivity (predict default when it does happen)
sum(truepos)/sum(ynew==1)

[1] 0.2857143

# Specificity (predict no default when it does not happen)
sum(trueneg)/sum(ynew==0)

[1] 0.9444444

## R macro for plotting the ROC curve
## plot the ROC curve for classification of y with p
roc <- function(p,y){
  y <- factor(y)
  n <- length(p)
  p <- as.vector(p)
  Q < - p > matrix(rep(seq(0,1,length=500),n),ncol=500,byrow=TRUE)
  fp < - colSums((y==levels(y)[1])*Q)/sum(y==levels(y)[1])
  tp < - colSums((y==levels(y)[2])*Q)/sum(y==levels(y)[2])
  plot(fp, tp, xlab="1-Specificity", ylab="Sensitivity")
  abline(a=0,b=1,lty=2,col=8)
}

## ROC for hold-out period
roc(p=ptest,y=ynew)

## ROC for all cases (in-sample)
credglmall <- glm(credit$Default ~ Xcred,family=binomial)
roc(p=credglmall$fitted, y=credglmall$y)

## using the ROCR package to graph the ROC curves
library(ROCR)
## input is a data frame consisting of two columns
## predictions in first column and actual outcomes in the second

## ROC for hold-out period
predictions=ptest
labels=ynew
data=data.frame(predictions,labels)
data
## pred: function to create prediction objects
pred <- prediction(data$predictions,data$labels)
pred
## perf: creates the input to be plotted
## sensitivity and one minus specificity (the false positive rate)
perf <- performance(pred, "sens", "fpr")
perf
plot(perf)

## ROC for all cases (in-sample)
credglmall <- glm(credit$Default ~ Xcred,family=binomial)
predictions=credglmall$fitted
labels=credglmall$y
```

```
data=data.frame(predictions,labels)
pred <- prediction(data$predictions,data$labels)
perf <- performance(pred, "sens", "fpr")
plot(perf)
```

The R program can be found on the webpage that accompanies this book.

CHAPTER 9

Classification Using a Nearest Neighbor Analysis

In classification, we start with a training set of objects with information on their group membership (an object can be from one of $g \geq 2$ possible groups) and a set of their measurable characteristics (also referred to as the *features*). The information we have on the training set is used to predict the unknown group membership of a new set of objects that are solely described by their measurable characteristics.

EXAMPLE 9.1 We have available a set of glass shards from six possible glass types ($g = 6$ groups):

WinF: float glass window,
WinNF: nonfloat glass window,
Veh: vehicle window,
Con: container (bottles),
Tabl: tableware,
Head: vehicle headlamp.

We know the group membership of each shard and we have available measurements on shard characteristics such as the refractive index (RI), and percentages of Na, Mg, Al, Si, K, Ca, Ba, and Fe. The measurable variables are often referred to as the *features*. Here we are dealing with nine features. Features of the various objects can be visualized as lying in a feature space of dimension $d = 9$.

We are presented with a new shard and its measured characteristics, but not its group membership. How can we infer (predict) the new shard's membership?

EXAMPLE 9.2 We have available the outcomes on several loans (no default/default; here there are just $g = 2$ groups) that are characterized by their

Data Mining and Business Analytics with R, First Edition. Johannes Ledolter.

features such as the amount of the loan borrowed, duration of the loan, interest rate of the loan, and characteristics of the borrower (income, whether borrower has a job, and so on)

A new customer walks through the door. We are told the characteristics of the loan and of the borrower. How can we predict the eventual outcome of the loan?

9.1 THE *K*-NEAREST NEIGHBOR ALGORITHM

One useful method for making the classification relies on the k-nearest neighbor algorithm. The *k-nearest neighbor (knn) algorithm* classifies new objects according to the outcome of the closest object or the outcomes of several closest objects in the feature space of the training set. The k-nearest neighbor algorithm is among the simplest of all machine learning algorithms: an object is classified by a majority vote of its neighbors, with the new object being assigned to the class that is most common among its k nearest neighbors (k is a positive integer, and typically small).

Here is a detailed explanation of how the k-nearest neighbor algorithm works. Take an object with d features (but unknown group membership). Calculate the distance of this object (distance is defined below) to every one of the objects in the training set (which have known group membership). Look at the k closest neighbors in the training set. Look at how the closest k neighbors are classified and determine the most frequent classification. This becomes the (predicted) classification of the object. If there are ties among the outcomes of its k-nearest neighbors, all tied candidates are included in a vote and ties are broken at random. If $k = 1$, the object is simply assigned to the class of its very nearest neighbor.

The neighbors are taken from a set of objects for which the correct classification is known. This can be thought of as the training set for the algorithm, although no explicit training step is required. The training samples are vectors in a multi-dimensional feature space, each with a specified class label. The training phase of the algorithm only consists of storing the feature vectors and class labels of the training samples.

In the classification phase, k is a user-defined constant, and a new object with given features (sometimes also referred to as a *query* or *test point*) is classified by assigning to it the label that is most frequent among the k training samples nearest to that new object.

Usually, with continuous features (such as income in multiples of $1000, or age in years), the Euclidean distance is used as the distance metric. Assume that the d features for case 1 are given by $x_{11}, x_{12}, \ldots, x_{1d}$ and that the features for case 2 are given by $x_{21}, x_{22}, \ldots, x_{2d}$. Then the Euclidean distance between cases 1 and 2 is defined as $\sqrt{(x_{11} - x_{21})^2 + (x_{12} - x_{22})^2 + \cdots + (x_{1d} - x_{2d})^2}$. If there is only one feature, the Euclidean distance is the absolute value of the difference. It is usually recommended to standardize the feature variables if their units are quite different. Otherwise, more weight would be given to feature variables with larger units.

For categorical features (gender, or age classified as young, middle-aged, and old) other metrics can be used. The overlap metric (also called the *Hamming distance*), for example, creates for two strings of equal length a sequence of mismatches and matches between corresponding positions of the two feature vectors, and then counts the number of mismatches. The distance is small if there are few mismatches.

A drawback of the basic "majority voting" classification of the k-nearest neighbor algorithm is that classes with more frequent outcomes tend to dominate the classification of the new object. Because of their large numbers, they tend to show up more often among the k-nearest neighbors when neighbors are computed. Ways of overcoming this problem are discussed in the literature, but are not included in our introductory discussion.

The k-nearest neighbor algorithm is sensitive to the local structure of the data, and its accuracy is affected by the presence of noisy or irrelevant features. The best choice of k depends upon the data. Larger values of k tend to reduce the effect of noise on the classification, but will make boundaries between classes less distinct. A good value of k can be selected by *cross-validation*. In binary (two-class) classification problems, it is helpful to choose k to be an odd number as this avoids tied votes.

The "naïve" (brute force) version of the algorithm is easy to implement. However, computing the distances from a test sample to all objects in the training set is computationally intensive, especially when the size of the training set is large. Many nearest neighbor search algorithms have been proposed, which seek to reduce the number of distance evaluations that need to be performed. Using an appropriate nearest neighbor search algorithm makes the procedure computationally tractable even for very large data sets. A reliable algorithm, *knn*, can be found in the R package **class**. In the following discussion we list R programs for two examples and we comment on the resulting output.

9.2 EXAMPLE 1: FORENSIC GLASS

We analyze the data set of 214 glass shards of six possible glass types ($g = 6$ groups):

WinF: float glass window,

WinNF: nonfloat window,

Veh: vehicle window,

Con: container (bottles),

Tabl: tableware,

Head: vehicle headlamp.

We know the group membership of each shard and we have available measurements on shard characteristics including the refractive index (RI) and the percentages of Na, Mg, Al, Si, K, Ca, Ba, and Fe.

The forensic glass data set is taken from the UCI (University of California at Irwin) Machine Learning Repository (Asuncion and Newman, 2007). The data set was provided in 1987 by Vina Spiehler and B. German of the Home Office Forensic Science Service, Reading, Berkshire RG7 4PN. It has been used as a test case for numerous classification algorithms.

Box plots of the nine feature variables, stratified for the six different shard groups, show how features vary across glass types. These graphs suggest that it may be feasible to classify a shard on the basis of its features. Next, we select 200 shards for the training set and the remaining 14 shards as the test (holdout) data set. The nearest neighbor method, with various values for the number of nearest neighbors k, is used to classify the 14 shards. Initially, our distance calculations use only the RI and the percentage of Al. With just two features it is quite easy to see graphically how nearest neighbor methods arrive at their classification. The scatter plots shown below represent the 14 shards of the test data set by solid symbols, while the 200 shards of the training set are represented by open symbols. The color of the symbol represents the type of glass. For example, in the nearest neighbor method with $k = 1$ (the scatter plot on the left side of the figure) the test shard is classified blue (container glass) if the closest shard in the training set is also blue (container glass). The most common glass type of the closest five neighbors ($k = 5$) determines the classification of each test case in the scatter plot on the right side. 11 of the 14 shards (or 78.6%) are correctly classified with $k = 1$, and 10 of the 14 shards (or 71.4%) are correctly classified with $k = 5$. These proportions will change, often considerably, with different random samples. Cross-validation, where each of the 214 shards is left out and classified one at a time, provides a better (more stable) assessment of how the number of nearest neighbors affects the proportion of correct classifications. The proportions of correct classification from cross-validation with varying numbers of nearest neighbors k range between 61% and 65% and change little with the number of nearest neighbors. Using all nine features improves the proportion of correct classification to about 71%. Using three nearest neighbors appears to work best, even though the proportion of correct classification for the single closest neighbor ($k = 1$) is quite similar.

```
#### ******* Forensic Glass ****** ####

library(textir) ## needed to standardize the data
library(MASS)   ## a library of example data sets

data(fgl)       ## loads the data into R; see help(fgl)
fgl
## data consists of 214 cases
## here are illustrative box plots of the features
## stratified by glass type
par(mfrow=c(3,3), mai=c(.3,.6,.1,.1))
plot(RI ~ type, data=fgl, col=c(grey(.2),2:6))
plot(Al ~ type, data=fgl, col=c(grey(.2),2:6))
```

```
plot(Na ~ type, data=fgl, col=c(grey(.2),2:6))
plot(Mg ~ type, data=fgl, col=c(grey(.2),2:6))
plot(Ba ~ type, data=fgl, col=c(grey(.2),2:6))
plot(Si ~ type, data=fgl, col=c(grey(.2),2:6))
plot(K ~ type, data=fgl, col=c(grey(.2),2:6))
plot(Ca ~ type, data=fgl, col=c(grey(.2),2:6))
plot(Fe ~ type, data=fgl, col=c(grey(.2),2:6))
```

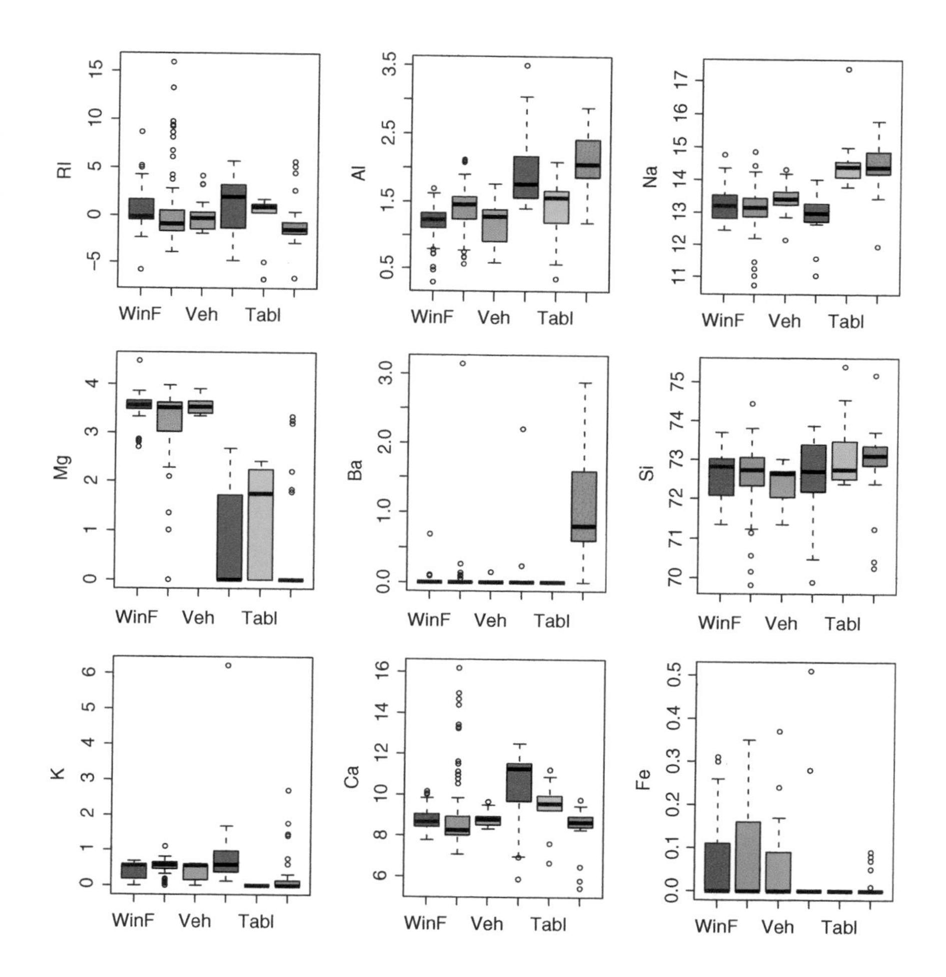


```
## for illustration, consider the RIxAl plane
## use nt=200 training cases to find the nearest neighbors for
## the remaining 14 cases. These 14 cases become the
## evaluation (test, hold-out) cases
```

```
n=length(fgl$type)
nt=200
set.seed(1)
## to make the calculations reproducible in repeated runs
train <- sample(1:n,nt)

## Standardization of the data is preferable, especially if
## units of the features are quite different
## could do this from scratch by calculating the mean and
## standard deviation of each feature, and use those to
## standardize.
## Even simpler, use the normalize function in the R-package
## textir; it converts data frame columns to mean 0 and sd 1
x <- normalize(fgl[,c(4,1)])
x[1:3,]

library(class)
nearest1 <- knn(train=x[train,],test=x[-train,],
+    cl=fgl$type[train],k=1)
nearest5 <- knn(train=x[train,],test=x[-train,],
+    cl=fgl$type[train],k=5)
data.frame(fgl$type[-train],nearest1,nearest5)

   fgl.type..train. nearest1 nearest5
1              WinF     WinF     WinF
2              WinF     WinF     WinF
3              WinF     WinF     WinF
4              WinF     WinF     WinF
5              WinF     WinF     WinF
6              WinF     WinF    WinNF
7              WinF     WinF     WinF
8             WinNF    WinNF    WinNF
9             WinNF     Head     Head
10            WinNF    WinNF    WinNF
11            WinNF    WinNF    WinNF
12            WinNF    WinNF    WinNF
13              Con     Head    WinNF
14             Tabl     Head     Head

## plot them to see how it worked on the training set
par(mfrow=c(1,2))
## plot for k=1 (single) nearest neighbor
plot(x[train,],col=fgl$type[train],
+    cex=.8,main="1-nearest neighbor")
points(x[-train,],bg=nearest1,pch=21,col=grey(.9),cex=1.25)
## plot for k=5 nearest neighbors
plot(x[train,],col=fgl$type[train],cex=.8,
+    main="5-nearest neighbors")
points(x[-train,],bg=nearest5,pch=21,col=grey(.9),cex=1.25)
```

```
legend("topright",legend=levels(fgl$type),
+    fill=1:6,bty="n",cex=.75)
```

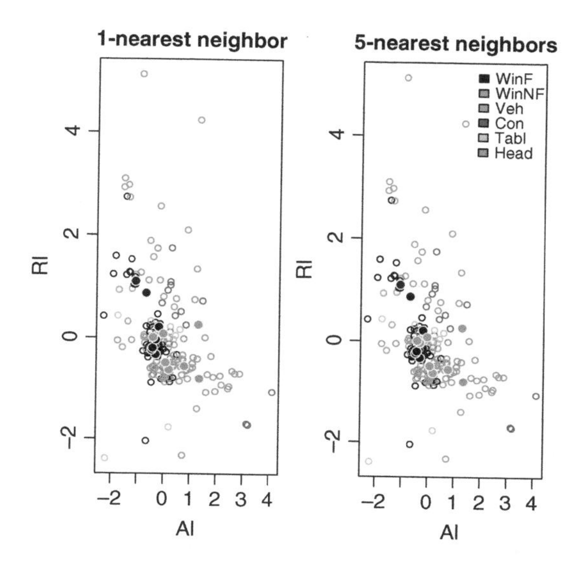


```
## calculate the proportion of correct classifications on this one
## training set

pcorrn1=100*sum(fgl$type[-train]==nearest1)/(n-nt)
pcorrn5=100*sum(fgl$type[-train]==nearest5)/(n-nt)
pcorrn1

[1] 78.57143

pcorrn5

[1] 71.42857

## cross-validation (leave one out)
pcorr=dim(10)
for (k in 1:10) {
pred=knn.cv(x,fgl$type,k)
pcorr[k]=100*sum(fgl$type==pred)/n
}
pcorr

[1] 62.61682 61.21495 61.68224 61.68224 64.48598 61.21495 62.14953 62.14953
[9] 63.55140 63.55140

## Note: Different runs may give you slightly different results as
## ties are broken at random

## using all nine dimensions (RI plus 8 chemical concentrations)

x <- normalize(fgl[,c(1:9)])
```

```
nearest1 <- knn(train=x[train,],test=x[-train,],
+    cl=fgl$type[train],k=1)
nearest5 <- knn(train=x[train,],test=x[-train,],
+    cl=fgl$type[train],k=5)
data.frame(fgl$type[-train],nearest1,nearest5)

   fgl.type..train. nearest1 nearest5
1              WinF     WinF     WinF
2              WinF      Veh     WinF
3              WinF     WinF     WinF
4              WinF    WinNF    WinNF
5              WinF     WinF     WinF
6              WinF     WinF     WinF
7              WinF     WinF     WinF
8             WinNF    WinNF    WinNF
9             WinNF    WinNF    WinNF
10            WinNF      Veh    WinNF
11            WinNF    WinNF    WinNF
12            WinNF     WinF    WinNF
13              Con      Con      Con
14             Tabl     Tabl     Tabl

## calculate the proportion of correct classifications on this one
## training set

pcorrn1=100*sum(fgl$type[-train]==nearest1)/(n-nt)
pcorrn5=100*sum(fgl$type[-train]==nearest5)/(n-nt)
pcorrn1

[1] 71.42857

pcorrn5

[1] 92.85714

## cross-validation (leave one out)

pcorr=dim(10)
for (k in 1:10) {
pred=knn.cv(x,fgl$type,k)
pcorr[k]=100*sum(fgl$type==pred)/n
}
pcorr

[1] 70.09346 70.09346 71.02804 68.22430 67.28972 64.95327 65.42056 66.35514
[9] 65.88785 66.35514
```

9.3 EXAMPLE 2: GERMAN CREDIT DATA

This data was considered in Chapter 7 on logistic regression (Section 7.9). The two outcomes are success (defaulting on the loan) and failure (not defaulting). The explanatory variables in the logistic regression included features of the loan and characteristics of the borrower. In the subsequent k-means classification we use the three continuous variables, duration, amount, and installment. Cross-validation shows that the method with $k = 5$ nearest neighbors correctly identifies about 65% of the outcomes.

```
#### ******* German Credit Data ******* ####
#### ******* data on 1000 loans ******* ####

library(textir) ## needed to standardize the data
library(class)  ## needed for knn

## read data and create some 'interesting' variables
credit <- read.csv("C:/DataMining/Data/germancredit.csv")
credit

credit$Default <- factor(credit$Default)

## re-level the credit history and a few other variables
credit$history = factor(credit$history, levels=c("A30","A31","A32",
+    "A33","A34"))
levels(credit$history) = c("good","good","poor","poor","terrible")
credit$foreign <- factor(credit$foreign, levels=c("A201","A202"),
+    labels=c("foreign","german"))
credit$rent <- factor(credit$housing=="A151")
credit$purpose <- factor(credit$purpose, levels=c("A40","A41","A42",
+    "A43","A44","A45","A46","A47","A48","A49","A410"))
levels(credit$purpose) <-c("newcar","usedcar",rep("goods/repair",4),
+    "edu",NA,"edu","biz","biz")

credit <- credit[,c("Default","duration","amount","installment","age",
+    "history", "purpose","foreign","rent")]
credit[1:3,]

  Default duration amount installment age  history      purpose foreign  rent
1       0        6   1169           4  67 terrible goods/repair foreign FALSE
2       1       48   5951           2  22     poor goods/repair foreign FALSE
3       0       12   2096           2  49 terrible          edu foreign FALSE

summary(credit) # check out the data

Default     duration         amount        installment         age
 0:700   Min.   : 4.0   Min.   :  250   Min.   :1.000   Min.   :19.00
 1:300   1st Qu.:12.0   1st Qu.: 1366   1st Qu.:2.000   1st Qu.:27.00
         Median :18.0   Median : 2320   Median :3.000   Median :33.00
         Mean   :20.9   Mean   : 3271   Mean   :2.973   Mean   :35.55
         3rd Qu.:24.0   3rd Qu.: 3972   3rd Qu.:4.000   3rd Qu.:42.00
         Max.   :72.0   Max.   :18424   Max.   :4.000   Max.   :75.00
    history              purpose          foreign         rent
 good    : 89   newcar      :234   foreign:963   FALSE:821
 poor    :618   usedcar     :103   german : 37   TRUE :179
 terrible:293   goods/repair:495
                edu         : 59
                biz         :109

## for illustration we consider just 3 loan characteristics:
## amount,duration,installment
## Standardization of the data is preferable, especially if
## units of the features are quite different
## We use the normalize function in the R-package textir;
## it converts data frame columns to mean-zero sd-one
```

```
x <- normalize(credit[,c(2,3,4)])
x[1:3,]

       duration     amount installment
[1,] -1.2358595 -0.7447588   0.9180178
[2,]  2.2470700  0.9493418  -0.8697481
[3,] -0.7382981 -0.4163541  -0.8697481

## training and prediction datasets
## training set of 900 borrowers; want to classify 100 new ones
set.seed(1)
train <- sample(1:1000,900) ## this is training set of 900 borrowers
xtrain <- x[train,]
xnew <- x[-train,]
ytrain <- credit$Default[train]
ynew <- credit$Default[-train]

## k-nearest neighbor method
library(class)
nearest1 <- knn(train=xtrain, test=xnew, cl=ytrain, k=1)
nearest3 <- knn(train=xtrain, test=xnew, cl=ytrain, k=3)
data.frame(ynew,nearest1,nearest3)[1:10,]

   ynew nearest1 nearest3
1     1        1        0
2     0        1        1
3     1        1        1
4     0        0        0
5     0        1        1
6     0        0        0
7     0        0        0
8     0        0        0
9     0        1        1
10    0        0        1

## calculate the proportion of correct classifications
pcorrn1=100*sum(ynew==nearest1)/100
pcorrn3=100*sum(ynew==nearest3)/100
pcorrn1

[1] 60

pcorrn3

[1] 61

## plot for 3nn
plot(xtrain[,c("amount","duration")],
+    col=c(4,3,6,2)[credit[train,"installment"]],
+    pch=c(1,2)[as.numeric(ytrain)],
+    main="Predicted default, by 3 nearest neighbors",cex.main=.95)
points(xnew[,c("amount","duration")],
+    bg=c(4,3,6,2)[credit[train,"installment"]],
+    pch=c(21,24)[as.numeric(nearest3)],cex=1.2,col=grey(.7))
legend("bottomright",pch=c(1,16,2,17),bg=c(1,1,1,1),
+    legend=c("data 0","pred 0","data 1","pred 1"),
+    title="default",bty="n",cex=.8)
```

```
legend("topleft",fill=c(4,3,6,2),legend=c(1,2,3,4),
+    title="installment %", horiz=TRUE,bty="n",col=grey(.7),cex=.8)
```

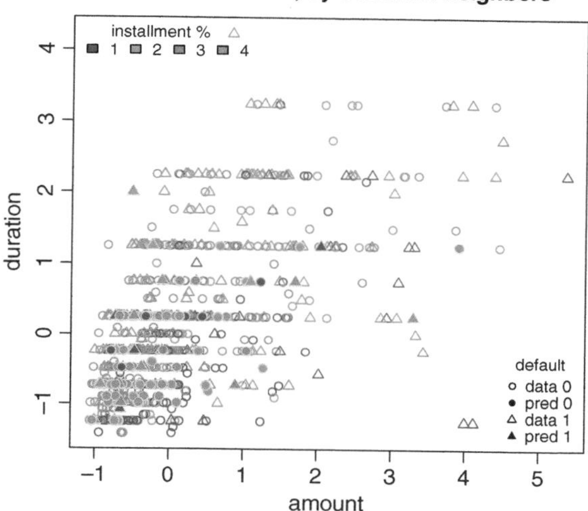


```
## above was for just one training set
## cross-validation (leave one out)
pcorr=dim(10)
for (k in 1:10) {
pred=knn.cv(x,cl=credit$Default,k)
pcorr[k]=100*sum(credit$Default==pred)/1000
        }
pcorr
```

```
[1] 61.0 61.1 64.2 62.6 64.4 65.9 65.6 67.6 66.0 67.0
```

REFERENCE

Asuncion, A. and Newman, D.J.: *UCI Machine Learning Repository*. Irvine, CA: University of California, School of Information and Computer Science, 2007. Available at http://www.ics.uci.edu/~mlearn/MLRepository.html. Accessed 2013 Jan 17.

CHAPTER 10

The Naïve Bayesian Analysis: a Model for Predicting a Categorical Response from Mostly Categorical Predictor Variables

The objective here is to infer (predict) an unknown binary response of a case from the information that is provided by its predictor or feature variables, which, in this application, are assumed categorical. One could search over the training set to find cases that match exactly the values of the predictor variables of the case in question, and then use the most frequent response of the matched cases for deciding whether the prediction for y should be 0 or 1. A common problem with this approach is that for categorical predictor variables (even when working with very large training sets), the data are sparse in the sense that there will be few cases (or even no cases) that match the values on the predictor variables of the case in question. The Bayesian approach can help.

We may not have data to make inferences about the conditional probability on the left-hand side of the equation shown below. But we can always write this probability as

$$P\left(y=1|x_1,x_2,\ldots,x_k\right)$$
$$=\frac{P\left(x_1,x_2,\ldots,x_k|y=1\right)P(y=1)}{P\left(x_1,x_2,\ldots,x_k|y=1\right)P\left(y=1\right)+P(x_1,x_2,\ldots,x_k|y=0)P\left(y=0\right)}.$$

This result is exact, and follows from basic conditional probability rules. It is often referred to as the *(general) Bayesian solution*. But also this solution is difficult to implement, because with a fine categorization of predictor variables it will be difficult to estimate the conditional joint probabilities $P(x_1,x_2,\ldots,x_k|y=1)$ and $P(x_1,x_2,\ldots,x_k|y=0)$ that are needed for its calculation. The prior probabilities,

Data Mining and Business Analytics with R, First Edition. Johannes Ledolter.

$P(y=1)$ and $P(y=0)$, on the other hand, are easy to estimate; one can use the relative frequencies from the training set.

The naïve Bayesian approach assumes that, conditional on the response, the predictors are independent. Under this assumption, we can write

$$P(y=1|x_1,x_2,\ldots,x_k) = \frac{\left[\prod_{i=1}^{k} P\left(x_i|y=1\right)\right] P(y=1)}{\left[\prod_{i=1}^{k} P\left(x_i|y=1\right)\right] P(y=1) + \left[\prod_{i=1}^{k} P\left(x_i|y=0\right)\right] P(y=0)}.$$

This can be implemented as usually there is enough information in the training set to estimate the marginal probabilities $P(x_i|y=1)$ and $P(x_i|y=0)$, for $i = 1,2,\ldots,k$.

10.1 EXAMPLE: DELAYED AIRPLANES

The data, taken from Shmueli et al. (2010), have been considered earlier in Chapter 7 on logistic regression. The data set consists of 2201 airplane flights in January 2004 from the Washington, DC area into NYC. The characteristic of interest (the response) is whether or not a flight has been delayed by more than 15 minutes (0 = no delay/1 = delay).

The explanatory variables include:

three different arrival airports (Kennedy, Newark, and LaGuardia; in that order),
three different departure airports (Reagan, Dulles, and Baltimore; in that order),
eight carriers,
a categorical variable for 16 different hours of departure (6 AM to 10 PM),
weather conditions (0/1),
day of week: 7 days with Monday = 1, ..., Sunday = 7.

In Section 7.7 (logistic regression), we selected 1320 cases (60%) as the training set. The success probability, with a delayed flight defined as "success" ($y = 1$), was 0.198 in the training set. The remaining 881 cases became the evaluation (test) data set.

The naïve rule that does not incorporate any covariate information and classifies every flight as being on time led to an overall error of 18.9% in the holdout data set. With this rule, we never make an error predicting a flight that is on time, but we make a 100% error predicting a flight that is delayed.

The logistic regression model in Section 7.7 with probability cutoff 0.5 was able to reduce (somewhat) the overall error in the holdout set to 17.6%. Among the 167 delayed flights, logistic regression identified correctly 14 delayed flights (8.4%), but it missed 153/167 delayed flights (92.6%). Furthermore, the logistic regression model classified 2 of the 714 on-time flights as being delayed. These improvements may appear minor, but the lift chart (which was also programmed) provided a quick and reliable method of identifying those flights that are most likely to be delayed.

We now use this data set, the 2201 flights from the Washington, DC area into NYC, to illustrate the naïve Bayesian method. The predictor variables are carrier (8 possibilities), day of week (7 possibilities; note this is different from logistic regression where we considered just two categories, Sunday/Monday vs remaining days), departure airport (3 possibilities), arrival airport (3 possibilities), weather (binary), and time of departure (16 one-hour time slots). These categorical predictor variables create (8)(7)(3)(3)(2)(16) = 16,128 groups, which is a very large number and even larger than the number of cases in the training data set (1320). Most of these groups contain no data.

The naïve Bayesian approach helps as it looks at each predictor variable separately. Estimates of the prior probabilities and of the conditional probabilities $P(x_i|y = 1)$ and $P(x_i|y = 0)$ are needed for the implementation of the naïve Bayesian approach. The probabilities are estimated from the training set consisting of the 60% randomly selected flights, and they are shown below. The remaining 40% of the 2201 flights serve as the holdout period.

Prior probabilities

$P(y = 0)$	$P(y = 1)$
0.802	0.198

Probabilities for scheduled time

For $y = 0$

1	2	3	4	5	6	7	8	9	10
0.061	0.068	0.072	0.059	0.050	0.034	0.064	0.077	0.095	0.063
11	12	13	14	15	16				
0.077	0.102	0.042	0.046	0.022	0.065				

For $y = 1$

1	2	3	4	5	6	7	8	9	10
0.038	0.046	0.065	0.027	0.027	0.008	0.065	0.050	0.153	0.092
11	12	13	14	15	16				
0.077	0.146	0.027	0.065	0.015	0.100				

Probabilities for scheduled carrier

For $y = 0$

1	2	3	4	5	6	7	8
0.041	0.228	0.185	0.122	0.015	0.180	0.015	0.214

For $y = 1$

1	2	3	4	5	6	7	8
0.069	0.337	0.092	0.176	0.011	0.222	0.015	0.077

Probabilities for scheduled destination

For $y = 0$

1	2	3
0.286	0.169	0.545

For $y = 1$

1	2	3
0.387	0.211	0.402

Probabilities for scheduled origin

For $y = 0$

1	2	3
0.064	0.653	0.283

For $y = 1$

1	2	3
0.096	0.490	0.414

Probabilities for weather

For $y = 0$

0	1
1	0

For $y = 1$

0	1
0.920	0.080

Probabilities for scheduled day of week

For $y = 0$

1	2	3	4	5	6	7
0.127	0.137	0.154	0.178	0.176	0.128	0.099

For $y = 1$

1	2	3	4	5	6	7
0.207	0.153	0.126	0.142	0.146	0.054	0.172

With these estimates, we can determine probabilities such as

$$\begin{aligned} P(y = 1|&\text{Carrier} = 7, \text{DOW} = 7, \text{DepTime} = 9\ \text{AM} - 10\ \text{AM}, \text{Dest} = \text{LGA}, \\ &\text{Origin} = \text{DCA}, \text{Weather} = 0) \\ &= \frac{[(0.015)(0.172)(0.027)(0.402)(0.490)(0.920)](0.198)}{[(0.015)(0.172)(0.027)(0.402)(0.490)(0.920)](0.198) + [(0.015)(0.099)(0.059)(0.545)(0.653)(1)](0.802)} \\ &= 0.09. \end{aligned}$$

Recall that in our coding, DepTime 9 AM − 10 AM = 4, LaGuardia = 3, and DCA = 2.

How well does this method fare? We apply this method to the cases of the holdout period. We score a case as success (delayed flight; $y = 1$) if its probability is 0.5 or larger, and as failure (on-time flight; $y = 0$), otherwise. From these data, we obtain a 2×2 table of classifications, and from that table we calculate the proportions of incorrect classifications (misclassifications). We are not doing better than with logistic regression. The misclassification proportion of the naïve Bayesian approach is 19.52%. We predict 30 delayed flights (out of 167) correctly, but fail to identify $137/(137 + 30)$, or 73%, of delayed flights. Furthermore, $35/(35 + 679)$, or 4.9%, of on-time flights are predicted as delayed.

```
## coding as 1 if probability 0.5 or larger
gg1=floor(gg+0.5)
ttt=table(response[-train],gg1)
ttt
       gg1
      0   1
  0 679  35
  1 137  30
error=(ttt[1,2]+ttt[2,1])/n2
error
[1] 0.1952327
```

The R program that we used to carry out the calculations can be found on the webpage that accompanies this book.

REFERENCE

Shmueli G., Patel, N.R., and Bruce, P.C.: *Data Mining for Business Intelligence*. Second edition. Hoboken, NJ: *John Wiley & Sons, Inc.*, 2010.

CHAPTER 11

Multinomial Logistic Regression

In this chapter we extend the logit link function of logistic regression in Chapter 7 to the multinomial situation where a categorical response variable can take on one of several (more than two) outcomes. We discuss an approach of estimating the class probabilities for a multicategory response, and we use these probabilities to classify new cases into one of several outcome groups.

Assume that we have g possible (unordered) outcomes $\{1, 2, \ldots, g\}$, with multinomial probabilities $P[y = k] = p_k$, $k = 1, 2, \ldots, g$. Similar to the logit-type expressions in the binary case, the probabilities are parameterized as

$$p_1 = P[y = 1] = \frac{1}{1 + \sum_{h=2}^{g} \exp(\alpha_h + x\beta_h)}$$

and

$$p_k = P[y = k] = \frac{\exp(\alpha_k + x\beta_k)}{1 + \sum_{h=2}^{g} \exp(\alpha_h + x\beta_h)}, \quad \text{for } k = 2, \ldots, g.$$

The probabilities sum to 1. Here we have adopted group 1 as the standard category, but any other group could be used instead.

The log-odds interpretation of the logistic regression model still applies, as

$$\log\left(\frac{p_k}{p_1}\right) = \alpha_k + x\beta_k, \quad \text{for } k = 2, \ldots, g.$$

Maximum likelihood estimates of the parameters and their standard errors are obtained by extending the analysis of the logistic regression model. The likelihood is the product of multinomial probabilities, with the g probabilities parameterized in the previous equations. The extensions involved are fairly straightforward mathematically. Usually an iterative reweighted least squares algorithm is used to carry out the estimation. For $g = 2$, we are back to the logistic regression model.

Data Mining and Business Analytics with R, First Edition. Johannes Ledolter.

However, there are several drawbacks to these models:

1. The models involve many parameters, which makes their interpretation tedious. Changing the explanatory variable x by one unit (from value x to $x + 1$) changes the odds of getting an outcome from group k relative to getting an outcome from group 1 (i.e., the ratio of the two group probabilities) by the factor $\exp(\beta_k)$. For $\beta_k = 1.46$, for example, a unit change in x increases the odds by the multiplicative factor $\exp(1.46) = 4.3$. For $\beta_k = 0$ and $\exp(0) = 1$, the change in x does not affect the ratio of these two probabilities. Similarly, a one-unit change in x changes the odds of getting an outcome from group k relative to getting an outcome from group r (i.e., now different from group 1) by the factor $\exp(\beta_k - \beta_r)$. Many such group comparisons need to be carried out if one wants to interpret the estimates, and even for a single covariate (as we have assumed here) the interpretation is not that easy.

   Fortunately, the main interest in data mining is not so much the interpretation of the parameter estimates, but the estimation of the class probabilities $p_1 = P[y = 1] = 1/[1 + \sum_{h=2}^{g} \exp(\alpha_h + x\beta_h)]$ and $p_k = P[y = k] = [\exp(\alpha_k + x\beta_k)]/[1 + \sum_{h=2}^{g} \exp(\alpha_h + x\beta_h)]$ (for $k = 2, \ldots, g$), and the classification of a new case with given covariate (feature) vector x. This is easily accomplished by substituting the parameter estimates into the probability equations. For visualization, one can graph the resulting probabilities, overlaid for each of the g categories, against the values of the covariate x.

   Note that here we consider just a single covariate. The extension to several covariates is again straightforward mathematically, but increases considerably the number of parameters (instead of a single parameter β_k, we now have a coefficient for each covariate and a vector of parameters), which makes the interpretation and the graphing of probabilities even more challenging.

2. Maximum-likelihood estimation (MLE) of logistic and multinomial logistic models encounters numerical problems if the data is separable and if the predicted probabilities are close to either 0 or 1. Imagine a scatter plot of two predictor variables coming from two groups where a line through the data can separate the two groups exactly. For example, assume that the observations of one group lie in the upper-right corner, while the observations of the second group are in the lower left corner, and that there is no overlap. Many different lines through the data achieve a perfect separation, and this explains why in such situations MLE of the coefficients in the logistic regression model will run into numerical difficulties. It will be impossible to come up with unique parameter estimates, although the classification will not be affected by such multicollinearity.

   Estimation difficulties also arise if the model includes too many parameters that need to be estimated, and this happens quite often as multinomial logistic regression models for responses with several outcomes have lots of parameters. An extreme case of this situation occurs in the analysis of text data (Chapter 19). But even in the absence of numerical estimation problems, it is not advisable to use that many estimated coefficients for the prediction of the probabilities and the subsequent classification of new cases. One needs to avoid model overfitting. We discussed earlier that overfitted models do not work well for predicting the outcomes of new cases. Shrinkage methods such as the LASSO (discussed in Chapter 6) add to the estimation criterion a regularization component that involves the L1 norm of the parameter vector. Bayesian estimation methods achieve similar shrinkage by putting Laplace priors on the coefficients. Such methods should be considered when estimating multinomial logistic regression models with many parameters, as they shrink the estimates and "zero out" many of the unneeded variables.

11.1 COMPUTER SOFTWARE

Several choices are available to estimate multinomial logistic regression models in R. For example, one can use the command *mlogit* in the package **mlogit**, the command *vglm* in the package **VGAM**, or the *mnlm* function in the package **textir**. These packages differ somewhat with respect to the data input, the model parameterization they adopt, and the estimation approach they use. For example, *mlogit* and *vglm* follow the parameterization outlined in this section that sets one group as the standard, and both packages obtain maximum likelihood estimates. The function *mnlm* in **textir** adopts a slightly different parameterization and implements a Bayesian estimation approach with a penalty function that regularizes the parameter estimates. The penalty is introduced via a prior distribution on the coefficients. For a detailed discussion of the estimation method, see Taddy (2012). The penalty estimation approach in *mnlm* allows for shrinkage, similar to the shrinkage of LASSO estimates that has been discussed in Chapter 6. This approach is especially useful for models with many parameters. It avoids many of the numerical problems that arise with MLE, and more importantly, it avoids overfitting.

11.2 EXAMPLE 1: FORENSIC GLASS

We have used the forensic glass data set containing 214 cases, six categories (WinF, WinNF, Veh, Con, Tabl, and Head) and nine predictor (explanatory, covariate, feature) variables in an earlier chapter (Chapter 9) on the classification with the nearest neighbor method. The box plots in Chapter 9 showed how each of the nine features varies across the six glass types. A feature that differs little across the six glass types is unlikely to help in classifying shards. On the other hand, features that differ across the six glass types such as the proportions of Na, Mg, and Al can be expected to aid in classification. In the following discussion, we use the three features, proportions of Na, Mg, and Al, to illustrate the multinomial logistic regression model. A multinomial logistic regression model with just three features makes the interpretation of the results somewhat easier, and it also avoids the numerical estimation difficulties that we encounter when fitting the model with all nine features. We use the *vglm* function from the R package **VGAM** to estimate the model. We standardize all feature variables to have zero means and standard deviations one; standardization helps with the estimation, especially if the features are of very different scales. The multinomial logistic regression model with all nine features will be estimated with *nmlm* of the R library **textir**. This will have the advantage of shrinking some of the many estimates toward zero.

```
## Forensic Glass
## you need to install the packages first
library(VGAM)       ## VGAM to estimate multinomial logistic regression
library(textir)     ## to standardize the features
library(MASS)       ## a library of example datasets
data(fgl)           ## loads the data into R; see help(fgl)
fgl

        RI    Na   Mg   Al    Si    K    Ca   Ba   Fe  type
1     3.01 13.64 4.49 1.10 71.78 0.06  8.75 0.00 0.00  WinF
2    -0.39 13.89 3.60 1.36 72.73 0.48  7.83 0.00 0.00  WinF
3    -1.82 13.53 3.55 1.54 72.99 0.39  7.78 0.00 0.00  WinF
4    -0.34 13.21 3.69 1.29 72.61 0.57  8.22 0.00 0.00  WinF
. . .
211  -1.15 14.92 0.00 1.99 73.06 0.00  8.40 1.59 0.00  Head
212   2.65 14.36 0.00 2.02 73.42 0.00  8.44 1.64 0.00  Head
213  -1.49 14.38 0.00 1.94 73.61 0.00  8.48 1.57 0.00  Head
214  -0.89 14.23 0.00 2.08 73.36 0.00  8.62 1.67 0.00  Head

## standardization, using the normalize function in the library
## textir
covars <- normalize(fgl[,1:9],s=sdev(fgl[,1:9]))
sd(covars) ## convince yourself that features are standardized
dd=data.frame(cbind(type=fgl$type,covars))
gg <- vglm(type ~ Na+Mg+Al,multinomial,data=dd)
summary(gg)
```

The R function *vglm* considers the last group (not the first) as the standard. The estimates of five intercepts ($\alpha_1, \ldots, \alpha_5$) and of five vectors of regression coefficients ($\beta_1, \ldots, \beta_5$) are shown below; each coefficient vector contains three elements, as there are three covariates in the model. The model contains 20 parameters, which are estimated on 214 cases.

```
Call:
vglm(formula = type ~ Na + Mg + Al, family = multinomial, data = dd)

Pearson Residuals:
                       Min       1Q    Median        3Q    Max
log(mu[,1]/mu[,6]) -8.9405 -0.40405 -0.007747  0.556957 1.7873
log(mu[,2]/mu[,6]) -8.9436 -0.48007 -0.148789  0.677820 4.0956
log(mu[,3]/mu[,6]) -8.3415 -0.25511 -0.177947 -0.009132 4.2084
log(mu[,4]/mu[,6]) -7.6165 -0.10943 -0.037835 -0.017610 1.6631
log(mu[,5]/mu[,6]) -6.7660 -0.11338 -0.041860 -0.019653 4.0316
```

```
Coefficients:
               Estimate Std. Error   z value
(Intercept):1  1.613703    0.84001  1.921057
(Intercept):2  3.444128    0.72131  4.774792
(Intercept):3  0.999448    0.93007  1.074594
(Intercept):4  0.067163    0.95554  0.070288
(Intercept):5  0.339579    0.89779  0.378239
Na:1          -2.483557    0.65323 -3.801955
Na:2          -2.031676    0.55399 -3.667326
Na:3          -1.409505    0.72721 -1.938243
Na:4          -2.382624    0.59434 -4.008837
Na:5           0.151459    0.53353  0.283879
Mg:1           3.842907    0.76674  5.012003
Mg:2           1.697162    0.47748  3.554387
Mg:3           3.291350    1.02370  3.215158
Mg:4           0.051466    0.50284  0.102351
Mg:5           0.699274    0.50346  1.388924
Al:1          -3.719793    0.68049 -5.466312
Al:2          -1.704689    0.54805 -3.110489
Al:3          -3.006102    0.75556 -3.978654
Al:4           0.263510    0.40013  0.658562
Al:5          -1.394559    0.51315 -2.717660

Number of linear predictors:  5

Names of linear predictors:

log(mu[,1]/mu[,6]), log(mu[,2]/mu[,6]), log(mu[,3]/mu[,6]), log(mu[,4]/mu[,6]),
log(mu[,5]/mu[,6])

Dispersion Parameter for multinomial family:   1

Residual deviance: 379.6956 on 1050 degrees of freedom

Log-likelihood: -189.8478 on 1050 degrees of freedom

Number of iterations: 7

predict(gg) ## obtain log-odds relative to last group
```

The *predict* command returns, for each case i with features x_i, the fitted logits $\widehat{\alpha}_k + x_i'\widehat{\beta}_k$. For each of the 214 cases we obtain five logits ($k = 1, \ldots, 5$) as there are six groups, with the last group being used as the standard.

```
   log(mu[,1]/mu[,6]) log(mu[,2]/mu[,6]) log(mu[,3]/mu[,6])
1           8.2875576         6.16853078          6.7952221
2           3.2189387         3.61161665          2.7674061
3           2.8395198         3.83386485          2.1909148
4           6.0483537         5.64832947          4.5679615

. . .
```

```
    log(mu[,4]/mu[,6]) log(mu[,5]/mu[,6])
1        -7.278033e-01        2.221312538
2        -1.351765e+00        1.109980963
3        -2.081657e-01        0.516194638
4         5.985536e-01        1.223013881
. . .
```

```
round(fitted(gg),2)   ## probabilities
```

We use the estimates of the coefficients to obtain, for each case i with features x_i, the fitted probability for each of the six possible outcomes (groups). These are obtained by executing the *fitted* command. The predicted probabilities for the first couple of cases, together with the actual classification, are shown as follows:

```
       1    2    3    4    5    6
1   0.74 0.09 0.17 0.00 0.00 0.00
2   0.30 0.45 0.19 0.00 0.04 0.01
3   0.23 0.61 0.12 0.01 0.02 0.01
4   0.52 0.35 0.12 0.00 0.00 0.00
. . .
```

```
cbind(round(fitted(gg),2),fgl$type)
```

```
       1    2    3    4    5    6
1   0.74 0.09 0.17 0.00 0.00 0.00 1
2   0.30 0.45 0.19 0.00 0.04 0.01 1
3   0.23 0.61 0.12 0.01 0.02 0.01 1
4   0.52 0.35 0.12 0.00 0.00 0.00 1
. . .
```

```
## boxplots of estimated probabilities against true group
dWinF=fgl$type=="WinF"
dWinNF=fgl$type=="WinNF"
dVeh=fgl$type=="Veh"
dCon=fgl$type=="Con"
dTable=fgl$type=="Tabl"
dHead=fgl$type=="Head"
yy1=c(fitted(gg)[dWinF,1],fitted(gg)[dWinNF,2],
+    fitted(gg)[dVeh,3],fitted(gg)[dCon,4],fitted(gg)[dTable,5],
+    fitted(gg)[dHead,6])
xx1=c(fgl$type[dWinF],fgl$type[dWinNF],fgl$type[dVeh],
+    fgl$type[dCon],fgl$type[dTable],fgl$type[dHead])
boxplot(yy1~xx1,ylim=c(0,1),xlab="1=WinF,2=WinNF,3=Veh,
+    4=Con,5=Table,6=Head")
```

Below, we plot the probabilities of group membership against group. For each group, we construct a box plot of its estimated probabilities, using the estimated probabilities of those shards that were known to come from that group. A good model should give us large probabilities in each of the six groups. The results show that we are doing poorly in identifying Vehicle and Table glass, but much better in identifying Head glass; most probabilities for WinF, WinNF, and Table glass are in the 40–60% range.

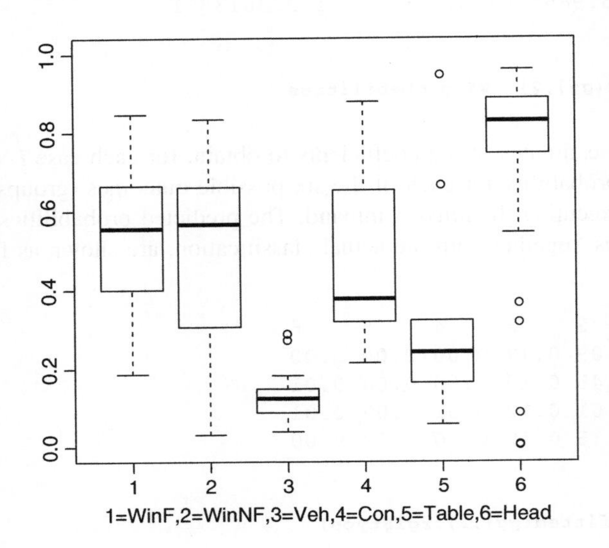

So far, the coefficients were estimated from the complete set of 214 shards. Next, we evaluate the performance of the multinomial logistic model in terms of its out-of-sample prediction. We select a training set of 194 cases at random, estimate the model on these 194 cases, and use the model to predict the probabilities of group membership for the 20 cases that have been withheld from the estimation. The predicted group membership probabilities are given below. Using the majority rule (which classifies an item into the group with the highest probability), we find five incorrect classifications (5/20, or 25%); they are labeled in bold-face type.

```
## performance in predicting a single set of 20 new cases
library(VGAM)
library(textir)
library(MASS) ## a library of example datasets
data(fgl)     ## loads the data into R; see help(fgl)
fgl
covars <- normalize(fgl[,1:9],s=sdev(fgl[,1:9]))
dd=data.frame(cbind(type=fgl$type,covars))
n=length(fgl$type)
nt=n-20
set.seed(1)
train <- sample(1:n,nt)
## predict
gg <- vglm(type ~ Na+Mg+Al,multinomial,data=dd[train,])
p1=predict(gg,newdata=dd[-train,])
p1=exp(p1)
## we calculate the probabilities from the predicted logits
```

```
sum=(1+p1[,1]+p1[,2]+p1[,3]+p1[,4]+p1[,5])
probWinF=round(p1[,1]/sum,2)    ## WinF
probWinNF=round(p1[,2]/sum,2)   ## WinNF
probVeh=round(p1[,3]/sum,2)     ## Veh
probCon=round(p1[,4]/sum,2)     ## Con
probTable=round(p1[,5]/sum,2)   ## Table
probHead=round(1/sum,2)         ## Head
ppp=data.frame(probWinF,probWinNF,probVeh,probCon,probTable,probHead,
+    fgl$type[-train])
ppp
```

```
    probWinF probWinNF probVeh probCon probTable probHead fgl.type..train.
1       0.73      0.08    0.19    0.00      0.00     0.00             WinF
23      0.58      0.32    0.09    0.00      0.00     0.00             WinF
26      0.59      0.30    0.11    0.00      0.00     0.00             WinF
33      0.58      0.32    0.09    0.00      0.00     0.00             WinF
46      0.35      0.47    0.15    0.01      0.02     0.01             WinF
55      0.25      0.63    0.09    0.01      0.02     0.01             WinF
58      0.52      0.38    0.09    0.00      0.00     0.00             WinF
65      0.74      0.11    0.15    0.00      0.00     0.00             WinF
77      0.25      0.59    0.13    0.01      0.02     0.01            WinNF
81      0.03      0.70    0.02    0.21      0.01     0.03            WinNF
84      0.27      0.61    0.09    0.02      0.01     0.01            WinNF
96      0.29      0.55    0.12    0.01      0.02     0.01            WinNF
103     0.73      0.21    0.06    0.00      0.00     0.00            WinNF
123     0.31      0.55    0.11    0.01      0.01     0.01            WinNF
136     0.58      0.28    0.14    0.00      0.00     0.00            WinNF
146     0.63      0.27    0.10    0.00      0.00     0.00            WinNF
169     0.00      0.16    0.00    0.66      0.01     0.16              Con
172     0.00      0.00    0.00    0.82      0.00     0.18              Con
183     0.00      0.02    0.00    0.12      0.03     0.83             Tabl
199     0.00      0.00    0.00    0.10      0.00     0.90             Head
```

The proportion of incorrect classifications (here 5 of 20, or 25%) depends on the selected test data set. In order to obtain more stable estimates, we replicate the analysis for 100 randomly selected test data sets. On average, about 60% of glass shards are identified correctly; about 40% are misclassified.

```
## performance from 100 replications predicting 20 new cases
library(VGAM)
library(textir)
library(MASS) ## a library of example datasets
data(fgl)     ## loads the data into R; see help(fgl)
fgl
covars <- normalize(fgl[,1:9],s=sdev(fgl[,1:9]))
dd=data.frame(cbind(type=fgl$type,covars))
## out-of-sample prediction
set.seed(1)
out=dim(20)
proportion=dim(100)
prob=matrix(nrow=20,ncol=6)
n=length(fgl$type)
nt=n-20
for (kkk in 1:100) {
train <- sample(1:n,nt)
## predict
```

```
gg <- vglm(type ~ Na+Mg+Al,multinomial,data=dd[train,])
p1=predict(gg,newdata=dd[-train,])
p1=exp(p1)
## we calculate the probabilities from the predicted logits
sum=(1+p1[,1]+p1[,2]+p1[,3]+p1[,4]+p1[,5])
prob[,1]=p1[,1]/sum    ## WinF
prob[,2]=p1[,2]/sum    ## WinNF
prob[,3]=p1[,3]/sum    ## Veh
prob[,4]=p1[,4]/sum    ## Con
prob[,5]=p1[,5]/sum    ## Table
prob[,6]=1/sum         ## Head
for (k in 1:20) {
pp=prob[k,]
out[k]=max(pp)==pp[fgl$type[-train]][k]
}
proportion[kkk]=sum(out)/20
}
## proportion of correct classification
proportion
```

```
 [1]  0.70  0.75  0.60  0.65  0.65  0.80  0.70  0.75  0.55  0.55
[11]  0.75  0.65  0.50  0.55  0.55  0.60  0.60  0.75  0.60  0.65
[21]  0.65  0.50  0.55  0.60  0.45  0.45  0.60  0.55  0.60  0.65
[31]  0.60  0.50  0.55  0.70  0.40  0.45  0.60  0.50  0.60  0.60
[41]  0.75  0.60  0.60  0.75  0.50  0.50  0.60  0.65  0.70  0.80
[51]  0.40  0.60  0.40  0.60  0.40  0.65  0.50  0.50  0.70  0.60
[61]  0.50  0.65  0.65  0.45  0.85  0.50  0.45  0.65  0.40  0.60
[71]  0.60  0.55  0.60  0.45  0.70  0.40  0.70  0.55  0.75  0.45
[81]  0.60  0.45  0.65  0.70  0.50  0.80  0.50  0.40  0.55  0.65
[91]  0.60  0.45  0.50  0.80  0.65  0.75  0.75  0.70  0.70  0.70
```

```
mean(proportion)
```

```
[1] 0.5965
```

```
boxplot(ylim=c(0,1),ylab="percent correct classification",proportion)
```

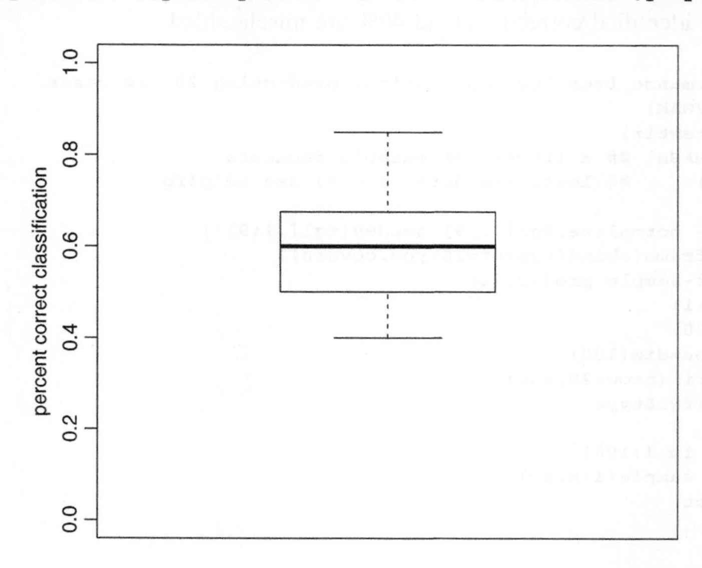

11.3 EXAMPLE 2: FORENSIC GLASS REVISITED

Until now the analysis considered only three features. Next, we consider the full data set, with all six classes and 214 cases. The number of parameters in the multinomial logistic regression model is quite large (with six groups and one group as the standard, there are 5 intercepts and $(5)(9) = 45$ slopes, for a total number of 50 parameters), considering that there are only 214 cases. It is not surprising that an MLE approach runs into difficulties. And this is exactly what we find when we estimate the model using *vglm*. The estimates of the parameters are unusually large and with both positive and negative signs, their standard errors are huge, and their t-ratios are close to zero, all diagnostics that point to serious multicollinearity and an overspecified model. The parameter estimates are meaningless, and the calculation of the multinomial logistic group probabilities with such large estimates is likely to lead to numerical problems.

Simplification of the model and shrinkage of the estimates would certainly be beneficial, which brings us to the Bayesian penalty estimation approach that is carried out by *mnlm* in the R library **textir**. For a description of this program and its current defaults, type `help(mnlm)` into your R session window. One penalty approach in *mnlm* puts a Laplace prior distribution on each nonintercept regression coefficient β and obtains the posterior modes of the multinomial logistic regression parameters. A large scale parameter $\lambda \geq 0$ in the Laplace prior density, $p(\beta) = (\lambda/2)\exp(-\lambda|\beta|)$, implies considerable shrinkage. A small scale parameter (or very small penalty) specifies a flat prior and induces little or no shrinkage. The scale parameter can be specified using the argument `penalty = lambda` in the mnlm calling statement, and the appropriate value of $\lambda \geq 0$ and the optimal amount of shrinkage can be determined by cross-validation. Since the same scale parameter is applied to all coefficients, it is recommended to first standardize the covariates, especially if the covariates are of very different magnitudes.

A second shrinkage approach in *mnlm* allows the scale parameters of the Laplace prior to vary across all nonintercept regression coefficients. It specifies a hyper prior distribution for the scale parameters, adopting a Gamma hyper prior distribution with shape parameter s and rate parameter r. The current default specification for the Laplace scale parameter (`penalty = c(1,1)`) specifies $s = 1$ and $r = 1$, a gamma distribution with mean 1 and standard deviation 1. The density of this distribution resembles an exponential decay; while giving considerable prior probability to small Laplace scales (which imply little shrinkage), the prior also allows for large Laplace scale parameters (which imply considerable shrinkage). If you want even more shrinkage, specify a smaller value of r such as $r = 0.2$ (`penalty = c(1,0.2)`), which results in a gamma prior with mean 5 and standard deviation 5. This prior density allows for even larger Laplace scale parameters and considerable shrinkage.

Below, we use the function *mnlm* of the **textir** R library to estimate the coefficients of the multinomial logistic regression model. We standardize the covariates, and we perform a cross-validation to determine the appropriate scale parameter

$\lambda \geq 0$ of the Laplace prior distribution and the appropriate shrinkage of the non-intercept regression coefficients in the multinomial logistic regression model. The results show that the proportion of correct classification varies very little with the scale (penalty) parameter $\lambda \geq 0$, and that the correct out-of-sample classification will not be better than about 65%. For extremely large shrinkage (when all regression parameters are shrunk to zero), a new observation is classified into the WinNF group as this is the most frequent group (76 out of 214 pieces are from this group), for a correct classification of $100(76/214) = 35.5\%$.

```
library(textir)
set.seed(1)
library(MASS) ## a library of example datasets
data(fgl)     ## loads the data into R; see help(fgl)

covars <- normalize(fgl[,1:9],s=sdev(fgl[,1:9]))
n=length(fgl$type)
prop=dim(30)
pen=dim(30)
out=dim(n)

for (j in 1:30) {
pen[j]=0.1*j
for (k in 1:n) {
train1=c(1:n)
train=train1[train1!=k]
glasslm <- mnlm(counts=fgl$type[train],penalty=pen[j],
+   covars=covars[train,])
prob=predict(glasslm,covars[-train,])
prob=round(prob,3)
out[k]=max(prob)==prob[fgl$type[-train]]
}
prop[j]=sum(out)/n
}
## proportion of correct classifications using Laplace scale
## penalty
output=cbind(pen,prop)
round(output,3)

      pen  prop
 [1,] 0.1 0.626
 [2,] 0.2 0.631
 [3,] 0.3 0.631
 [4,] 0.4 0.640
 [5,] 0.5 0.645
 [6,] 0.6 0.640
 [7,] 0.7 0.640
 [8,] 0.8 0.654
 [9,] 0.9 0.640
```

```
[10,] 1.0 0.645
[11,] 1.1 0.650
[12,] 1.2 0.645
[13,] 1.3 0.645
[14,] 1.4 0.640
[15,] 1.5 0.636
[16,] 1.6 0.640
[17,] 1.7 0.640
[18,] 1.8 0.640
[19,] 1.9 0.640
[20,] 2.0 0.645
[21,] 2.1 0.645
[22,] 2.2 0.640
[23,] 2.3 0.636
[24,] 2.4 0.636
[25,] 2.5 0.631
[26,] 2.6 0.631
[27,] 2.7 0.631
[28,] 2.8 0.631
[29,] 2.9 0.631
[30,] 3.0 0.631
. . .
[1000,] 100 0.355
```

The annotated output of the analysis with Laplace prior $\lambda = 1.0$ is shown below:

```
library(textir)
library(MASS) ## a library of example datasets
data(fgl)      ## loads the data into R; see help(fgl)
fgl$type
covars <- normalize(fgl[,1:9],s=sdev(fgl[,1:9]))
glasslm <- mnlm(counts=fgl$type,penalty=1.0,covars=covars)
glasslm$intercept

category  intercept
   WinF    0.5787068
   WinNF   2.0459126
   Veh    -0.1046593
   Con    -0.7854798
   Tabl   -1.1089260
   Head   -0.4823157

glasslm$loadings
round(as.matrix(glasslm$loadings)[,],2)
```

```
         covariate
category     RI  Na     Mg    Al    Si     K    Ca    Ba    Fe
   WinF    0.00 0.0   1.90 -2.04  0.00  0.00  0.00  0.00  0.00
   WinNF   0.00 0.0   0.01 -0.23 -0.50  0.00 -0.13  0.00  0.29
   Veh    -1.79 0.0   0.08 -2.12 -1.80 -0.64  0.00  0.00  0.00
   Con     0.00 0.0  -0.88  1.54  0.00  0.28  0.34  0.00  0.00
   Tabl    0.00 1.9  -0.06  0.13  0.33 -2.16  0.00 -0.32 -0.40
   Head    0.69 1.6  -1.03  0.96  0.87  0.27 -0.55  1.02 -0.10
```

Observe that quite a few of the estimated coefficients are zero and many of the variables are not entering the logit equations.

```
fitted(glasslm)
as.matrix(fitted(glasslm)[1,])
```

```
                     response
            WinF WinNF Veh Con Tabl Head
  [1,] 0.7470912     0   0   0    0    0
```

This command calculates the fitted count expectations. For a binomial or multinomial response, they are stored in a simple triplet matrix (simple triplet matrices are explained in Appendix 11.A) with empty entries for zero count observations. For example, consider the first row with response 1 in the first column (WinF). The estimated probability for WinF is shown in the first column of the fitted count expectations, with zeros in the other columns as they have zero observed counts.

```
round(predict(glasslm,covars),2)
```

```
               probability
      WinF WinNF  Veh  Con Tabl Head
  1   0.75   0.13 0.10 0.00 0.02 0.00
  2   0.42   0.36 0.09 0.01 0.07 0.04
  3   0.31   0.46 0.09 0.02 0.08 0.04
  4   0.53   0.33 0.12 0.01 0.01 0.01
  . . .

  211 0.00   0.00 0.00 0.00 0.01 0.99
  212 0.00   0.00 0.00 0.00 0.00 1.00
  213 0.00   0.00 0.00 0.00 0.01 0.99
  214 0.00   0.00 0.00 0.00 0.00 0.99
```

```
plot(glasslm)
```

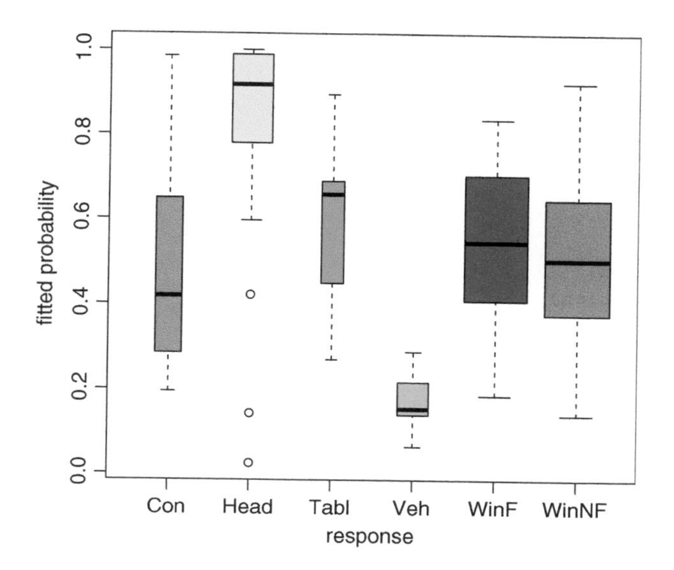

Here we plot the probabilities of group membership against group. The results show that we are still doing poorly in identifying Vehicle glass, but somewhat better in identifying Head glass. The results are only a modest improvement over those of the model with just Na, Mg, and Al.

Fitting the multinomial logistic regression model with penalty $\lambda = 1.0$ to the 194 shards that had been randomly selected from the 214 shards, predicting the probabilities and classifying the remaining 20 items, and repeating the out-of-sample prediction/classification 100 times, leads to 65% correct classification. This is not much of an improvement over the results of the multinomial logistic model with just three covariates (which was 60%).

```
library(textir)
library(MASS) ## a library of example datasets
data(fgl)     ## loads the data into R; see help(fgl)
covars <- normalize(fgl[,1:9],s=sdev(fgl[,1:9]))
sd(covars)
set.seed(1)
pp=dim(6)
out=dim(20)
proportion=dim(100)
n=length(fgl$type)
nt=n-20
for (kkk in 1:100) {
train <- sample(1:n,nt)
glasslm=mnlm(counts=fgl$type[train],penalty=1,
+    covars=covars[train,])
prob=predict(glasslm,covars[-train,])
```

```
for (k in 1:20) {
pp=prob[k,]
out[k]=max(pp)==pp[fgl$type[-train]][k]
}
proportion[kkk]=sum(out)/20
}

proportion
mean(proportion)
boxplot(proportion)
```

Finally, we enlarge our model by including all interaction effects. This new model includes 1 intercept, 9 linear coefficients, and $(9)(8)/2 = 36$ cross-products, for a total of 46 terms and $(5)(46) = 230$ coefficients as there are six possible outcomes. One could not fit such a model with maximum likelihood. And even if one could, it would not be advisable as many of the terms (perhaps even most) should probably be set zero. One certainly needs a penalty component in the estimation that shrinks (and "zeros out") unneeded components. Fitting this extremely overparameterized multinomial logistic regression model with penalty $\lambda = 1.0$ to the 194 shards that had been randomly selected from the 214, predicting the probabilities and classifying the remaining 20 items, and repeating the out-of-sample prediction/classification 100 times, leads to 68% correct classification, a slight improvement over the 65% that results from the multinomial logistic model that models the logits as linear functions of the six covariates.

```
library(textir)
library(MASS) ## a library of example datasets
data(fgl)     ## loads the data into R; see help(fgl)
X <- model.matrix(~.+.^2, data=fgl[,1:9])[,-1]
X[1:3,]       ## to see the contents
## -1 removes the intercept
dim(X)        ## X has 45 columns
covars <- normalize(X,s=sdev(X))
sd(covars)
set.seed(1)
pp=dim(6)
out=dim(20)
proportion=dim(100)
n=length(fgl$type)
nt=n-20

for (kkk in 1:100) {
train <- sample(1:n,nt)
glasslm=mnlm(counts=fgl$type[train],penalty=1,
+    covars=covars[train,])
prob=predict(glasslm,covars[-train,])
```

```
for (k in 1:20) {
pp=prob[k,]
out[k]=max(pp)==pp[fgl$type[-train]][k]
}
proportion[kkk]=sum(out)/20
}

proportion
mean(proportion)
boxplot(proportion)
```

All R programs can be found on the webpage that accompanies this book.

APPENDIX 11.A SPECIFICATION OF A SIMPLE TRIPLET MATRIX

The R function *mnlm* makes use of simple triplet matrices. A *simple triplet matrix* is a very efficient representation of a large, but sparse matrix. Consider an n by m matrix of counts, where n represents the number of cases (observations) and m is the number of categories. Each row has a one for that observation's category, and zeros elsewhere. You can specify this matrix as a standard matrix, or specify it as a "simple triplet matrix" object using just three elements: The row "i," column "j," and entry value "v." Everything else in the matrix is assumed to be zero. This is a useful way to store sparse matrices without using too much memory.

Here are some examples:

```
i=c(1,2,3,4,5,6)
j=c(1,1,1,2,2,2)
v=c(5,5,5,6,6,6)
b=simple_triplet_matrix(i,j,v)
b

    A 6x2 simple triplet matrix.

as.matrix(b)[,]

     [,1] [,2]
[1,]    5    0
[2,]    5    0
[3,]    5    0
[4,]    0    6
[5,]    0    6
[6,]    0    6

v=c(11,12,22,33,44,55)
b=simple_triplet_matrix(i,j,v)
```

```
as.matrix(b)[,]

     [,1] [,2]
[1,]   11    0
[2,]   12    0
[3,]   22    0
[4,]    0   33
[5,]    0   44
[6,]    0   55

i=c(1,2,3,4,5,6)
j=c(1,2,3,4,5,6)
v=c(5,5,5,6,6,6)
b=simple_triplet_matrix(i,j,v)
b

    A 6x6 simple triplet matrix.

as.matrix(b)[,]

     [,1] [,2] [,3] [,4] [,5] [,6]
[1,]   5    0    0    0    0    0
[2,]   0    5    0    0    0    0
[3,]   0    0    5    0    0    0
[4,]   0    0    0    6    0    0
[5,]   0    0    0    0    6    0
[6,]   0    0    0    0    0    6

i=c(1,2,3,4,5,6,7,8,9,10,11,12,13,14,15,16)
j=c(1,1,2,3,3,4,4,4,4,5,5,6,6,6,6,6)
v=c(1,1,1,1,1,1,1,1,1,1,1,1,1,1,1,1)
b=simple_triplet_matrix(i,j,v)
b

    A 16x6 simple triplet matrix.

as.matrix(b)[,]

      [,1] [,2] [,3] [,4] [,5] [,6]
 [1,]    1    0    0    0    0    0
 [2,]    1    0    0    0    0    0
 [3,]    0    1    0    0    0    0
 [4,]    0    0    1    0    0    0
 [5,]    0    0    1    0    0    0
 [6,]    0    0    0    1    0    0
 [7,]    0    0    0    1    0    0
 [8,]    0    0    0    1    0    0
 [9,]    0    0    0    1    0    0
[10,]    0    0    0    0    1    0
```

```
[11,]    0    0    0    0    1    0
[12,]    0    0    0    0    0    1
[13,]    0    0    0    0    0    1
[14,]    0    0    0    0    0    1
[15,]    0    0    0    0    0    1
[16,]    0    0    0    0    0    1
```

REFERENCES

Agresti, A.: *Categorical Data Analysis*. Second edition. New York: John Wiley & Sons, Inc., 2002.

Abraham, B. and Ledolter, J.: *Introduction to Regression Modeling*. Belmont, CA: Duxbury Press, 2006 (Chapter 11).

Taddy, M.: Multinomial inverse regression for text analysis. 2012. Available at http://arxiv.org/abs/1012.2098. Accessed 2013 Feb 10. To appear in *Journal of the American Statistical Association*, Vol. 108 (2013).

CHAPTER 12

More on Classification and a Discussion on Discriminant Analysis

Assume that we are given an object with known feature vector with k components $x = (x_1, x_2, \ldots, x_k)'$ and that we wish to classify this object into one of g mutually exclusive groups $G1, G2, \ldots, Gg$. Let us start the discussion with two groups, $G1$ and $G2$.

Let us assume that there are certain costs of misclassification. Let $c(2|1)$ be the cost of misclassifying the object into group $G2$ if it actually belongs to $G1$, and let $c(1|2)$ be the cost of misclassifying the object into group $G1$ if it actually belongs to $G2$. We wish to allow for asymmetric cost structure in general, even though in some cases the costs of misclassification may be the same (i.e., $c(2|1) = c(1|2)$).

Assume that the features $x = (x_1, x_2, \ldots, x_k)'$ vary among objects from the same group, and that the distributions that describe their variability are different for the two groups. Let $f_1(x)$ be the that probability density of the feature vector in group $G1$, and let $f_2(x)$ be the probability density of the feature vector in group $G2$. Let $P(2|1)$ be the probability that we classify an object from $G1$ into $G2$, and let $P(1|2)$ be the probability that we classify an object from $G2$ into $G1$. For any given classification region, these probabilities can be calculated from the densities $f_1(x)$ and $f_2(x)$. Furthermore, assume that we have prior probabilities for the group association of an object. The prior probability that the object is from group $G1$ is p_1; the prior that it is from group $G2$ is $p_2 = 1 - p_1$. Often there is no strong prior, and then we assume that $p_1 = p_2 = 0.5$.

A sensible classification rule minimizes the *expected cost of misclassification*,

$$\text{EMC} = c(2|1)P(2|1)p_1 + c(1|2)P(1|2)p_2. \tag{12.1}$$

$P(2|1)p_1$ in the earlier equation represents the probability of getting an object from group $G1$ *and* misclassifying it into group $G2$. This incurs cost $c(2|1)$. Similarly, $P(1|2)p_2$ represents the probability of getting an object from group $G2$ and misclassifying it into group $G1$. This incurs cost $c(1|2)$.

Data Mining and Business Analytics with R, First Edition. Johannes Ledolter.

A straightforward proof (involving probability and integral calculus, however), gives us a rule that minimizes the expected cost of misclassification. It specifies that we should classify an object with feature vector x into group $G2$ when

$$\frac{f_1(x)}{f_2(x)} < \frac{c(1|2)}{c(2|1)} \frac{p_2}{p_1}; \qquad (12.2)$$

otherwise, we classify the object into group $G1$.

An alternative way to solve the classification problem is to look at the posterior probability (after having observed the feature vector x) of the object being in $G2$. That is,

$$P(G2|x) = \frac{P(x|G2)P(G2)}{P(x|G2)P(G2) + P(x|G1)P(G1)} = \frac{f_2(x)p_2}{f_2(x)p_2 + f_1(x)p_1}. \qquad (12.3)$$

The rule that classifies an object with feature vector x into $G2$ for posterior probability

$$P(G2|x) = \frac{f_2(x)p_2}{f_2(x)p_2 + f_1(x)p_1} > c^* = \frac{c(2|1)}{c(2|1) + c(1|2)} \qquad (12.4)$$

is equivalent to the rule that minimizes the expected cost of misclassification. Note that for symmetric cost, the posterior probability is compared to $c^* = 0.5$. It classifies the object with features x into the group with the highest posterior probability and follows the majority rule assignment.

Simple algebra shows the earlier equivalence:

$$\frac{f_2(x)p_2}{f_2(x)p_2 + f_1(x)p_1} > c^*$$

is equivalent to

$$f_2(x)p_2 + f_1(x)p_1 < \frac{f_2(x)p_2}{c^*}$$

and

$$\frac{f_1(x)}{f_2(x)} < \frac{p_2}{p_1} \frac{1-c^*}{c^*} = \frac{p_2}{p_1} \frac{c(1|2)}{c(2|1)}.$$

In order to make the general classification rule operational, we need to make further assumptions about the densities $f_1(x)$ and $f_2(x)$. A reasonable assumption for quantitative features (perhaps after some suitable transformation) is normality. Thus, we assume that $f_1(x)$ is the density of a multivariate normal distribution (k-dimensional, as there are k features) with mean vector $E(x) = \mu_1$ and covariance matrix $V(x) = \Sigma_1$, and that $f_2(x)$ is the density of a multivariate normal distribution with mean vector $E(x) = \mu_2$ and covariance matrix $V(x) = \Sigma_2$. For small samples, this is probably a reasonable assumption, as usually there would not be enough information in the data to suggest generalizations. For large samples (such as they arise in data mining applications), we certainly can relax this parametric assumption and use nonparametric procedures such as the nearest neighbor classification. We discussed this procedure in Chapter 9.

Under the normal distribution assumption we can simplify the ratio $f_1(x)/f_2(x)$ in the classification rule in Equation 12.2. After some simplifications (which involve matrix algebra and knowledge of the multivariate normal distribution), we can rewrite the earlier classification rule. Accordingly, we classify an object with feature vector $x = (x_1, x_2, \ldots, x_k)'$ into group $G2$ when

$$-\left(\frac{1}{2}\right)x'(\Sigma_1^{-1} - \Sigma_2^{-1})x + (\mu_1'\Sigma_1^{-1} - \mu_2'\Sigma_2^{-1})x - w^* < \ln\left[\frac{c(1|2)\,p_2}{c(2|1)\,p_1}\right], \quad (12.5)$$

where

$$w^* = \frac{1}{2}\ln\frac{|\Sigma_1|}{|\Sigma_2|} + \frac{1}{2}(\mu_1'\Sigma_1^{-1}\mu_1 - \mu_2'\Sigma_2^{-1}\mu_2)$$

is a scalar that depends on the parameters of the two normal distributions. We have been consistent in denoting the feature vector as a $k \times 1$ column vector and using the symbol $'$ for the transpose in our vector/matrix calculations.

Note that the classification rule in Equation 12.5 defines a quadratic region in the feature (vector) space, and one speaks of a *quadratic discriminant function*. How would one use this rule? From a training set (which consists of the observed features of n_1 objects from group $G1$ and n_2 objects from group $G2$), we calculate vectors of sample averages ($\overline{x}_1$ and $\overline{x}_2$) and sample covariance matrices (S_1 and S_2), and use them in place of the mean vectors μ_1 and μ_2, and matrices Σ_1 and Σ_2. For each new object to be classified, we calculate the left-hand side of Equation 12.5 for given values of the feature vector x. This calculated value for the left-hand side is compared to the expression on the right-hand side. For equal priors and symmetric misclassification costs, the right-hand side in Equation 12.5 is equal to 0. If the left-hand side is negative, we classify the object into group $G2$; otherwise we classify the object into group $G1$.

Calculations simplify considerably if we assume equal covariance matrices; that is, $\Sigma_1 = \Sigma_2 = \Sigma$. Then we classify the object into group $G2$ if

$$(\mu_1 - \mu_2)'\Sigma^{-1}x - \frac{1}{2}(\mu_1 - \mu_2)'\Sigma^{-1}(\mu_1 + \mu_2) < \ln\left[\frac{c(1|2)\,p_2}{c(2|1)\,p_1}\right]. \quad (12.6)$$

This equation represents a *linear discriminant function*; the quadratic function involving the feature vector has now disappeared, and all that remains are the linear terms. How would one use this rule? Again, we calculate the sample mean vectors of the features for groups $G1$ and $G2$. The common covariance matrix Σ can be estimated from the pooled covariance matrix $[(n_1 - 1)S_1 + (n_2 - 1)S_2]/(n_1 + n_2 - 2)$.

These classification procedures can be extended to $g > 2$ groups. If all misclassification costs are the same, the rule that minimizes the expected cost of misclassification allocates an object with feature vector x to group j if

$$p_j f_j(x) > p_i f_i(x), \quad \text{for all } i \neq j. \tag{12.7}$$

This rule is identical to the one that allocates x to the group with the largest posterior probability.

Again, under normal distribution assumptions these rules simplify. Under the assumption of equal covariance matrices (i.e., with g groups, $\Sigma_1 = \Sigma_2 = \cdots = \Sigma_g = \Sigma$), the procedure involves the calculation of g (linear) discriminant scores (which now depend on the g means, $\mu_1, \mu_2, \ldots, \mu_g$, the common covariance matrix Σ, and the g prior probabilities). We assign an object with feature vector x to the group with the largest discriminant score.

12.1 FISHER'S LINEAR DISCRIMINANT FUNCTION

Fisher solved the two-group classification problem from another vantage point. Consider two groups, $G1$ and $G2$, feature vector $x = (x_1, x_2, \ldots, x_k)'$, mean vectors $\mu_1 = E(x|G1)$ and $\mu_2 = E(x|G2)$ for groups $G1$ and $G2$, and common covariance matrix $V(x|G1) = V(x|G2) = \Sigma$. Fisher was looking for a linear transformation $y = \ell' x$ that transforms the k-dimensional feature vector x into a scalar variable y such that in the transformed space the two groups are separated as far as possible. He calls this function the *linear discriminant function*. The (scalar) means of the transformed feature vector are given by $\mu_{1y} = \ell'\mu_1$ for group $G1$ and $\mu_{2y} = \ell'\mu_2$ for group $G2$. The variance of the linear combination, $V(y) = \ell'\Sigma\ell$, is the same for both groups. Fisher's goal was to separate the means of the linear transforms, $\mu_{1y} = \ell'\mu_1$ and $\mu_{2y} = \ell'\mu_2$, as much as possible. Of course, this had to be done relative to the variability of the transform $V(y) = \ell'\Sigma\ell$ as the transform also changes the scale. Fisher found that the transformation that achieves maximal separation is given by $\ell = \Sigma^{-1}(\mu_1 - \mu_2)$. He then used this transformation in a classification procedure. For a new object with feature vector x he calculated the value of the discriminant function $\ell' x = (\mu_1 - \mu_2)'\Sigma^{-1}x$, and compared it to the midpoint of the two (univariate) populations means, $m = (1/2)[\mu_{1y} + \mu_{2y}] = (1/2)(\mu_1 - \mu_2)'\Sigma^{-1}(\mu_1 + \mu_2)$. He showed that the expected value of $(\mu_1 - \mu_2)'\Sigma^{-1}x$ is greater than m for objects from $G1$, and less than m for objects from $G2$. This then led him to classify the object with feature vector x into group $G2$ when

$$(\mu_1 - \mu_2)'\Sigma^{-1}x < \frac{1}{2}(\mu_1 - \mu_2)'\Sigma^{-1}(\mu_1 + \mu_2). \tag{12.8}$$

Look back to the optimal classification rule in Equation 12.6 that we found earlier in the context of normal distributions with equal covariance matrices, equal prior probabilities, and equal misclassification costs. We see that this rule and Fisher's linear discriminant procedure are identical.

Fisher extended this approach to $g > 2$ groups. His extension involves the determination of two or more discriminant functions that maximize the separation between groups and, furthermore, are orthogonal to each other. With two discriminant functions, the observations (and also their group means) are transformed into a two-dimensional discriminant space, and in that transformed space new observations are assigned to groups by minimizing their Euclidean distances to the group averages. Fisher's approach reduces the dimensions from a usually large number of characteristics (k) to a relatively small number of linear combinations, giving an excellent graphical display of possible groupings. Fisher's approach with $r = \min(g - 1, k)$ discriminant functions is equivalent to the approach that classifies observations into the group with the largest linear discriminant score that is calculated assuming equal priors (Johnson and Wichern, 1988, p. 524).

We consider four examples. We analyze Fisher's iris data and data on MBA applications that have been taken from Johnson and Wichern (1988). The latter example should be of interest to a business audience. In addition, we provide another analysis of the forensic glass data set and the German default data that have been used in previous chapters. The findings from discriminant analysis can be compared to earlier results on (i) nearest neighbor analysis and (ii) logistic regression. The functions *lda* and *qda* in the R library **MASS** are used for linear and quadratic discriminant analysis. After listing the R programs (which are also given on the webpage that accompanies this text), we interpret the cross-validation performance of these discriminant procedures.

12.2 EXAMPLE 1: GERMAN CREDIT DATA

```
#### ******* German Credit Data ******* ####
#### ******* data on 1000 loans ******* ####

library(MASS)
## MASS includes lda and qda for discriminant analysis
set.seed(1)

## read data and create some 'interesting' variables
credit <- read.csv("C:/DataMining/Data/germancredit.csv")
credit

credit$Default <- factor(credit$Default)

## re-level the credit history and a few other variables
credit$history = factor(credit$history, levels=c("A30","A31",
+    "A32","A33","A34"))
levels(credit$history) = c("good","good","poor","poor",
+    " terrible")
credit$foreign <- factor(credit$foreign,levels=c("A201",
+    "A202"),labels=c("foreign","german"))
```

```
credit$rent <- factor(credit$housing=="A151")
credit$purpose <- factor(credit$purpose,
levels=c("A40","A41","A42","A43","A44","A45","A46","A47",
+    "A48","A49","A410"))
levels(credit$purpose) <-c("newcar","usedcar",rep("goods/repair",4),
+    "edu",NA,"edu","biz","biz")

## take the continuous variables duration, amount,
## installment, age. With indicators the assumptions of a
## normal distribution would be tenuous at best; hence these
## variables are not considered here

cred1=credit[, c("Default","duration","amount","installment",
+    "age")]
cred1
summary(cred1)

hist(cred1$duration)
hist(cred1$amount)
hist(cred1$installment)
hist(cred1$age)
cred1$Default
cred1=data.frame(cred1)

## linear discriminant analysis
## class proportions of the training set used as prior
## probabilities
zlin=lda(Default~.,cred1)
predict(zlin,newdata=data.frame(duration=6,amount=1100,
+    installment=4,age=67))
predict(zlin,newdata=data.frame(duration=6,amount=1100,
+    installment=4,age=67))$class
zqua=qda(Default~.,cred1)
predict(zqua,newdata=data.frame(duration=6,amount=1100,
+    installment=4,age=67))
predict(zqua,newdata=data.frame(duration=6,amount=1100,
+    installment=4,age=67))$class

n=1000
neval=1
errlin=dim(n)
errqua=dim(n)
## leave-one-out evaluation
for (k in 1:n) {
train1=c(1:n)
train=train1[train1!=k]
```

```
## linear discriminant analysis
zlin=lda(Default~.,cred1[train,])
predict(zlin,cred1[-train,])$class
tablin=table(cred1$Default[-train],predict(zlin,
+    cred1[-train,])$class)
errlin[k]=(neval-sum(diag(tablin)))/neval
## quadratic discriminant analysis
zqua=qda(Default~.,cred1[train,])
predict(zqua,cred1[-train,])$class
tablin=table(cred1$Default[-train],predict(zqua,
+    cred1[-train,])$class)
errqua[k]=(neval-sum(diag(tablin)))/neval
}
merrlin=mean(errlin)
merrlin
merrqua=mean(errqua)
merrqua
```

Crossvalidation leads to a 29.0% misclassification rate for linear discriminant analysis and 29.5% for quadratic discriminant analysis.

12.3 EXAMPLE 2: FISHER IRIS DATA

We consider three different species of irises (50 each). Four characteristics are measured on each iris: Sepal.Length, Sepal.Width, Petal.Length, and Petal.Width. The objective is to classify each iris on the basis of these four features.

```
library(MASS)
## MASS includes lda and qda for discriminant analysis
set.seed(1)
Iris=data.frame(rbind(iris3[,,1],iris3[,,2],iris3[,,3]),
+    Sp=rep(c("s","c","v"),rep(50,3)))
Iris
## linear discriminant analysis
## equal prior probabilities as same number from each species
zlin=lda(Sp~.,Iris,prior=c(1,1,1)/3)
predict(zlin,newdata=data.frame(Sepal.L.=5.1,Sepal.W.=3.5,
+    Petal.L.=1.4, Petal.W.=0.2))
predict(zlin,newdata=data.frame(Sepal.L.=5.1,Sepal.W.=3.5,
+    Petal.L.=1.4, Petal.W.=0.2))$class
## quadratic discriminant analysis
zqua=lda(Sp~.,Iris,prior=c(1,1,1)/3)
predict(zqua,newdata=data.frame(Sepal.L.=5.1,Sepal.W.=3.5,
+    Petal.L.=1.4, Petal.W.=0.2))
predict(zqua,newdata=data.frame(Sepal.L.=5.1,Sepal.W.=3.5,
+    Petal.L.=1.4, Petal.W.=0.2))$class
```

```
n=150
nt=100
neval=n-nt
rep=1000
errlin=dim(rep)
errqua=dim(rep)
for (k in 1:rep) {
train=sample(1:n,nt)
## linear discriminant analysis
m1=lda(Sp~.,Iris[train,],prior=c(1,1,1)/3)
predict(m1,Iris[-train,])$class
tablin=table(Iris$Sp[-train],predict(m1,Iris[-train,])$class)
errlin[k]=(neval-sum(diag(tablin)))/neval
## quadratic discriminant analysis
m2=qda(Sp~.,Iris[train,],prior=c(1,1,1)/3)
predict(m2,Iris[-train,])$class
tablin=table(Iris$Sp[-train],predict(m2,Iris[-train,])$class)
errqua[k]=(neval-sum(diag(tablin)))/neval
}
merrlin=mean(errlin)
merrlin
merrqua=mean(errqua)
merrqua
```

We evaluate linear and quadratic discriminant analyses by randomly selecting 100 of 150 plants, estimating the parameters from the training data, and classifying the remaining 50 plants of the holdout sample. We repeat this 1000 times. We achieve a 2.16% misclassification rate for linear discriminant analysis (and 2.61% for the quadratic version).

12.4 EXAMPLE 3: FORENSIC GLASS DATA

```
library(MASS)
## MASS includes lda and qda for discriminant analysis
set.seed(1)
data(fgl)
glass=data.frame(fgl)
glass

## linear discriminant analysis
m1=lda(type~.,glass)
m1
predict(m1,newdata=data.frame(RI=3.0,Na=13,Mg=4,Al=1,Si=70,
+    K=0.06,Ca=9,Ba=0,Fe=0))
predict(m1,newdata=data.frame(RI=3.0,Na=13,Mg=4,Al=1,Si=70,
+    K=0.06,Ca=9,Ba=0,Fe=0))$class
```

```
## quadratic discriminant analysis: Not enough data as
## each 9x9 covariance matrix includes (9)(10)/2 = 45
## unknown coefficients

n=length(fgl$type)
nt=200
neval=n-nt

rep=100
errlin=dim(rep)

for (k in 1:rep) {
train=sample(1:n,nt)
glass[train,]
## linear discriminant analysis
m1=lda(type~.,glass[train,])
predict(m1,glass[-train,])$class
tablin=table(glass$type[-train],predict(m1,
+    glass[-train,])$class)
errlin[k]=(neval-sum(diag(tablin)))/neval
}
merrlin=mean(errlin)
merrlin

n=214
neval=1
errlin=dim(n)
errqua=dim(n)

for (k in 1:n) {
train1=c(1:n)
train=train1[train1!=k]
## linear discriminant analysis
m1=lda(type~.,glass[train,])
predict(m1,glass[-train,])$class
tablin=table(glass$type[-train],predict(m1,glass[-train,])
+    $class)
errlin[k]=(neval-sum(diag(tablin)))/neval
}
merrlin=mean(errlin)
merrlin
```

We evaluate the linear discriminant analysis by randomly selecting 200 of 214 shards, estimating the parameters on the training data, and classifying the remaining 14 shards of the holdout sample. We repeat this 100 times. We achieve a 37.2% misclassification rate.

A crossvalidation (leave-one-out) analysis is carried out below. Its misclassification error rate of 35.5% is similar to that of the nearest neighbor algorithm.

```
n=214
neval=1
errlin=dim(n)
errqua=dim(n)

for (k in 1:n) {
train1=c(1:n)
train=train1[train1!=k]
## linear discriminant analysis
m1=lda(type~.,glass[train,])
predict(m1,glass[-train,])$class
tablin=table(glass$type[-train],predict(m1,glass[-train,])
+     $class)
errlin[k]=(neval-sum(diag(tablin)))/neval
}

merrlin=mean(errlin)
merrlin
```

12.5 EXAMPLE 4: MBA ADMISSION DATA

Johnson and Wichern (1988, Chapter 11) provide admission data for applicants to graduate schools in business. The objective is to use the GPA and GMAT scores to predict the likelihood of admission (admit, notadmit, and borderline).

```
library(MASS)
set.seed(1)
## reading the data
admit <- read.csv("C:/DataMining/Data/admission.csv")
adm=data.frame(admit)
adm
plot(adm$GPA,adm$GMAT,col=adm$De)

## linear discriminant analysis
m1=lda(De~.,adm)
m1
predict(m1,newdata=data.frame(GPA=3.21,GMAT=497))

## quadratic discriminant analysis
m2=qda(De~.,adm)
m2
predict(m2,newdata=data.frame(GPA=3.21,GMAT=497))
```

```
n=85
nt=60
neval=n-nt

rep=100
errlin=dim(rep)

for (k in 1:rep) {
train=sample(1:n,nt)
## linear discriminant analysis
m1=lda(De~.,adm[train,])
predict(m1,adm[-train,])$class

tablin=table(adm$De[-train],predict(m1,adm[-train,])$class)
errlin[k]=(neval-sum(diag(tablin)))/neval
}

merrlin=mean(errlin)
merrlin
```

We evaluate the linear discriminant analysis by randomly selecting 60 of 85 students, estimating the parameters on the training data, and classifying the remaining 25 students of the holdout sample. We repeat this 100 times. We achieve a 9.4% misclassification rate.

REFERENCE

Johnson, R.A. and Wichern, D.W.: *Applied Multivariate Statistical Analysis.* Second edition. Englewood Cliffs, NJ: Prentice Hall, 1988.

CHAPTER 13

Decision Trees

Ordinary regression, logistic regression, and multinomial logistic regression are discussed in Chapters 3, 7, and 11 of this book. These methods relate a response variable (a continuous variable in ordinary regression, and a categorical variable with two or more outcome groups in logistic and multinomial logistic regression) to a set of explanatory variables. A common feature of regression models is their *parametric nature*. Most often, linear relationships are being considered, which means that the effect on the response of a change in the explanatory variable from x to $x + 1$ is the same for any value x. Furthermore, it is usually assumed that the effect does not depend on the levels of other explanatory variables. Of course, these assumptions can be relaxed with the introduction of quadratic (or higher order) terms and components that allow for interaction. While it is then true that the effect of a change in x depends on the value of x and those of other covariates, the dependence is still parametric (e.g., as implied by the quadratic function). With the decision trees and regression/classification trees that are discussed in this and the next chapter, the models become truly nonparametric and they allow for very flexible representations. Decision trees provide a convenient and efficient representation of knowledge.

Decision trees use a tree-logic to make predictions. Classification and regression trees are known under their acronym CART. We talk about classification trees if the response is categorical; we talk about regression trees if the response is continuous. Regression trees try to predict a (numeric) mean response at the leaves of the tree, such as the expected amount of rain in inches, or the expected default rate on loans. Classification trees try to predict the class probabilities at the leaves, such as the probability that there will be rain, the probability of defaulting on a loan, or the preference for one of five different types of movie genres.

Let us start the discussion with a very simple example of a decision tree. Here, the decision is whether or not to take an umbrella. The decision depends on the weather, on the predicted rain probability, and on whether it is sunny or cloudy when you leave the house. You can go through the various decision nodes of the diagram and determine whether you should take an umbrella or leave it at home.

Data Mining and Business Analytics with R, First Edition. Johannes Ledolter.

For example, if the forecast predicts rain with a probability between 30% and 70% and if it is cloudy when you leave the house, you better take an umbrella. Tree-logic uses a series of steps to come to a conclusion. The trick is to combine several mini-decisions such that they result in good choices.

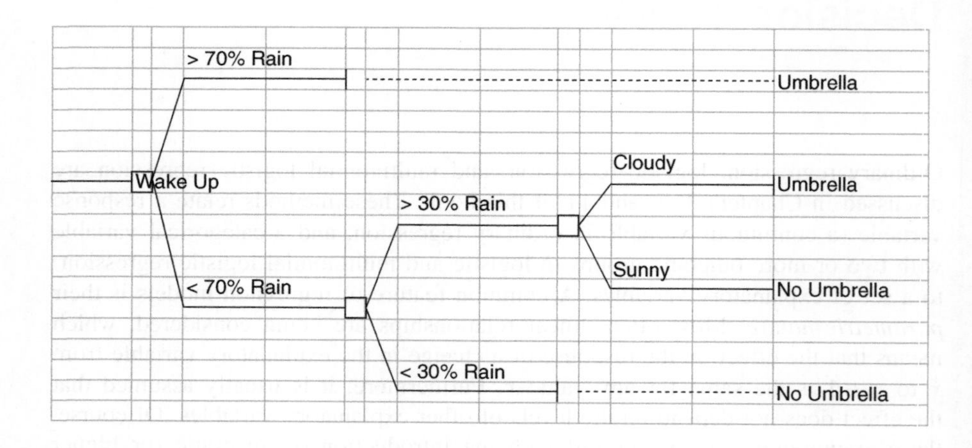

Each decision defines a *node*. The final prediction is made at a *leaf* or *terminal node*. A given set of covariates x dictates your path through the tree nodes, eventually leading to a leaf (or terminal) node at the end of the tree.

In our illustrative example, the path through the tree and the splits at the various nodes are assumed given. In actual applications, one has a training set with known outcomes, and the path through the tree and the splits need to be estimated. We need a way to estimate the sequence of decisions. How many decision nodes should there be? What is the order of making the decisions? Obviously, there is a huge set of possible tree configurations. We solve the problem by thinking recursively and by always splitting a node into just two branches. We start by splitting the complete data into two branches that make different decisions about the response y. Then we take each new partition and split it again into two, and again, and again, and so on.

Growing a tree with the CART algorithm involves the following steps. Find the split location in a covariate x that minimizes a certain measure of node impurity of the tree (node impurity is explained below). If, for example, the set of covariates comprises five variables, we try each variable one at a time, find the best two-way split on each variable, and select the variable and the split that minimize the node impurity of the tree. This certainly involves many computations and queries, but fast and efficient computer software is available. In the regression tree, the predictions at the nodes are given by the mean responses. Class proportions characterize the nodes in the classification tree, with the most frequent category representing the predicted classification.

Each child node of the split tree contains a subset of the data, with its own predicted $\widehat{y}$ and $\widehat{p}$. View each child as a new data set, and grow the tree from there. Stop growing when no split reduces the node impurity of the tree. Instead of being based on (regression) coefficients, the predictions (the $\widehat{y}$ in the regression tree, and the predicted probabilities $\widehat{p}$ in the classification tree) are functions of decisions at the decision nodes (which depend on the covariates in some nonparametric way).

Programs are available for regression and classification trees. We use the R package **tree**, with syntax

```
mytree = tree(y ~ x1 + x2 + x3..., data=mydata)
```

There are a few additional arguments in this function, all of which dictate the splitting of the tree. Requirements on minimum leaf size (number of observations in a leaf), minimum improvement in node impurity, and maximum depth of the tree (number of splits within the longest branch) avoid trees that are split too often. One can print, summarize, and plot the tree.

Regression and classification trees partition a data set into exhaustive and nonoverlapping nodes. The data partitioning is nonoverlapping (every element of the data set is part of just one node) and exhaustive (every element of the data set must fall into one of the nodes). At each node of the tree, the response of interest is summarized by an average response (for regression trees) or a frequency distribution (for classification trees).

Tree construction uses a recursive partitioning of the data set; the approach is also referred to as *divide-and-conquer partitioning*. At each stage, the data set is split into two nonoverlapping, smaller data sets. The objective of a split is to increase the homogeneity of the resulting smaller data sets with respect to the target variable (alternatively, we can think of this as decreasing the impurity, or the disorder). We continually divide the data set by creating node splits of smaller and smaller data sets.

We need a measure of node impurity that guides our splitting of the data set and that tells us whether it makes sense to split a certain node even further. For regression trees, which predict a numerical response y, we use the *regression deviance*. The regression deviance at a given node of the regression tree is the sum of squares $D_{\text{node}} = \sum_{i=1}^{n} (y_i - \overline{y})^2$. Here, $y_1, y_2, \ldots, y_n$ are the responses of the data elements that make up the node, and $\overline{y}$ is their average. The regression deviance measures the node impurity (disorder) and assesses the homogeneity of the responses within the node. The deviance is simply the sum of the squared residuals, with the residual defined as the difference between the observation and the group (node) mean. Here we have defined the deviance of a single node in a tree. The deviance of a regression tree T, described by an exhaustive collection of nonoverlapping nodes, is obtained by adding the deviances of its nodes; that is, $D_{\text{T}} = \sum D_{\text{node}}$. Small values of the deviance reflect homogeneity and node purity (or node order). Consider an initial regression tree T1 and a regression tree T2 that splits one of the nodes of T1 into two and represents each of the child nodes by their average. Splitting a node of a tree into two can never increase the deviance of the tree, as any split allows

Table 13.1 Splitting a Regression Tree

x	y		Mean $\bar{y}$ Left	Mean $\bar{y}$ Right	Deviance Left	Deviance Right	Deviance Tree	
		No split					17.2	
1	10	Split at 1	10	8	0	14	14	
2	8	Split at 2	9	8	2	14	16	
4	11	Split at 4	29/3	6.5	4.667	0.5	5.167	Best split
6	7	Split at 6	9	6	10	0	10	
8	6	No split					17.2	

for more flexibility (in the worst case, the averages of the two split nodes are the same). Hence the difference between the deviance of the initial tree T1 (D_{T1}) and the deviance of tree T2 (D_{T2}), $D_{T1} - D_{T2} \geq 0$, reflects the benefit of the splitting. We want this difference to be as large as possible. Computer programs refer to the regression deviance also as the analysis of variance (ANOVA) deviance, as the analysis of variance in statistics deals with partitioning sums of squares.

For splitting a node on a continuous variable, we order the continuous variable from the smallest to the largest value and evaluate the impurity measures for splits at the order statistics, subject to satisfying the requirements on the minimum number of units in each node. For simplicity, take a small data set with five observations on a single attribute x and a continuous response y. The data in Table 13.1 are already ordered on the values of the attribute. The mean of all five observations is 8.4, and the ANOVA deviance of the data is

$$(10 - 8.4)^2 + (8 - 8.4)^2 + (11 - 8.4)^2 + (7 - 8.4)^2 + (6 - 8.4)^2 = 17.2.$$

For the first possible split, which puts units with $x \leq 1$ into the lower node and units with $x > 1$ into the upper node, the ANOVA deviance for the left node is $(10 - 10)^2 = 0$, and the ANOVA deviance for the right node is $(8 - 8)^2 + (11 - 8)^2 + (7 - 8)^2 + (6 - 8)^2 = 14$. The deviance of the split tree is 14. For the second split, which puts units with $x \leq 2$ into the lower node and units with $x > 2$ into the upper node, the ANOVA deviance for the lower (left) node is $(10 - 9)^2 + (8 - 9)^2 = 2$, and the ANOVA deviance for the upper (right) node is $(11 - 8)^2 + (7 - 8)^2 + (6 - 8)^2 = 14$. The deviance of the split tree is 16. For the third split, which puts units with $x \leq 4$ into the lower node and units with $x > 4$ into the upper node, the ANOVA deviance for the left node is $(10 - 29/3)^2 + (8 - 29/3)^2 + (11 - 29/3)^2 = 42/9 = 4.667$, and the ANOVA deviance for the right node is $(7 - 6.5)^2 + (6 - 6.5)^2 = 0.50$. The deviance of the split tree is 5.167. For the fourth and final split, which puts units with $x \leq 6$ into the lower node and units with $x > 6$ into the upper node, the ANOVA deviance for the left node is $(10 - 9)^2 + (8 - 9)^2 + (11 - 9)^2 + (7 - 9)^2 = 10$, and the ANOVA deviance for the right node is $(6 - 6)^2 = 0$. The deviance of the split tree is 10.

The best split is the third split, which puts units with $x \leq 4$ into the lower node and units with $x > 4$ into the upper node. This example illustrates the calculations that need to be carried out. Remember that the software for constructing trees has to perform these tasks not just for small data sets of five observations, but for large data sets; also, it has to do this for several variables and not just one; and it has to do this repeatedly as the data set is being divided into smaller and smaller splits. Fortunately, efficient and fast procedures for doing this are readily available.

How does the search change if one has a categorical attribute x with, say, three possible groups (e.g., three different sales districts, A, B, and C)? Then we need to evaluate the deviance of the following $2^{(3-1)} - 1 = 3$ splits: (A) versus (B,C); (B) versus (A,C); and (C) versus (A,B). What if there are four possible groups? Then we need to evaluate $2^{(4-1)} - 1 = 7$ splits: (A) versus (B,C,D); (B) versus (A,C,D); (C) versus (A,B,D); (D) versus (A,B,C); (A,B) versus (C,D); (A,C) versus (B,D); and (A,D) versus (B,C). In general, for categorical variables with k possible outcomes, one has to evaluate the deviance of $2^{(k-1)} - 1$ splits.

In the classification setting, the response is a class assignment with m different possible outcomes. Consider a node with n data elements and number of occurrences $n_1, n_2, \ldots, n_m$ for the m possible outcomes ($n = n_1 + n_2 + \cdots + n_m$). We need a measure that assesses the impurity of the classification within that node. A node that classifies all its items into a single category (i.e., achieves a classification with 100% certainty) needs no additional help (additional information) to classify an observation. It represents maximal homogeneity, maximal information, and minimal node impurity. On the other hand, a node that assigns its elements evenly among two or more outcomes (for two outcomes, half of the observations of the node are assigned to one outcome group, while the other half is assigned to the other) describes a situation of minimal homogeneity, minimal information, and maximal impurity. A measure of impurity for a node with number of occurrences $n_1, n_2, \ldots, n_m$ for the m possible outcomes ($n = n_1 + n_2 + \cdots + n_m$) is provided by the *classification deviance*,

$$D_{\text{node}} = -2\sum_{k=1}^{m} n_k \log\left(\frac{n_k}{n}\right) = -2\left[\sum_{k=1}^{m} n_k \log(n_k) - n\log(n)\right],$$

with $0\log(0) = 0$. For a perfect classification (and no impurity) at that node, $D_{Node} = 0$. For n observations assigned evenly to the m possible outcomes and maximum impurity, $D_{Node} = 2n\log(m)$. Here we have defined the deviance of a single node in a tree. The deviance of a classification tree T, described by an exhaustive collection of nonoverlapping nodes, is obtained by adding the deviances of its nodes; that is, $D_{\text{T}} = \sum D_{\text{node}}$. Splitting a node of a tree into two by minimizing the sum of deviances from the two resulting nodes can never increase the tree deviance; in the worst case, the sum of the deviances equals the deviance of the unsplit node. The difference between the deviance of the initial tree T1 (D_{T1}) and the deviance of tree T2 (D_{T2}), $D_{\text{T1}} - D_{\text{T2}} \geq 0$, reflects the benefit of the splitting. Again, we want to find the variables on which to split and the values for the splits such that this difference is as large as possible.

Let us describe the classification deviance with a simple example. Consider the classification into just two groups. Suppose the data set of 100 observations has 60

YES and 40 NO. The node impurity is expressed by the classification deviance

$$-2[60 \log(60) + 40 \log(40) - 100 \log(100)] = 134.60.$$

Assume that a certain covariate splits the data set of 100 into two groups of sizes 30 and 70. In scenario 1, the proportions of YES and NO stay the same, that is G1: (YES = 18, NO = 12) and G2: (YES = 42, NO = 28). The classification deviance of the split tree is

$$-2[18 \log(18) + 12 \log(12) - 30 \log(30)] - 2[42 \log(42) + 28 \log(28) - 70 \log(70)]$$
$$= 40.38 + 94.22 = 134.60.$$

We see that the change in deviance is 134.60 − 134.60 = 0, which tells us that there is no benefit in splitting the tree. The split has not increased the homogeneity of the tree.

Under scenario 2, the proportions of YES and NO exhibit greater homogeneity. Suppose G1: (YES = 25, NO = 5) and G2: (YES = 35, NO = 35), as the numbers of YES need to add up to 60. The classification deviance of the split tree is

$$-2[25 \log(25) + 5 \log(5) - 30 \log(30)] - 2[35 \log(35) + 35 \log(35) - 70 \log(70)]$$
$$= 27.03 + 97.04 = 124.07.$$

The change in deviance is 134.60 − 124.07 = 10.53, which tells us that there is a benefit in splitting the tree. This particular split has increased the homogeneity of the tree. However, there may be other splits that are even better.

The literature and computer programs refer to the classification deviance as the *entropy* (or the *information*) criterion. Why is it referred to as entropy? The entropy of a discrete random variable with m possible outcome values $o_1, o_2, \ldots, o_m$ and associated probabilities $p(o_1), p(o_2), \ldots, p(o_m)$ is given by $E = -\sum_{k=1}^{m} p(o_k) \log(p(o_k))$. It measures the information content of a random variable, with small values of entropy reflecting homogeneity or little impurity (disorder). For two groups ($m = 2$) and $p(o_1) = p$ and $p(o_2) = 1 - p$, the entropy is given by $E = -p \log(p) - (1 - p) \log(1 - p)$. The entropy is 0 for $p = p(o_1) = 1$ or $p = p(o_1) = 0$, and it is largest (it is $\log(2)$) for $p = p(o_1) = p(o_2) = 0.5$. Replacing the unknown class probabilities $p(o_k)$ with relative frequencies n_k/n provides an estimate of the entropy of a single realization. This must be multiplied by n, as the node contains n data elements, resulting in the node entropy $-\sum_{k=1}^{m} n_k \log(n_k/n)$. Apart from the factor 2, this is the classification deviance. But the multiplication by 2 makes no difference to how splits are carried out, as we compare the difference of the entropies before and after a split. The factor 2 in our deviance definition can be justified from maximum likelihood ratio tests that compare twice the difference of the logarithms of the likelihoods of two competing models: in our case, the trees before and after a split. Assuming a multinomial distribution for the occurrences in m groups leads to our classification deviance, which includes the multiple 2.

Other criteria for splitting a classification tree are available, such as the *Gini index* or the *number of misclassifications* that are made when assigning a node element to the most common category in that node. The Gini index,

$$\text{Gini} = \sum_{k=1}^{m} p(o_k)[1 - p(o_k)] = 1 - \sum_{k=1}^{m} [p(o_k)]^2,$$

leads to the node similarity measure $\text{Gini}_{\text{node}} = \sum_{k=1}^{m} n_k[1 - (n_k/n)]$. It is quite similar to the deviance (or entropy), and it usually leads to very similar splits. To see the similarity, take the binary classification ($m = 2$) as an example, with $p(o_1) = p$ and $p(o_2) = 1 - p$. The Gini index is $2p(1 - p)$, and we saw earlier that the entropy is $-p\log(p) - (1 - p)\log(1 - p)$. The misclassification error is also easy to determine; it is given by $\min(p, 1 - p)$. The graph of these three measures against p in Figure 13.1 shows their close similarity. The entropy and the Gini index are differentiable for all p, which makes them more amenable to numeric optimization.

13.1 EXAMPLE 1: PROSTATE CANCER

Let us consider the example of prostate cancer that we analyzed previously in the context of our discussion of penalized variable selection (LASSO) in Chapter 6. Biopsy results are available for $n = 97$ men of various ages. The information includes

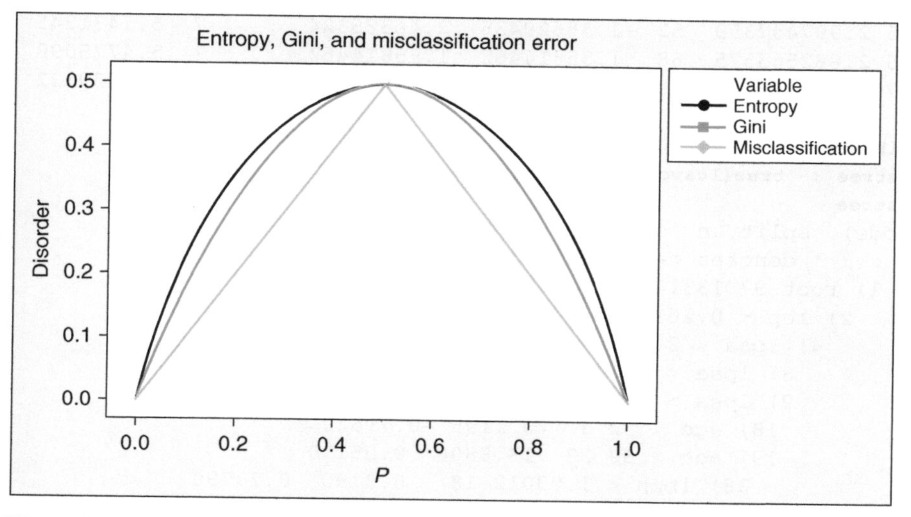

Figure 13.1 Node impurity measures in the two-class classification as a function of the proportion of falling into class 1. The entropy measure has been scaled to pass through the point (0.5,0.5).

Gleason Score (gleason): scores are assigned to the two most common tumor patterns, ranging from 2 to 10; in this data set, the range is from 6 to 9.

Prostate-specific antigen (psa): laboratory results on protein production.

Capsular penetration (cp): reach of cancer into gland lining.

Benign prostatic hyperplasia amount (bph): size of prostate.

The goal is to predict tumor log volume (which measures the tumor's size/spread). The predicted tumor size affects the patients' treatment options, which include chemotherapy, radiation treatment, and surgical removal of the prostate.

The data on the 97 prostate cancer patients is given below. The response is log volume (lcavol). We try to predict this variable from five covariates (age; logarithms of bph, cp, and psa; and the Gleason score). Here the response is a continuous measurement variable, and we are dealing with a regression tree. We use the sum of squared residuals as the impurity (fitting) criterion.

```
prostate <- read.csv("C:/DataMining/Data/prostate.csv")
prostate
        lcavol  age        lbph         lcp   gleason   lpsa
1 -0.579818495   50 -1.38629436 -1.38629436         6 -0.4307829
2 -0.994252273   58 -1.38629436 -1.38629436         6 -0.1625189
3 -0.510825624   74 -1.38629436 -1.38629436         7 -0.1625189
4 -1.203972804   58 -1.38629436 -1.38629436         6 -0.1625189
.
.
.
94 3.821003607   44 -1.38629436  2.16905370         7  4.6844434
95 2.907447359   52 -1.38629436  2.46385324         7  5.1431245
96 2.882563575   68  1.55814462  1.55814462         7  5.4775090
97 3.471966453   68  0.43825493  2.90416508         7  5.5829322

library(tree)
pstree <- tree(lcavol ~., data=prostate, mincut=1)
pstree
node), split, n, deviance, yval
      * denotes terminal node
  1) root 97 133.4000  1.35000
    2) lcp < 0.261624 63  64.1100  0.79250
      4) lpsa < 2.30257 35  24.7200  0.27870
        8) lpsa < 0.104522 4   0.3311 -0.82220 *
        9) lpsa > 0.104522 31  18.9200  0.42070
         18) age < 52 3   0.1195 -0.79620 *
         19) age > 52 28  13.8800  0.55110
           38) lbph < 1.09012 18   6.3190  0.73790
             76) age < 65.5 14   4.0670  0.55550
              152) lcp < -0.698172 11   2.1200  0.37820 *
              153) lcp > -0.698172 3   0.3329  1.20600 *
             77) age > 65.5 4   0.1552  1.37600 *
```

```
        39) lbph > 1.09012 10   5.8010  0.21490
          78) lpsa < 1.96623 7   2.8370 -0.08817 *
          79) lpsa > 1.96623 3   0.8212  0.92200 *
    5) lpsa > 2.30257 28  18.6000  1.43500
     10) lpsa < 3.24598 23  11.6100  1.23300 *
     11) lpsa > 3.24598 5   1.7560  2.36200 *
  3) lcp > 0.261624 34  13.3900  2.38300
    6) lcp < 2.13963 25   6.6620  2.14700
     12) age < 62.5 7   0.7253  1.68600 *
     13) age > 62.5 18   3.8700  2.32600 *
    7) lcp > 2.13963 9   1.4750  3.03800 *
```

```
plot(pstree, col=8)
text(pstree, digits=2)
```

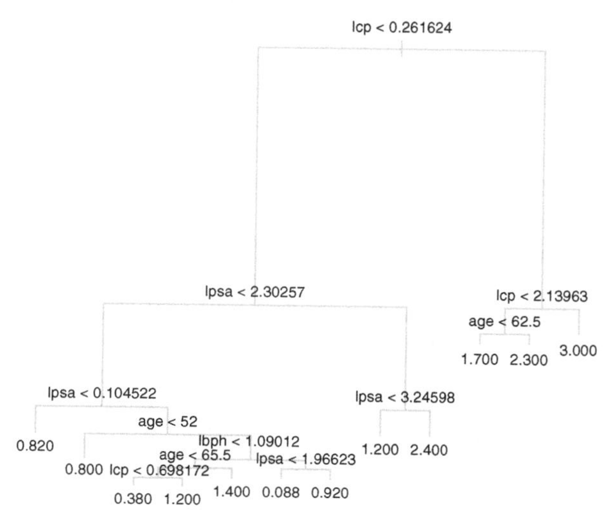

We obtain a tree with 12 leaf nodes (just count up the number of terminal nodes). The size of our tree is 12. The graphs, with and without text, show the nodes and how they are split. The y-axis on the graph reflects the deviances of the various trees that are fit to the data. We start on top at the root of the tree. The deviance (the sum of squares from the single mean 1.35) is 133.4. The first split is on log capsular penetration, with lcp < 0.261624 (node 2) and lcp > 0.261624 (node 3). The deviance of the tree with just this one split is $64.11 + 13.39 = 77.5$. The reduction from 133.4 to 77.5 is reflected on the y-axis of the graph. The next split (under lcp < 0.261624) is on lpsa, creating node 4 (lpsa < 2.30257) and node 5 (lpsa > 2.30257), and so on. The reductions in deviances become smaller and smaller, which unfortunately often creates an overlap in the printed labels of the tree.

How does one use this tree for prediction? As illustration, take a patient, 65 years old and with lcp $= 2$ (log capsular penetration). What is his prediction of log tumor size? Looking at the splits on the tree, you go right on the first split (lcp > 0.261624), left on the second split (lcp < 2.13963), and right on the third split (age > 62.5): the model predicts 2.30 for his log volume. What about a patient with lcp $= 0.20$ and lpsa $= 2.40$? His prediction for log volume is 1.20, and so on.

CART represents a pretty powerful nonparametric technique that generalizes parametric regression models. It allows for nonlinearity and variable interactions without having to specify the structure in advance. Moreover, violations of constant variance (a critical assumption in regression) are no problem. But the biggest challenge with such flexible models is how to avoid *overfitting*. If the splitting algorithm is not stopped, the tree algorithm will ultimately "extract" all information from the data, including information that is not and cannot be predicted in the population with the current set of predictors. In other words, it extracts random or noise variation. The first defense against overfitting is to stop generating new split nodes when subsequent splits result only in very little overall improvement of the prediction. For example, if we can predict 90% of all cases correctly from 10 splits, and 90.1% of all cases from 11 splits, then it obviously makes little sense to add that 11th split to the tree. There are several criteria for automatically stopping the splitting (tree-building) process. The basic constraints (mincut, mindev) lead to a full tree fit with a certain number of terminal nodes. In the prostate example, we specified mincut $= 1$ (minimum number of observations to include in a child node is 1) and obtained a tree of size 12. If we specify mincut $= 5$, the final tree ends up with fewer leaf nodes and a smaller size. Check that its size is 9.

Once the tree-building algorithm has stopped, it is always useful to evaluate the quality of the prediction of the current tree in samples of observations that did not participate in the fitting computations. Such *cross-validation* subjects the tree computed from one set of observations (the learning sample) to another independent set of observations (the evaluation/test sample). If most or all of the splits determined by the analysis of the learning sample are essentially based on "random noise," then the prediction for the test sample will be poor. If one sees a big difference between the in-sample and the out-of-sample performance, one can infer that the selected tree is not very useful, and not of the "right size."

Pruning back the tree is another good defense against overfitting. There, we select a simpler tree than the tree that resulted from the tree-building (i.e., the growing) algorithm. The hope is that the simpler tree does better when it is used to predict or classify "new" observations. We prune the tree by removing splits from the bottom up. At each step, we remove the split that contributes least to deviance reduction, thus reversing CART's growth process. Instead of deviance reduction, one can also look at the change in the number of misclassification, if this is felt to be a more appropriate criterion. Each prune step produces a candidate tree model, now of reduced size (number of terminal nodes).

Let us explain this in more detail. Assume that our criterion for pruning the tree is the minimization of the deviance of the tree. With this objective function, the in-sample (training set) performance will only get worse with pruning as every pruned tree will have a larger deviance; remember that we grew the tree by minimizing the deviance, and so, taking away a split that we found useful in growing the tree will make things worse. To achieve some pruning, we must modify the objective function and introduce a penalty for the complexity of the tree. This is reasonable, as we know that overfitting the data will hurt our out-of-sample predictions. Instead of minimizing the deviance of the tree, $D(\mathrm{T})$, the pruning step minimizes the *cost complexity* of the tree, $D(\mathrm{T}) + \alpha\mathrm{Size}(\mathrm{T})$. The size of the tree T is the number of its terminal nodes, and α is a penalty term. We already saw that with $\alpha = 0$ (i.e., no penalty for the size of the tree), our final grown tree cannot be simplified (pruned). For illustration, look at terminal nodes 152 and 153 at the very bottom of the tree. Pruning that split increases the deviance by $4.0670 - (2.1200 + 0.3329) = 1.6151$; for $\alpha = 0$, no pruning step is executed and we keep our final tree of size 12. What if the penalty is $\alpha = 1.7$? Then the increase in the deviance (1.6151) is less than the penalty that is associated with the one extra node (1.7), and in this case, the tree could be pruned back. Let us check whether additional pruning among other terminal nodes can be done. What about terminal nodes 78 and 79? They cannot be pruned as the increase in deviance $5.8010 - (2.8370 + 0.8212) = 2.1428 > 1.70$. What about the five terminal nodes 152, 153, 77, 78, and 79? They cannot be pruned, as the increase in deviance $13.8800 - (2.1200 + 0.3329 + 0.1552 + 2.8370 + 0.8212) = 7.6137$ is larger than four times the penalty, $(4)(1.7) = 6.80$, and so on. It turns out that nodes 152 and 153 are the only nodes that can be pruned. The simplified tree has size 11. This is what you see by executing the prune step with $\alpha = 1.7$ (note that the R function *prune.tree* uses the letter k for the penalty term).

```
pstcut <- prune.tree(pstree,k=1.7)
plot(pstcut)
pstcut
```

```
node), split, n, deviance, yval
      * denotes terminal node
```

```
1) root 97 133.4000  1.35000
  2) lcp < 0.261624 63  64.1100  0.79250
    4) lpsa < 2.30257 35  24.7200  0.27870
      8) lpsa < 0.104522 4   0.3311 -0.82220 *
      9) lpsa > 0.104522 31  18.9200  0.42070
       18) age < 52 3   0.1195 -0.79620 *
       19) age > 52 28  13.8800  0.55110
         38) lbph < 1.09012 18   6.3190  0.73790
           76) age < 65.5 14   4.0670  0.55550 *
           77) age > 65.5 4   0.1552  1.37600 *
         39) lbph > 1.09012 10   5.8010  0.21490
           78) lpsa < 1.96623 7   2.8370 -0.08817 *
           79) lpsa > 1.96623 3   0.8212  0.92200 *
    5) lpsa > 2.30257 28  18.6000  1.43500
     10) lpsa < 3.24598 23  11.6100  1.23300 *
     11) lpsa > 3.24598 5   1.7560  2.36200 *
  3) lcp > 0.261624 34  13.3900  2.38300
    6) lcp < 2.13963 25   6.6620  2.14700
     12) age < 62.5 7   0.7253  1.68600 *
     13) age > 62.5 18   3.8700  2.32600 *
    7) lcp > 2.13963 9   1.4750  3.03800 *
```

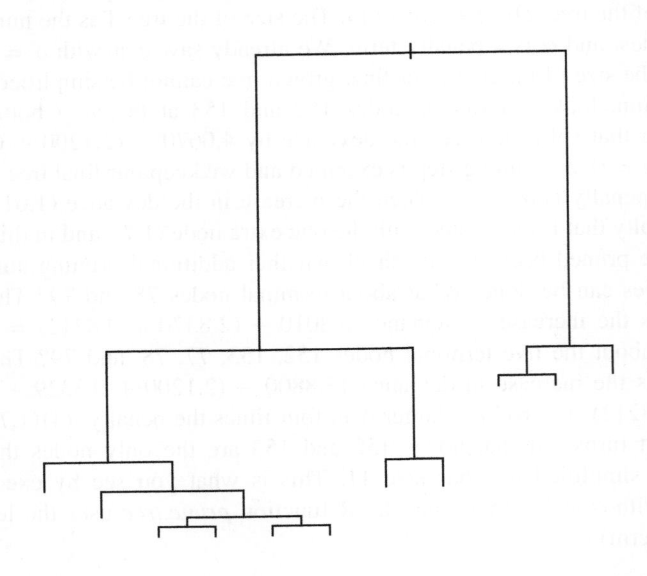

Now let us consider $\alpha = 2.05$, and investigate whether further pruning can be done. Terminal nodes 12 and 13 cannot be pruned, as the increase in deviance $6.6620 - (0.7253 + 3.8700) = 2.0667 > 2.05$, the penalty for that one extra terminal node. Terminal nodes 10 and 11 cannot be pruned either, as the increase in deviance $18.6000 - (11.6100 + 1.7560) = 5.2340 > 2.05$, the penalty for that one extra terminal node. However, the four terminal nodes 76, 77, 78, and 79 can be pruned at the same time. The increase in deviance $13.8800 - (6.3190 + 5.8010) = 1.7600$ is smaller than the penalty $(3)(2.05) = 6.15$; here the pruning collapses four leaf nodes into one and the difference in the sizes of the tree is 3. No other pruning is possible. The pruned tree (given below) has size 8.

```
pstcut <- prune.tree(pstree,k=2.05)
plot(pstcut)
pstcut
```

```
1) root 97 133.4000  1.3500
  2) lcp < 0.261624 63  64.1100  0.7925
    4) lpsa < 2.30257 35  24.7200  0.2787
      8) lpsa < 0.104522 4   0.3311 -0.8222 *
      9) lpsa > 0.104522 31  18.9200  0.4207
       18) age < 52 3   0.1195 -0.7962 *
       19) age > 52 28  13.8800  0.5511 *
    5) lpsa > 2.30257 28  18.6000  1.4350
     10) lpsa < 3.24598 23  11.6100  1.2330 *
     11) lpsa > 3.24598 5   1.7560  2.3620 *
  3) lcp > 0.261624 34  13.3900  2.3830
    6) lcp < 2.13963 25   6.6620  2.1470
     12) age < 62.5 7   0.7253  1.6860 *
     13) age > 62.5 18   3.8700  2.3260 *
    7) lcp > 2.13963 9   1.4750  3.0380 *
```

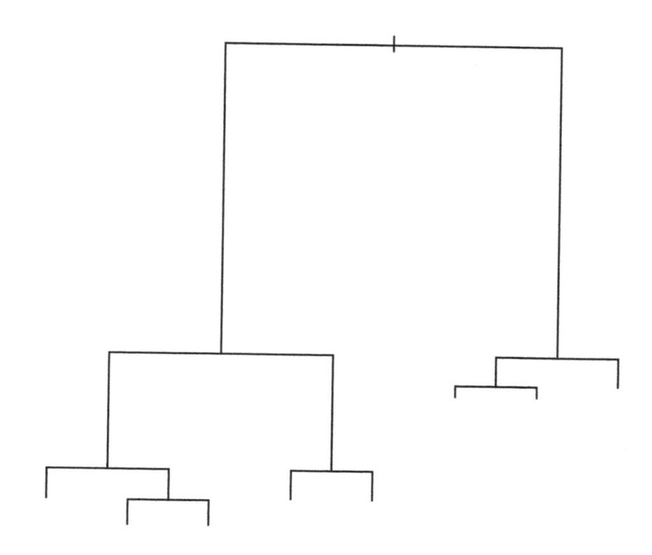

Next, consider the somewhat larger $\alpha = 3.0$, and investigate whether further pruning can be done. Nodes 12 and 13 can be pruned. The increase in deviance $6.6620 - (0.7253 + 3.8700) = 2.0667 < 3.0$ is smaller than its penalty (note that with $\alpha = 2.05$ these two nodes could not be pruned). Nothing else can be pruned. For example, terminal nodes 18 and 19 cannot be pruned, as the resulting increase in deviance $18.9200 - (0.1195 + 13.8800) = 4.9205 > 3.0$. The pruned tree (given below) has size 7.

```
pstcut <- prune.tree(pstree,k=3)
plot(pstcut)
pstcut
```

```
1) root 97 133.4000  1.3500
  2) lcp < 0.261624 63  64.1100  0.7925
    4) lpsa < 2.30257 35  24.7200  0.2787
      8) lpsa < 0.104522 4   0.3311 -0.8222 *
      9) lpsa > 0.104522 31  18.9200  0.4207
       18) age < 52 3   0.1195 -0.7962 *
       19) age > 52 28  13.8800  0.5511 *
    5) lpsa > 2.30257 28  18.6000  1.4350
     10) lpsa < 3.24598 23  11.6100  1.2330 *
     11) lpsa > 3.24598 5   1.7560  2.3620 *
  3) lcp > 0.261624 34  13.3900  2.3830
    6) lcp < 2.13963 25   6.6620  2.1470 *
    7) lcp > 2.13963 9  1.4750  3.0380 *
```

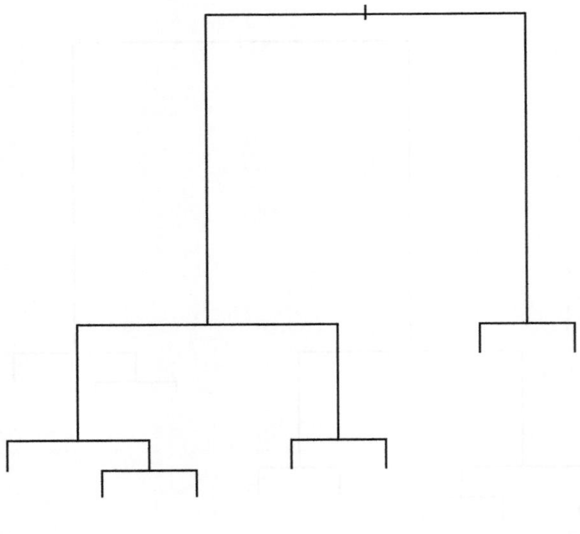

The R command *prune.tree*, without specifying the cost complexity parameter α (labeled as argument k), traces out the pruned trees for changing α (and implied size of the tree). Tree deviances are also listed. You get those by adding the deviances across all leaf nodes. For example, the deviance of the pruned tree with $\alpha = 1.7$ and 11 terminal nodes is $0.3311 + 0.1195 + 4.0670 + 0.1552 + 2.8370 + 0.8212 + 11.6100 + 1.7560 + 0.7253 + 3.8700 + 1.4750 = 27.77$.

```
pstcut <- prune.tree(pstree)
pstcut

$size
 [1] 12 11 8 7 6 5 4 3 2 1
$dev
 [1]  26.15491  27.76888  33.76664  35.83388  40.75225  45.98251  51.23381
 [8]  56.70719  77.50140 133.35903
$k
 [1]      -Inf  1.613972  1.999253  2.067239  4.918373  5.230262  5.251294
 [8]  5.473378 20.794213 55.857635
$method
[1] "deviance"
attr(,"class")
[1] "prune"         "tree.sequence"

plot(pstcut)
```

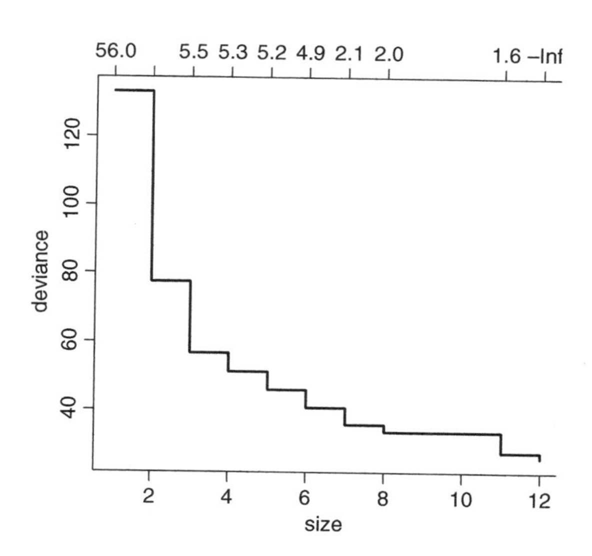

Assume that you want a pruned tree of size 7. You could get this tree with $\alpha = 3.0$ (in fact, any value that is between 2.067 and 4.918). The deviance of that tree is 35.83. You can check that a value $\alpha = 5.0$ leads to a tree of size 6, with the deviance being 40.75. You can use the earlier table and the associated graph to trade off the size of the tree and the deviance. The graph shows the tree deviance as a function of the penalty (x-scale on top) and the size of the tree (x-scale at bottom). Even a tree of size 3 has fairly acceptable deviance (56.7); it is for trees of even lower size that the deviance goes up rapidly. You can learn about the structure of the tree with three terminal nodes by using the *prune.tree* command with option "best = 3". The deviance of this tree is 24.72 + 18.60 + 13.39 = 56.71.

```
pstcut <- prune.tree(pstree,best=3)
pstcut

1) root 97 133.40 1.3500
  2) lcp < 0.261624 63   64.11 0.7925
    4) lpsa < 2.30257 35   24.72 0.2787 *
    5) lpsa > 2.30257 28   18.60 1.4350 *
  3) lcp > 0.261624 34   13.39 2.3830 *
```

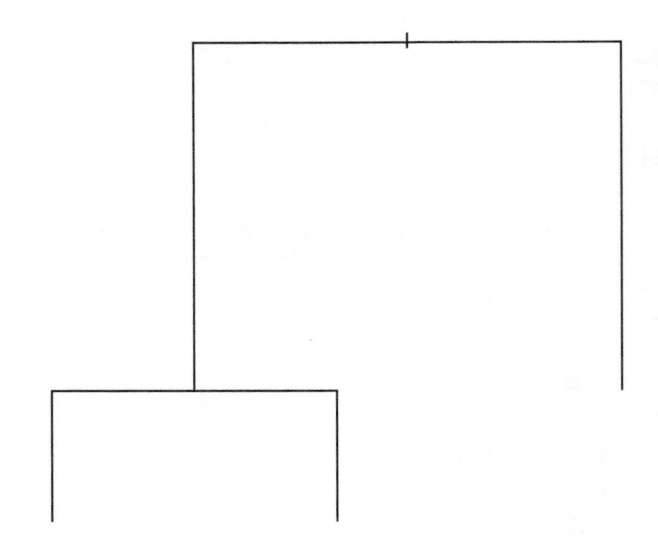

Note that the in-sample (training set) performance is bound to get worse by pruning as the tree gets less complicated. However, if a complicated tree is just "chasing after noise", then a simpler tree, which does worse on the training set, may actually predict better on a holdout data set. So an even better way to determine the size of the tree is to evaluate the performance of the pruned trees on a holdout sample. There we can compare the out-of-sample prediction performances of the candidate trees of varying sizes and select the size of tree that performs best.

This method of selecting the right size of a tree from a holdout sample is an essential step for generating tree models that hold up well for prediction. In order to perform evaluations to measure the tree performance as a function of tree size, it is necessary to have evaluation (test) data samples that are independent of the learning data set that has been used to build the tree. However, independent evaluation data is often difficult or expensive to obtain, and sometimes it is also undesirable to hold back data from the learning data set to use for a separate test because that weakens the learning data set. We need independent data that was not used to build the tree to measure the performance, but, on the other hand, we want to use all data to build the tree. *V-fold cross-validation*, described below, is a technique for performing independent tree size tests without requiring separate data sets and without reducing the data that is used to build the tree.

All of the rows in the learning data set are used to build the tree. This tree is intentionally allowed to grow larger than is likely to be optimal. This is called the *reference*, or unpruned tree. The reference tree is the best tree that fits the learning data set. Next, the learning data set is partitioned into a certain number of groups called *folds*. The partitioning is done using stratification methods so that the distributions of the target variable are approximately the same in the partitioned groups. The number of groups that the rows are partitioned into is the "V" in "V-fold cross classification." Research has shown that little is gained by using more than 10 partitions; so 10 is the recommended default number of partitions.

Let us assume 10 partitions are created. We combine the rows in nine of the partitions into a new pseudo-learning data set. A test tree is constructed and subsequently pruned, resulting in a set of terminal nodes for each possible tree size. The 10% (1 out of 10 partitions) of the data that is held back from the test tree is independent of the test tree, and it is used for an out-of-sample evaluation of the test tree. The 10% of the data that is held back is run through the test tree and the tree's performance is evaluated at its various class sizes. The evaluation is usually carried out in terms of deviance or in terms of classification errors if classification trees are involved. Next, a different set of nine partitions is collected into a new pseudo-learning data set. The partition being held back this time is different than the partition that was held back for the first test tree. A second test tree is built and its performance measures are computed using the data that was held back when it was built. This process is repeated 10 times, building 10 separate test trees. In each case, 90% of the data is used to build a test tree and 10% is held back for independent testing. A different 10% is held back for each test tree.

Once the 10 test trees have been built, their out-of-sample performances, for given tree size, are averaged. The averaged error rate for a particular tree size is known as the *cross-validation cost* (or CV cost). The tree size that produces the minimum CV cost is found. The reference tree is then pruned back to the number of nodes matching the size that produces the minimum CV cost. Pruning is done in a stepwise bottom-up manner, as has been explained previously, removing the least important nodes during each pruning cycle. It is important to note that the test trees built during the CV process are used only to find the optimal tree size. Their structure (which may be different in each test tree) has no bearing on the structure of the reference tree that is constructed using the full learning data set. The reference tree that is pruned back to the optimal size determined by CV is the best tree to use for predicting/classifying future data sets.

V-fold CV is carried out with the R command *cv.tree*. The graph of the CV deviances shown in the following indicates that, for the prostate example, a tree of size 3 is appropriate. The deviance is smallest for trees of size 3. The reference tree that was obtained from all the data is now being pruned back to size 3. CV chooses capsular penetration and PSA as the deciding variables. Note the interaction: the effect of capsular penetration on the response (log volume) depends on PSA. The final picture shows that CART divides up the space of the explanatory variables into rectangles, with each rectangle leading to a different prediction. The size of the circles of the data points in the respective rectangles reflects the magnitude of the response. This graph confirms that the tree splits are quite reasonable.

```
set.seed(2)
cvpst <- cv.tree(pstree, K=10)
cvpst$size
 [1] 12 11  8  7  6  5  4  3 2 1
cvpst$dev
 [1]  73.22055  70.18426  70.79439  70.24455  65.40004  65.30020  65.30020
 [8]  64.93791  90.18312 134.09981
plot(cvpst, pch=21, bg=8, type="p", cex=1.5, ylim=c(65,100))
```

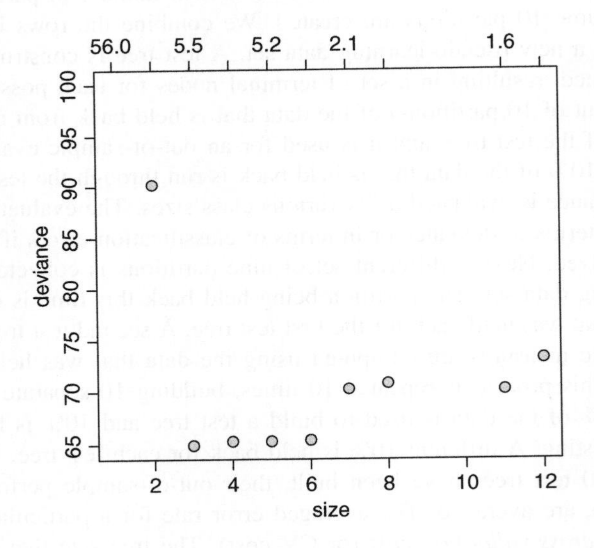


```
pstcut <- prune.tree(pstree, best=3)
pstcut

node), split, n, deviance, yval
      * denotes terminal node
1) root 97 133.40 1.3500
  2) lcp < 0.261624 63  64.11 0.7925
    4) lpsa < 2.30257 35  24.72 0.2787 *
    5) lpsa > 2.30257 28  18.60 1.4350 *
  3) lcp > 0.261624 34  13.39 2.3830 *

plot(pstcut, col=8)
text(pstcut)
```

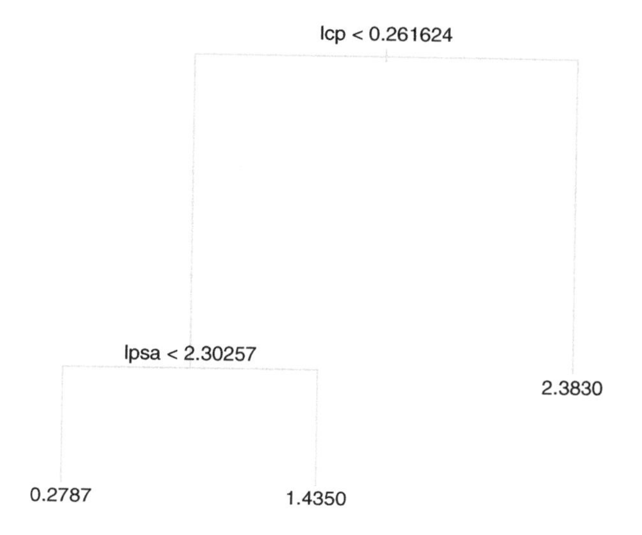


```
plot(prostate[,c("lcp","lpsa")],cex=0.2*exp(prostate$lcavol))
abline(v=.261624, col=4, lwd=2)
lines(x=c(-2,.261624), y=c(2.30257,2.30257), col=4, lwd=2)
```

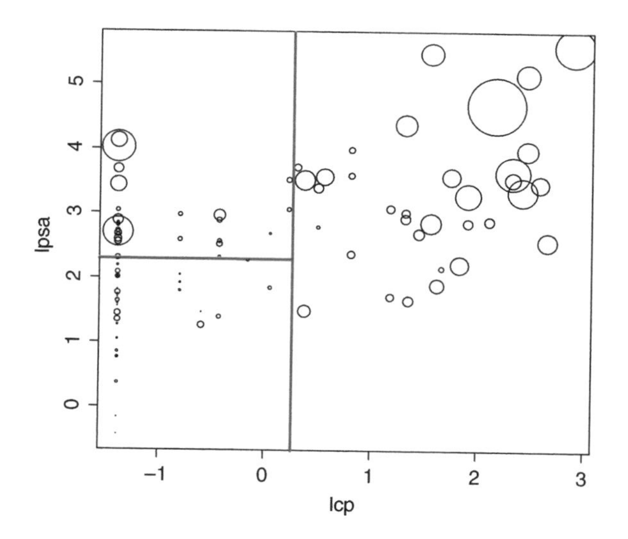

13.2 EXAMPLE 2: MOTORCYCLE ACCELERATION

The motorcycle accident data set consists of the observed accelerations on a motorcycle rider's helmet at 133 different time points after a simulated impact. The data consists of times (in milliseconds after impact) and acceleration (in g). The objective here is to predict acceleration as a function of time. The relationship between acceleration and time is rather complicated. The scatter plot of acceleration against time shows that it would be quite difficult to specify a parametric regression model. Smoothing the time series of accelerations would be one approach. One could predict the acceleration for a certain given time (say 10 ms) by the average of all accelerations recorded within a certain window around that time (e.g., within ± 3 ms, for a window extending from 7 to 13 ms). The smoothness of the resulting fitted curve depends on the length of the window, with longer windows improving the smoothness but in danger of missing changes of the true underlying function. We used such an approach in Chapter 4 when we discussed nonparametric regression models. Here we do not pursue this approach (you may want to apply, as an exercise, a local polynomial regression to this dataset), but fit a *regression tree* instead. The resulting fitted curve is added to the scatter plot of acceleration against time.

```
library(MASS)
library(tree)
data(mcycle)
mcycle
```

```
      times  accel
1      2.4    0.0
2      2.6   -1.3
3      3.2   -2.7
4      3.6    0.0
.
.
.
130   55.0   -2.7
131   55.0   10.7
132   55.4   -2.7
133   57.6   10.7
```

```
plot(accel~times,data=mcycle)
```

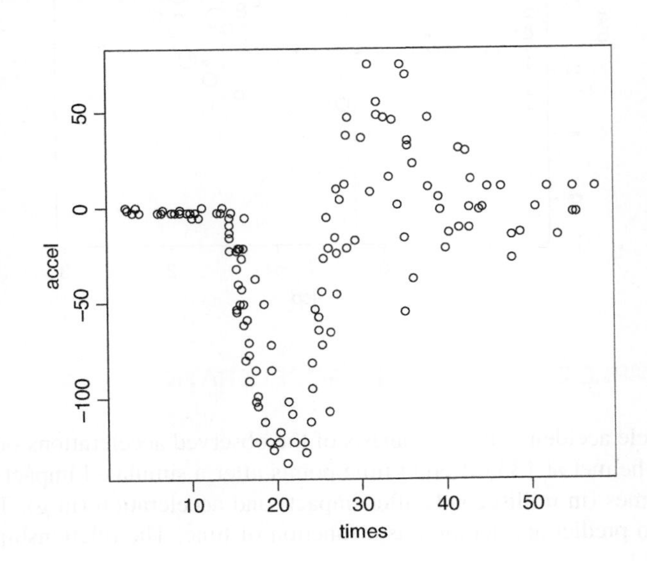

```
mct <- tree(accel ~ times, data=mcycle)
mct

node), split, n, deviance, yval
      * denotes terminal node
 1) root 133 308200.0  -25.550
   2) times < 27.4 84 160500.0  -47.320
     4) times < 16.5 43  18020.0  -16.480
       8) times < 15.1 28    724.1   -4.357 *
       9) times > 15.1 15   5494.0  -39.120 *
     5) times > 16.5 41  58660.0  -79.660
      10) times < 24.4 27  17040.0  -98.940
        20) times < 19.5 15   9045.0  -86.310 *
        21) times > 19.5 12   2616.0 -114.700 *
      11) times > 24.4 14  12240.0  -42.490 *
   3) times > 27.4 49  39670.0   11.780
     6) times < 35 16  13300.0   29.290
      12) times < 29.8 6   3900.0   10.250 *
      13) times > 29.8 10   5919.0   40.720 *
     7) times > 35 33  19080.0    3.291 *

plot(mct, col=8)
text(mct, cex=.75)
## we use different font size to avoid overlap
```

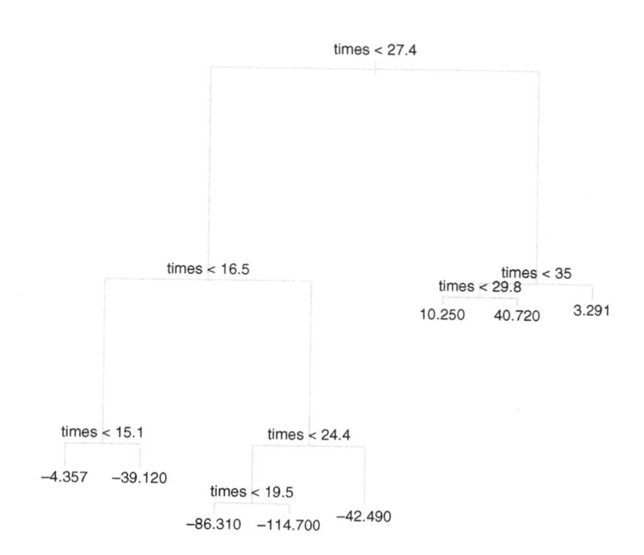


```
## scatter plot of data with overlay of fitted function
x=c(1:6000)
x=x/100
y1=seq(-4.357,-4.357,length.out=1510)
```

```
y2=seq(-39.120,-39.120,length.out=140)
y3=seq(-86.31,-86.31,length.out=300)
y4=seq(-114.7,-114.7,length.out=490)
y5=seq(-42.49,-42.49,length.out=300)
y6=seq(10.25,10.25,length.out=240)
y7=seq(40.72,40.72,length.out=520)
y8=seq(3.291,3.291,length.out=2500)
y=c(y1,y2,y3,y4,y5,y6,y7,y8)
plot(accel~times,data=mcycle)
lines(y~x)
```

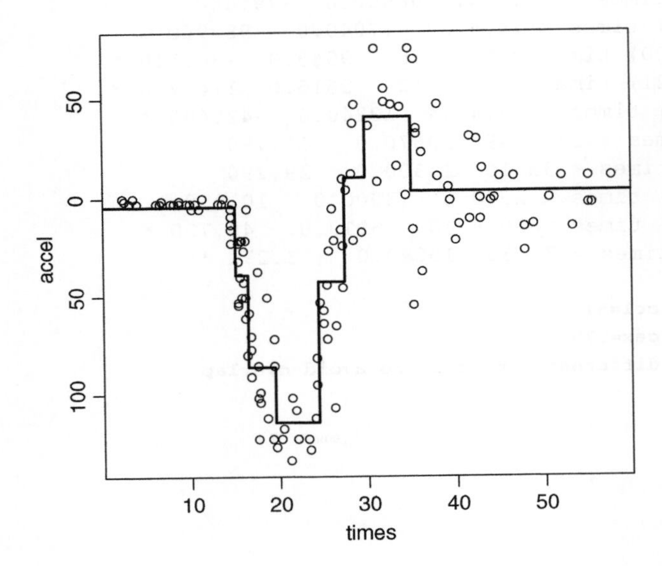

13.3 EXAMPLE 3: FISHER IRIS DATA REVISITED

We consider the Fisher iris data that we analyzed in Section 12.3 on discriminant analysis. Here we use the data to construct a *classification tree* for predicting the species of the iris (there are three classes: setosa, versicolor, and virginica) from their known characteristics (four characteristics are measured on each iris: Sepal.Length, Sepal.Width, Petal.Length, and Petal.Width). This is an example of a classification tree. The two preceding examples dealt with regression trees.

```
library(MASS)
library(tree)
## read in the iris data
iris
```

```
  Sepal.Length Sepal.Width Petal.Length Petal.Width   Species
1          5.1         3.5          1.4         0.2    setosa
2          4.9         3.0          1.4         0.2    setosa
3          4.7         3.2          1.3         0.2    setosa
```

```
. . .
. . .
148          6.5        3.0        5.2        2.0  virginica
149          6.2        3.4        5.4        2.3  virginica
150          5.9        3.0        5.1        1.8  virginica
```

```
iristree <- tree(Species~.,data=iris)
iristree
```

```
node), split, n, deviance, yval, (yprob)
      * denotes terminal node
 1) root 150 329.600 setosa ( 0.33333 0.33333 0.33333 )
   2) Petal.Length < 2.45 50   0.000 setosa ( 1.00000 0.00000 0.00000 ) *
   3) Petal.Length > 2.45 100 138.600 versicolor ( 0.00000 0.50000 0.50000 )
     6) Petal.Width < 1.75 54  33.320 versicolor ( 0.00000 0.90741 0.09259 )
      12) Petal.Length < 4.95 48   9.721 versicolor ( 0.00000 0.97917 0.02083 )
        24) Sepal.Length < 5.15 5   5.004 versicolor ( 0.00000 0.80000 0.20000 ) *
        25) Sepal.Length > 5.15 43   0.000 versicolor ( 0.00000 1.00000 0.00000 ) *
      13) Petal.Length > 4.95 6   7.638 virginica ( 0.00000 0.33333 0.66667 ) *
     7) Petal.Width > 1.75 46   9.635 virginica ( 0.00000 0.02174 0.97826 )
      14) Petal.Length < 4.95 6   5.407 virginica ( 0.00000 0.16667 0.83333 ) *
      15) Petal.Length > 4.95 40   0.000 virginica ( 0.00000 0.00000 1.00000 ) *
```

```
plot(iristree)
plot(iristree,col=8)
text(iristree,digits=2)
```

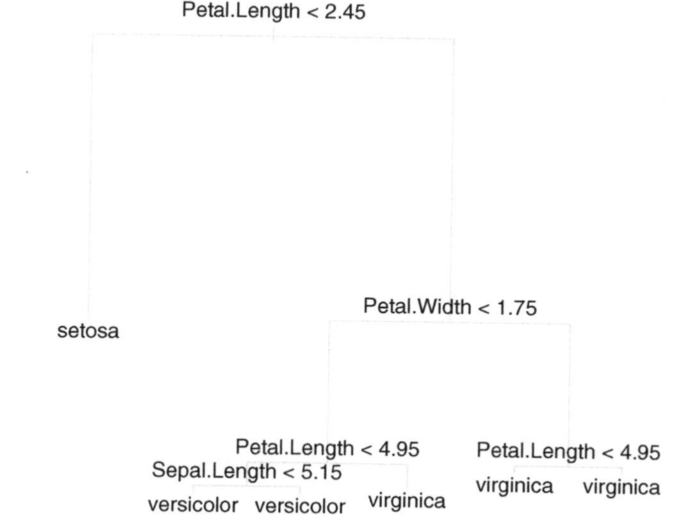


```
summary(iristree)
```

```
Classification tree:
tree(formula = Species ~ ., data = iris)
Variables actually used in tree construction:
[1] "Petal.Length" "Petal.Width"  "Sepal.Length"
```

```
Number of terminal nodes:  6
Residual mean deviance:  0.1253 = 18.05 / 144
Misclassification error rate: 0.02667 = 4 / 150
```

Consider terminal node 13, for example, with Petal.Width $<$ 1.75 and Petal.Length $>$ 4.95; four of six irises (i.e., the majority) in that group are classified as virginica. Since this is a terminal node that could not be split any further (as the default for the minimum number of observations in a node is 5), the most common category, virginica, is listed as the predicted classification. Four of 150 irises are misclassified by this tree, for a misclassification rate of 4/150, or 2.667%.

The tree with the six terminal nodes has two splits that lead to identical classifications. Nodes 24 and 25 (with a split on Sepal.Length) and nodes 14 and 15 (with a split on Petal.Length) lead to identical results, and these nodes and the trees below them can be snipped off. This shows that we can classify an iris on its Petal.Length and Petal.Width; the information on Sepal.Width and Sepal.Length does not enter the classification at all. An iris with Petal.Length $<$ 2.45 is classified as setosa. An iris with Petal.Length $>$ 2.45 and Petal.Width $>$ 1.75 is classified as virginica; so is an iris with Petal.Width $<$ 1.75 and Petal.Length $>$ 4.95. An iris with with Petal.Width $<$ 1.75 and Petal.Length between 2.45 and 4.95 is classified as versicolor.

```
irissnip=snip.tree(iristree,nodes=c(7,12))
plot(irissnip)
text(irissnip)
```

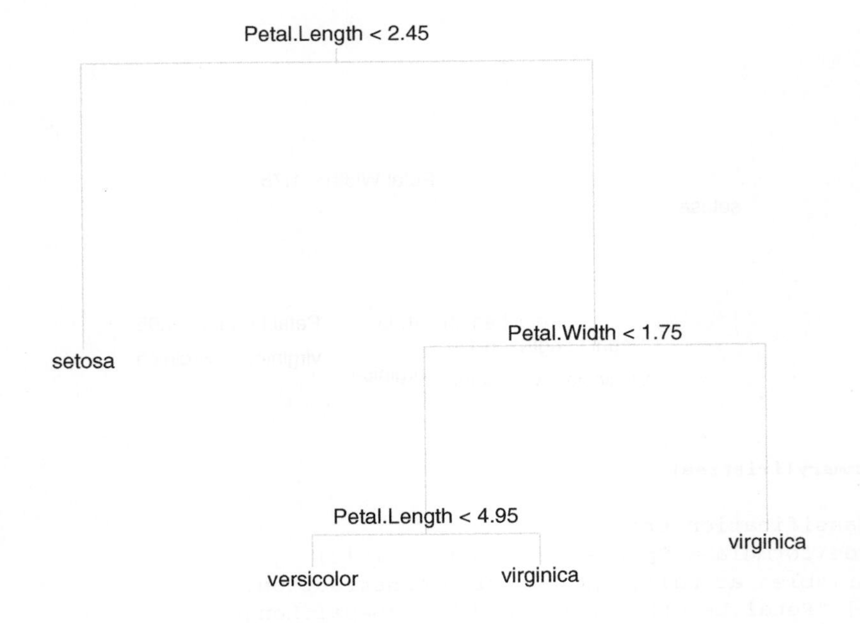

CHAPTER 14

Further Discussion on Regression and Classification Trees, Computer Software, and Other Useful Classification Methods

14.1 R PACKAGES FOR TREE CONSTRUCTION

R includes several excellent packages for tree construction, of which the foremost are **tree** and **rpart**. We have explained the package **tree** in the previous chapter. Alternatively, one can use the package **rpart** (rpart stands for recursive partitioning). The methodologies and the outputs of the two routines are very similar, and so are the solutions that are found by them. You can experiment with either package and compare the output. The following discussion may help you understand the slight differences in the results.

The deviance node impurity is used as the default in tree(); it coincides exactly with the criteria that have been covered in Chapter 13: the sum of squares criterion for regression trees and the classification deviance criterion for classification trees. The parameters in the tree.control() function—mincut (minimum number of items in a node, default 5), minsize (minimum number of items in a node before it is being considered for a cut; default 10), and mindev (the required change in deviance before a split is being carried out)—determine the forward motion of growing a tree. Forward growth stops when one of these stopping parameters is reached. The cross-validation routine cv.tree() and the pruning routine prune.tree() in the package **tree** use the deviance measure in the calculation of the complexity parameters. The cost complexity tells us about the required change in the deviance per number of cut leaves that is needed before a subtree can be pruned; if the cost complexity is smaller than the change in deviance, we simplify (i.e., prune) the tree. Note that the R package **tree** considers absolute, not relative changes. The output of the cross-validation routine cv.tree() lists the average cross-validation deviance for various tree sizes. One locates the size of tree that minimizes the average deviance. But one should not always insist on the absolute minimum. One usually scans the average deviances to find the

Data Mining and Business Analytics with R, First Edition. Johannes Ledolter.

smallest tree with an average deviance that is not much larger than the minimum deviance.

The Gini node impurity measure is used as the default in rpart(); the deviance impurity measure is used if it is specified through parms=list(split="information"). The forward growth of the tree is controlled through the parameters in the rpart.control function: minsplit (same as minsize in tree; the minimum number of observations that must exist in a node for a split to be attempted with default 20), minbucket (same as mincut in tree; number of observations in a terminal node with default round(minsplit/3)), and cp (the required percentage change in node impurity before a split is being carried out with default 0.01). Forward growth stops whenever one of the growth thresholds is reached. The complexity parameters in rtree() are defined proportionally relative to the root node (this is different from tree()), and they are defined in terms of the reduction in the number of misclassifications, and not in terms of the reduction of node impurity (the criterion used to grow the tree). Also, rpart() uses the number of misclassifications for the calculation of the errors (which are expressed relative to the number of errors at the root node). For given size of the tree (defined as number of splits, which is the size of the tree minus 1), rpart() calculates the number of errors in the training set, as well as the average and the standard deviation of the errors from the cross-validation of a tree of that size. Again one looks for the number of splits for which the average error from cross-validation has flattened out; it has been recommended that one adds one standard deviation to the minimal error and uses that as the cutoff. Also note that the cross-validation output from rpart() may list results for #splits=0 and #splits=2, but not for #splits=1. This happens when the reduction in the number of misclassification errors from #splits=0 to #splits=1 is smaller than half of the reduction in the number of misclassification errors from #splits=0 to #splits=2.

14.2 CHI-SQUARE AUTOMATIC INTERACTION DETECTION (CHAID)

CHAID or *chi-square automatic interaction detection* is a tree-structured classification procedure developed by Kass (1980) and further improved by Biggs et al. (1991). It is part of the SAS Enterprise Miner software and the SPSS data mining software AnswerTree.

Our previous discussion emphasized that classification and regression trees are useful for extracting structure from large multivariate data sets. These techniques partition a data set into mutually exclusive, exhaustive subsets that "best" describe a given response variable. Classification/regression trees are hierarchical displays that result from a series of questions about the outcomes of the predictor variables on each unit in the data set. At the end of the process, one knows the most likely response value or class membership of each unit. One speaks of a classification tree if the response variable is categorical, and of a regression tree if the response variable is continuous.

It is called a *tree* because the resulting display resembles an upside down tree, with a root on top (the entire data set), a series of branches connecting nodes, and leaves at the bottom. At each node (initially starting with the entire data set), a question about one of the predictor variables is posed; the branch taken at that node depends on the answer to this question. The order in which questions are asked and the rules about node splits are important, as they determine the structure of the tree. Many different algorithms have been proposed, and a detailed discussion of algorithms that split nodes of regression and classification trees on deviances has been given in the previous chapter. A general principle in tree construction focuses on "node purity" at each node-splitting juncture. The tree is built by repeatedly splitting subsets of the data to create further subsets, which are as homogeneous as possible with respect to the response variable. CART, the popular tree-based approach for continuous and categorical response variables that has been discussed in the previous chapter, generates binary trees. At each stage, it splits subsets of the data into just two child nodes of one of the predictor variables. See Breiman et al. (1984) for further discussion.

CHAID is designed to handle categorical response and categorical predictor variables. For example, the response may be the degree of success of an organization. Degree of success may be measured by a dichotomous (successful/not successful) variable, or success may be given as a nominal-scaled categorical variable with more than two outcomes. The predictor variables may be the size of the organization (small, medium, large), whether the organization is a multinational corporation or not, and the degree to which new management techniques are being adopted (on an integer scale from 1 to 10). A CHAID decision tree is constructed by splitting the units on one of the feature (explanatory) variables into two *or more* nodes; splitting the data set into more than two nodes makes this technique different from CART, which considers binary splits exclusively.

The basic idea behind CHAID is as follows. Assume a nominal response variable with d possible outcomes and a nominal explanatory variable with c outcomes; for the time being, consider just one explanatory variable; more explanatory variables will be introduced shortly. The Pearson chi-square statistic is used to assess the relationship among the two categorical variables. CHAID maximizes the Pearson chi-square statistic between the d outcomes on the response variable and all possible groupings of the c outcomes on the explanatory variable. For an explanatory variable with say $c = 7$ outcomes, the grouped table of frequencies with outcomes 1 and 2 on the explanatory variable in one group, outcomes 3 and 6 in a second group, and outcomes 4, 5, and 7 in the third group may maximize the Pearson chi-square statistics. Assuming that the chi-square statistic for this 3 by d table is significant, CHAID would then split the data into these three groups. The tree would show three leaves, with each leaf listing proportions of the d response categories calculated from the units in that particular leaf. This is straightforward, but the search over all possible groupings of the categorical explanatory variable is time consuming (especially as this has to be repeated for several categorical variables and for many nodes). Hence, the following "merge" strategy is suggested.

For a given node (at the beginning we start with all data) and for a given predictor variable, the procedure merges a pair of predictor variable outcomes if there is no statistically significant association between them and the outcomes on the response variable; the probability value of the chi-square statistic from the 2 by d table of frequencies is compared to a given alpha-to-merge value. The merge step is then repeated, and the procedure finds the next pair of categories to merge, which may now include previously merged categories. The merging step stops if the statistical significance for the last pair of predictor categories to be merged is significant. At that step the procedure will have found the best split for that predictor variable.

The next step in CHAID is to select the split variable. The predictor variable with the smallest p-value among the chi-square statistics testing the association between the response outcomes and the optimally merged outcomes on the predictor variables is selected for the split. That is, the predictor variable that yields the most significant split is selected for the split. If the smallest p-value is greater than some alpha-to-split value, then no further splits from the present node will be performed, and the present node is a terminal node. The process is repeated at each node until one of the stopping rules is triggered; that is, either a further split would violate user-specified minimum data requirements, or no more significant splits can be found.

Of course, such a search procedure may not find the split that is best at each node. Only an exhaustive search procedure will do that. The modifications by Biggs et al. (1991) enhance the search procedure.

14.3 ENSEMBLE METHODS: BAGGING, BOOSTING, AND RANDOM FORESTS

Do not put all your eggs into one basket. Because if all your eggs are in one basket and you drop the basket, you lose everything. Similarly, do not put all your trust in a single classification or prediction method. Use the information from several alternative methods if more than one method is available.

The advantage of aggregating several methods tends to be greatest if the methods are unrelated (independent). The advantage of aggregation disappears if the methods are identical (i.e., strongly correlated). In forecasting, there is a large literature on the benefits of combining forecasts of quantitative information such as sales, going back to the paper by Granger and Bates (1969). Suppose the head of a forecasting division has two sources of forecasts for the company's sales, one source being the forecasts developed by the division's econometrics group using a time series model and the other source being the aggregated forecasts of the regional sales managers of the company. Suppose that the forecast horizon is $h = 1$. Represent the one-step-ahead forecast of the econometrics group made at time t of the observation y_{t+1} by f_{t+1}^1 and that of the managers by f_{t+1}^2, and consider the combination forecast $f_{t+1}^c = \omega f_{t+1}^1 + (1 - \omega)\ f_{t+1}^2$. Assume that either forecast is unbiased (i.e., $E(y_{t+1} - f_{t+1}^1) = E(y_{t+1} - f_{t+1}^2) = 0$) and assume that the one-step-ahead forecast errors have variances σ_{11} and σ_{22}, and covariance σ_{12}. It is straightforward to show that the combined forecast is also unbiased, and that its variance is given by $E(y_{t+1} - f_{t+1}^c)^2 = \omega^2\sigma_{11} + (1-\omega)^2\sigma_{22} + 2\omega(1-\omega)\sigma_{12}$. This variance is smallest for $\omega = (\sigma_{22} - \sigma_{12})/(\sigma_{11} + \sigma_{22} - 2\sigma_{12})$. We can show this by taking the first derivative with respect to ω and setting it equal to zero. For forecasts with the same precision ($\sigma^2 = \sigma_{11} = \sigma_{22}$), the individual forecasts should be averaged ($\omega = 0.5$), and this is true for any value of the covariance σ_{12}. Then the forecast error of the resulting combined (averaged) forecast has variance $E(y_{t+1} - f_{t+1}^c)^2 = \sigma^2(1+\rho)/2 \leq \sigma^2$, which is never larger than the variance of an individual method. The benefit of averaging is actually largest if the correlation $\rho = \sigma_{12}/\sigma^2$ between the two unbiased forecast errors is negative, as then the two errors compensate each other.

The same idea can be applied to the classification of a new case into one of $m \geq 2$ different outcome categories. Assume that results from an ensemble of k different base classifiers are available. The ensemble returns a class prediction that is based on the votes of the base classifiers. The combined (ensemble) classifier is given by the outcome that occurs most often. An ensemble classifier tends to be more accurate than its components. Each base classifier in a binary classification may make mistakes, but the ensemble classifier will lead to an error only if over half of the base classifiers are incorrect. Ensemble methods work well if there is little correlation among the classifiers and if each classifier is better than random guessing. Combining classifiers that are all alike brings little benefit. Combining classifiers that resemble random guessing does not lead to an advantage either, as the distribution of the classifiers of the ensemble is again uniform across the possible outcome groups.

There are various ways of creating ensembles. The classifiers could come from different methods, such as logistic regression, nearest neighbor methods, classification trees, naïve Bayesian methods, or discriminant methods. Or, the classifiers can come from different subsets of the training part of the data. *Bootstrap* aggregation, or *bagging*, combines classifiers across different subsets of the training data. Assume that the training data set consists of d cases, each case being described by its attributes and the true classification. Bagging selects a new training set of d cases by selecting d of the cases through sampling *with* replacement (a certain case could come up multiple times in the new training set), constructs the model from this new training set, and uses the fitted model to classify the new case. This process is repeated k times, resulting in k classifications of the new test case. The ensemble classifier assigns to the new test case the most frequent classification. The bootstrap (or bagged) classifier is often considerably better than a single classifier that is derived from the original training set. It certainly will not be much worse, and it has been shown to be robust to overfitting and to noisy data.

Boosting is similar, except that there the ensemble classifier assigns weights to the individual base classifiers when combining them. *Adaptive boosting*, for example, works as follows. We start, as in bagging, constructing a new training set of d cases from the original training set by selecting d of the cases through sampling *with* replacement, constructing the model (classifier) from this new training set, and using the fitted model to classify the new case. In the beginning, each case of the original training set is assigned weight $1/d$. But now, we assess the internal performance of the resulting classifier on the new training set that has been used for its construction, and we revise the weights for drawing the next training sample. If a case in the training sample has been classified incorrectly, we increase its weight; if the case has been classified correctly, we reduce its weight. The revised weights reflect how difficult it is to classify each case. The revised weights are used to draw the next bootstrap sample, construct a new classifier, obtain the classification of the test case, assess the internal performance of the classifier, and revise the weights for the draw of the next training sample, and so on. Note that each new training sample increases the focus on misclassified cases. The boosting strategy tends to build a series of classifiers that complement each other. Once we have obtained k training sets and k classifiers, adaptive boosting combines the results, and it does so by assigning a weight to each classifier's vote that reflects how well this classifier has performed internally. Strategies for revising the weights from one draw to the next and for weighting the votes of the classifiers are discussed in the literature, and programs for implementing this approach are available. In general, boosting tends to increase the accuracy.

The *randomForest* method is another way to combine information across an ensemble. Assume that each of the k classifiers in the ensemble is a decision tree for classifying a new element into one of m possible outcome groups. At each node, an individual decision tree determines the split on the basis of a smaller, random selection of attributes, and not from the set of all attributes. Each tree in the forest of trees (hence the method's name) then votes on the classification of a new item, and the most popular class is returned as the ensemble solution.

Random attribute selection can be combined with bagging. There a training sample of the d elements from the initial training set of d elements is selected with replacement. At each node of each tree, a randomly selected set of A attributes (selected from the usually much larger set of all attributes) is used to determine the split. Each tree is grown to maximum size and is not pruned. Each tree votes, and the most popular class becomes the ensemble classification. Random forests formed with random attribute input selection are referred to as *Forest-RI*. Several modifications of this basic strategy have been developed, and computer software for their implementation is available.

When constructing a single tree one notices quite often that there is little difference between choosing one splitting variable or several others. One or more attribute variables usually have the same ability to partition the training data set into homogeneous groups, and in such cases, luck or small changes in the training data set determine whether the algorithm prefers one splitting variable over the others. Random forest techniques tend to do better. The randomness of variable selection delivers robustness to noise and the presence of attribute variables that have weak relationships with the target variable, to outliers, and to small changes in the training data set. These conditions usually have little impact on the final decisions made by the ensemble classifier. A random forest technique also handles underrepresented classes quite well. In addition, it has computational efficiencies as it restricts the search.

Software for carrying out this methodology is available in R; the package **randomForest** a good place to start. Typically, one constructs 500 random trees, with each tree constructed on training data that has been randomly selected from the original training data set (say 70% of the complete data). Randomization enters twice: through the randomly selected training data set (each tree uses different training data) and through the random selection of a few of the attributes (four or so) at each of the node splits. These two random selections make random forests robust to noisy observations and weak attributes. The randomForest() command, through its control parameter replace=TRUE/FALSE, achieves the randomization of the tree-specific training sets through two very similar sampling procedures. Under replace=TRUE, the training set is sampled with replacement from the number of cases that have been established as the training set; typically the sample size is the same as the size of the training set. Because of sampling with replacement, not all elements of the original training set are selected (on average, roughly 1/3 of the elements are not chosen) and a few elements are selected more than once. Under replace=FALSE, the sampling is without replacement, and in this case, the program samples 63.2% of the elements from the training set. Again, roughly one-third of the cases are not selected.

The randomForest package evaluates the classification procedure internally by computing classification errors on only those elements of the training set that have been left out of the training data for at least one of the constructed trees. Elements that are included in the training data for all of the constructed trees are not part of this evaluation. The program refers to the resulting error as the "out-of-bag" misclassification error; the expression "out of bag" comes from the bootstrap aggregation or bagging of the training data. Misclassification error rates for evaluation and test data sets can also be obtained, and they tend to give a more accurate picture of the out-of-sample performance.

The randomForest package also allows the user to specify the number of elements that should be included in each tree's training data set (typically the sample size is the same as the size of the training set). Furthermore, the program allows the user to specify the number of elements that should come from each of the outcome groups; for example, the statement sampsize=c(30,30) forces the selection of 30 samples from each of two groups. This can be an advantage in situations where the groups are not balanced, and one group is considerably more prevalent than the other. Without the sampsize argument, the program selects the cases from the training set at random, and trees constructed on unbalanced training data will have difficulties identifying members of the rare group. By forcing the program to select the same number of training elements from each strata, the trees are given a better chance to identify the rare group. The randomForest software package also includes several ways of assessing the importance of the attributes. The simplest way to determine importance is to count the number of trees in which a certain attribute is present for making a decision.

14.4 SUPPORT VECTOR MACHINES (SVM)

Support vector machines (SVM) comprise another very general group of classification methods, and they cover both linear and nonlinear classifiers. Assume that we want to classify two-dimensional training data into two groups. The information we have available can be shown through a scatter plot of the two attributes, and the plotting positions can be represented with two different labels or colors that identify the two classes. Suppose we are fortunate enough that the data are linearly separable, which means that a straight line through the data set separates the two classes exactly. Quite often there are several lines with different intercepts and slopes that can achieve the separation, and the objective is to find the line that achieves the best possible one. What do we mean by the best separation? Find a separation line and add two parallel lines of equal distance to that line such that each parallel line passes through the data point that is closest to the line of separation. The distance between the two parallel lines, also called the *margin*, measures how good the separation really is. If the distance between these two parallel lines is large, then the two groups are far apart and we have found an excellent classifier. The objective is to find the classifier with the best margin of separation. In two-dimensional space where data can be graphed on a scatter plot, it is fairly straightforward to find the separation line, the two parallel lines, and the margin. In high dimensional space, with more than two attributes and more than two groups, this becomes a complicated problem. Nevertheless mathematical tools are readily available to find the hyperplane that separates the groups with the largest margin. SVM achieves this by finding "support vectors" that identify the data points on the hyperplanes with the largest margins.

But, what if we are not so fortunate and our data is not linearly separable? Then SVM methods transform the data into a higher dimensional space in which such complete separation is possible. With an appropriate nonlinear mapping into a sufficiently high dimension, such separation is always possible, and the methods that we outlined for linearly separable data can be applied to the transforms. SVM tries several nonlinear transformations, including the one that considers powers and cross products of the attributes.

SVMs are useful, and programs are available in R. However, in this text, we do not discuss SVM in detail, as its full explanation requires advanced tools from mathematical optimization. For reference, you may want to look at the article "Support vector machines in R," by Karatzoglou et al. (2006), the command ksvm() in the R package **kernlab**, and the command svm() in the R package **e1071**.

14.5 NEURAL NETWORKS

Neural networks comprise yet another group of nonlinear procedures for prediction and classification. Neural networks are computational analogs of the processes that describe the working of neurons. Neural nets consist of connected input and output layers, each containing numerous units, where the connections among the units of the layers have weights associated with them. A multilayer neural network consists of an input layer of distinct units (usually input units represent the attributes of the case that needs to be classified), one or more hidden layers, and an output layer with several units that represent the class that needs to be predicted. Each layer is made up of several units. The inputs from the units of the first layer are weighted and fed simultaneously to a second layer of neuron-like units, known as the *hidden layer*. Each of its units takes as its input a weighted sum of the outputs of the previous layer and applies a nonlinear activation function to determine its output. The activation function is usually a logistic function that transforms the output to a number that is between 0 and 1; for this reason, it is often referred to as the *squashing* function. The output units may already be the actual output of the system (in which case we talk about a single hidden layer), or the outputs of the first hidden layer can be fed as inputs to another hidden layer (in which case we have two hidden layers), and so on if additional hidden layers are involved.

One needs to decide on the number of hidden layers and the number of units in the various layers, as well as the weights that connect the inputs and outputs of these layers. Neural nets are very general, and because of the nonlinear activation functions neural nets are able to approximate most nonlinear functional relationships very well. Of course, the weights in these systems have to be estimated by training the system on actual data. The iterative method of back-propagation can be used to determine the empirical weights that lead to the best fit on a given training sample. For that, one needs to specify a learning parameter that controls the speed of convergence of this iterative estimation method.

Neural nets are very general and can approximate complicated relationships. But they also come with disadvantages. One disadvantage of neural nets is that the approximating models relating inputs and outputs are purely "black box" models, and they provide very little insight into what these models really do. Also, the user of neural nets must make many modeling assumptions, such as the number of hidden layers and the number of units in each hidden layer, and usually there is little guidance on how to do this. It takes considerable experience to find the most appropriate representation. Furthermore, back-propagation can be quite slow if the learning constant is not chosen correctly.

In this chapter, we do not emphasize neural networks. Software for carrying out this methodology is available in R; the packages **nnet** and **neuralnet** are good places to start from.

14.6 THE R PACKAGE RATTLE: A USEFUL GRAPHICAL USER INTERFACE FOR DATA MINING

A graphical user interface (GUI) allows people to interact with software in more ways than just by typing text. The Microsoft Office products are well-known GUIs. Take Microsoft Excel, for example, which allows users to simply point a cursor at a certain icon and click. Excel and other spreadsheet programs, such as Minitab and SPSS for statistical applications, have become popular because they rely on a simple "point and click" interface. R, on the other hand, is not that user-friendly because one must type commands into an R console. However, more and more R GUIs are being developed. **Rcmdr** (R commander) is one of them, and it provides an excellent GUI especially for statistical analysis. Typing onto the R console the statement "library(Rcmdr)" opens up a graphical interface, and from there numerous statistical functions are just a mouse-click away.

The R library **rattle**, developed by Williams and discussed in his book "Data Mining with Rattle and R: the Art of Excavating Data for Knowledge Discovery" (Williams, 2011), provides a very useful GUI to important data-mining libraries in R. Typing onto the R console the statement "library(rattle)" and then "rattle()" into the next line opens up a graphical interface to R. Once the data is loaded into rattle, the analysis can be carried out by entering answers to simple queries and mouse clicks. This makes the R analysis simple and easy, as data mining is now carried out in a spreadsheet environment that is familiar to users of programs such as Excel or Minitab. By clicking on various window tabs, the user of rattle can execute data-mining tasks such as constructing regression/classification trees discussed in this and the previous chapter, clustering (forthcoming Chapter 15), and association analysis (forthcoming Chapter 16), and can do so without having to write out detailed R instructions. In addition, rattle makes it convenient to pre-process the data. Once the data has been loaded into rattle, it is easy to specify the data type and declare the data as continuous or categorical (rattle infers the data type automatically; while it is actually very good at this, the user can always over-write its selection). It is also easy to get information on missing data and identify outliers, and simple methods for adjusting outliers and imputing missing observations are available. Rattle provides useful graphical displays for one or more variables and calculates standard summary statistics such as means, standard deviations, and correlation coefficients, and it does so without the user having to write out R instructions. Rattle reduces much of the programming pain, as it becomes unnecessary to write out R instructions. The user does not need to remember the R code; information is simply entered into various windows and the information is being translated into R code. An added benefit of rattle is that the generated code is visible to the user, who then can change the code for more sophisticated analyses. Experiment with this R package by loading the library rattle, and start it by typing rattle().

Rattle makes it very easy to split the data into training, evaluation, and test data sets, and to evaluate the models that are fitted to one set of data on other data sets. The evaluation of the predictive performance of a model on the very same data that has been used for its estimation is most likely too optimistic. Just think of regression models that can be made to fit better and better just by increasing the number of estimated coefficients, but at the expense of their forecasting performance. It is important that the predictive performance of models and methods is being assessed on new observations that are independent of those that were used at the model estimation/fitting stage. Typically, one divides the data set into two, or sometimes three nonoverlapping parts: the training, evaluation, and test data sets. The training data set is used to build and estimate the model, and the test data set is used to evaluate the out-of-sample performance of the model (which can result in a prediction of a response variable or a classification of a new case). Sometimes a third data set, called the *evaluation data set*, is present. Consider LASSO regression for predicting the outcome of a certain response. The implementation of LASSO regression for prediction requires the knowledge of a penalty parameter λ, which usually is determined through cross-validation on an independent data set. This is where the evaluation data set comes in. For each of several choices of λ, penalized regression estimates are obtained from the training set; these estimates are then used to determine the predictions for cases in the evaluation data set. The resulting prediction errors are calculated, and the penalty parameter is estimated by minimizing the sum of the squared prediction errors from the evaluation data set. Once the penalty parameter is found, one can use the LASSO estimates from the training data, using the penalty that was found on the evaluation data set, to evaluate the model performance on the test data set. Rattle has easy ways of splitting the data into these three parts, and they use splits, such as 70/15/15. In certain other modeling situations, just two independent data sets are needed: the training data set and the test data set. Then we use the terms "test" and "evaluation" data sets interchangeably to refer to the same data set. With just two data sets, it is common to split the complete data set into two equal parts (50/50 split).

Rattle covers several important data-mining tasks, but not all methods discussed in this book are covered; it does not address regression and logistic regression, LASSO and false discovery rates, nearest neighbor methods, and principal components and partial least squares, network analysis and the analysis of text information (topics that are being discussed in subsequent chapters).

REFERENCES

Biggs, D., DeVille, B., and Suen, E.: A method of choosing multiway partitions for classification and decision trees. *Journal of Applied Statistics*, Vol. 18 (1991), 49–62.

Breiman, L., Friedman, J., Ohlsen, R., and Stone, C.: *Classification and Regression Trees*. Pacific Grove: Wadsworth, 1984.

Granger, C.W.J and Bates, J.: The combination of forecasts. *Operations Research Quarterly*, Vol. 20 (1969), 451–468.

Karatzoglou, A., Meyer, D., and Hornik, K: Support vector machines in R. *Journal of Statistical Software*, Vol. 15 (2006), No 1, 1–28.

Kass, G.V.: An exploratory technique for investigating large quantities of categorical data. *Applied Statistics*, Vol. 29 (1980), 119–127.

Williams, G.: *Data Mining with Rattle and R: the Art of Excavating Data for Knowledge Discovery*. New York: Springer, 2011.

CHAPTER 15

Clustering

15.1 *k*-MEANS CLUSTERING

We are given observations on n units, $(x_1, x_2, \ldots, x_n)$, with the observation on unit i representing a p-dimensional vector of features (attributes). The *k-means clustering* method partitions the n units into $k \leq n$ distinct clusters, $S = \{S_1, S_2, \ldots, S_k\}$, so as to minimize the within-cluster sum of squares,

$$\arg\min_S \sum_{j=1}^{k} \sum_{x_i \in S_j} \|x_i - \overline{m}_j\|^2.$$

The cluster mean $\overline{m}_j$ is the mean vector of the p attributes averaged over all units in cluster S_j. The norm $\|x - m\|^2 = \sum_{r=1}^{p}(x_r - m_r)^2$ sums the squared differences over the p attributes; it represents the (square of the) Euclidian distance between the p-dimensional vectors x and m. The norm assumes equal scales of the p attributes and it does not incorporate correlations among the features.

The number of clusters k must be given. Clustering is purely descriptive. Clustering groups the items according to how similar they are, and it does so in an unguided (unsupervised) manner. If we had a training set with known clusters, we could use a nearest neighbor analysis (discussed in Chapter 9) to assign the units of the evaluation (hold-out; test) data set to the known clusters; but here we assume that such training sets are not available.

The question is how the minimization should be carried out. While this is a difficult and time-intensive problem (it is known as an *NP hard* problem), several good heuristic algorithms for its solution do exist. The most common algorithms use an iterative refinement technique; see Hartigan (1975), and Hartigan and Wong (1979).

Given an initial set of k cluster means $\overline{m}_1^1, \overline{m}_2^1, \ldots, \overline{m}_k^1$ (details of the initialization are described below), the algorithm proceeds by alternating between the following two steps:

Data Mining and Business Analytics with R, First Edition. Johannes Ledolter.

An assignment step. We run through all units and assign each unit to the cluster with the closest mean. This creates the jth cluster assignment at iteration time (t):

$$S_j^{(t)} = \left\{ x_i : \|x_i - \overline{m}_j^{(t)}\| \leq \|x_i - \overline{m}_{j*}^{(t)}\|, \text{ for all } j* = 1, 2, \ldots, k \right\}.$$

An update step. We calculate k new cluster means as the centroids of the units in the clusters that have been created in the assignment step,

$$\overline{m}_j^{(t+1)} = \frac{1}{nu(S_j^{(t)})} \sum_{x_i \in S_j^{(t)}} x_i.$$

The algorithm has converged when the assignments no longer change.

Commonly Used Initialization Methods

- Randomly choose k units from the data set and use these as the initial cluster means;
- Randomly assign one of the k clusters to each unit and then proceed to the update step, computing the initial means as the centroids of the clusters' randomly assigned units. In general, this random partition method is thought to be preferable.

We use the function *k-means* in the R package **stats**; this algorithm is known to be fast and reliable. But as with all heuristic algorithms, there is no guarantee that it converges to the global optimum; its final result may depend on how the algorithm has been started. The parameter nstart in the function k-means can be used to reduce the sensitivity of the algorithm to the random selection of the initial clusters/cluster means. Note that the random initialization may result in slightly different cluster assignments if the same function k-means is executed repeatedly. However, this change in the cluster assignment should not cause undue worry, as in such cases the solutions are fairly close and most likely any of the proposed clustering assignments can be justified by the data. Also note that the cluster labels (1, 2, and so on) usually change from one run to the other. This only has to do with the labeling of the clusters, but not with the assignment of the elements to the clusters.

EXAMPLE 15.1 EUROPEAN PROTEIN CONSUMPTION

For our first example, we consider 25 European countries ($n = 25$ units) and their protein intakes (in percent) from nine major food sources ($p = 9$). The data are listed below. For example, Austria gets 8.9% of its protein from red meat, 19.9% from milk, and so on. It is of interest to learn whether the 25 countries can be separated into a smaller number of clusters. It may well be that Mediterranean countries get their protein intake from certain food categories, which are different from the food staples that are favored by North European and German-speaking countries.

The data, originally reported by Weber (1973), can be found in Hand et al. (1994, p. 297).

Country	Red Meat	White Meat	Eggs	Milk	Fish	Cereals	Starch	Nuts	Fr&Veg
Albania	10.1	1.4	0.5	8.9	0.2	42.3	0.6	5.5	1.7
Austria	8.9	14	4.3	19.9	2.1	28	3.6	1.3	4.3
Belgium	13.5	9.3	4.1	17.5	4.5	26.6	5.7	2.1	4
Bulgaria	7.8	6	1.6	8.3	1.2	56.7	1.1	3.7	4.2
Czechoslovakia	9.7	11.4	2.8	12.5	2	34.3	5	1.1	4
Denmark	10.6	10.8	3.7	25	9.9	21.9	4.8	0.7	2.4
E Germany	8.4	11.6	3.7	11.1	5.4	24.6	6.5	0.8	3.6
Finland	9.5	4.9	2.7	33.7	5.8	26.3	5.1	1	1.4
France	18	9.9	3.3	19.5	5.7	28.1	4.8	2.4	6.5
Greece	10.2	3	2.8	17.6	5.9	41.7	2.2	7.8	6.5
Hungary	5.3	12.4	2.9	9.7	0.3	40.1	4	5.4	4.2
Ireland	13.9	10	4.7	25.8	2.2	24	6.2	1.6	2.9
Italy	9	5.1	2.9	13.7	3.4	36.8	2.1	4.3	6.7
Netherlands	9.5	13.6	3.6	23.4	2.5	22.4	4.2	1.8	3.7
Norway	9.4	4.7	2.7	23.3	9.7	23	4.6	1.6	2.7
Poland	6.9	10.2	2.7	19.3	3	36.1	5.9	2	6.6
Portugal	6.2	3.7	1.1	4.9	14.2	27	5.9	4.7	7.9
Romania	6.2	6.3	1.5	11.1	1	49.6	3.1	5.3	2.8
Spain	7.1	3.4	3.1	8.6	7	29.2	5.7	5.9	7.2
Sweden	9.9	7.8	3.5	24.7	7.5	19.5	3.7	1.4	2
Switzerland	13.1	10.1	3.1	23.8	2.3	25.6	2.8	2.4	4.9
United Kingdom	17.4	5.7	4.7	20.6	4.3	24.3	4.7	3.4	3.3
USSR	9.3	4.6	2.1	16.6	3	43.6	6.4	3.4	2.9
W Germany	11.4	12.5	4.1	18.8	3.4	18.6	5.2	1.5	3.8
Yugoslavia	4.4	5	1.2	9.5	0.6	55.9	3	5.7	3.2

We start by clustering on the first two features, protein intake from red and white meat. With just two features and in two dimensions it is easy to get an appreciation for what clustering is trying to achieve. We program the software to cluster the 25 countries into three groups. The following scatter plot of protein intake from white meat against protein intake from red meat labels the k-means cluster associations of the countries in color. This gives us a good visualization of the three clusters. Clusters are formed by minimizing the Euclidean distance to the respective cluster centroids. The graph indicates that there exist several other reasonable cluster assignments that would not be much worse.

```
### *** European Protein Consumption, in grams/person-day *** ###

## read in the data
food <- read.csv("C:/DataMining/Data/protein.csv")
```

```
food[1:3,]
```

```
  Country RedMeat WhiteMeat Eggs Milk Fish Cereals Starch Nuts Fr.Veg
1 Albania    10.1       1.4  0.5  8.9  0.2    42.3    0.6  5.5    1.7
2 Austria     8.9      14.0  4.3 19.9  2.1    28.0    3.6  1.3    4.3
3 Belgium    13.5       9.3  4.1 17.5  4.5    26.6    5.7  2.1    4.0
```

```
## first, clustering on just Red and White meat (p=2) and k=3
## clusters
set.seed(1) ## to fix the random starting clusters
grpMeat <- kmeans(food[,c("WhiteMeat","RedMeat")], centers=3,
+    nstart=10)
grpMeat
## list of cluster assignments
o=order(grpMeat$cluster)
data.frame(food$Country[o],grpMeat$cluster[o])
```

```
   food.Country.o. grpMeat.cluster.o.
1          Albania                  1
2         Bulgaria                  1
3          Finland                  1
4           Greece                  1
5            Italy                  1
6           Norway                  1
7         Portugal                  1
8          Romania                  1
9            Spain                  1
10          Sweden                  1
11            USSR                  1
12      Yugoslavia                  1
13         Belgium                  2
14          France                  2
15         Ireland                  2
16     Switzerland                  2
17              UK                  2
18         Austria                  3
19  Czechoslovakia                  3
20         Denmark                  3
21       E Germany                  3
22         Hungary                  3
23     Netherlands                  3
24          Poland                  3
25       W Germany                  3
```

```
## plotting cluster assignments on Red and White meat scatter plot
plot(food$Red, food$White, type="n", xlim=c(3,19), xlab="Red Meat",
+    ylab="White Meat")
text(x=food$Red, y=food$White, labels=food$Country,
+    col=grpMeat$cluster+1)
```

Next, we cluster on all nine protein groups and prepare the program to create seven clusters. The resulting clusters, shown in color on a scatter plot of white meat against red meat (any other pair of features could be selected), actually make a lot of sense. Countries in close geographic proximity tend to be clustered into the same group.

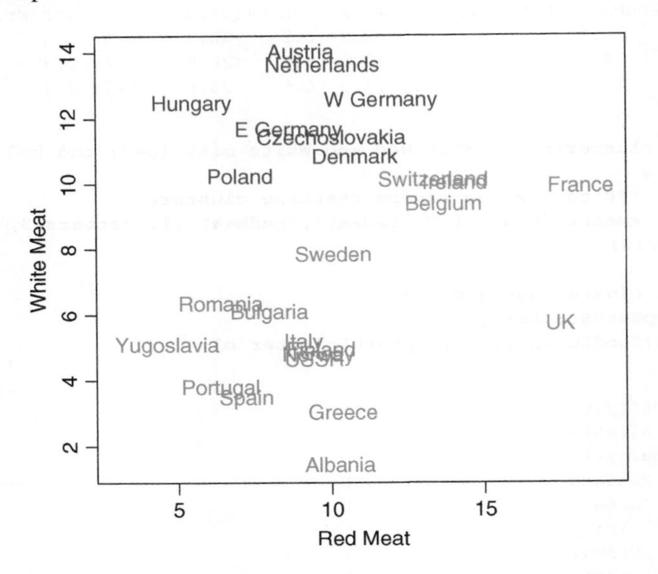


```
## same analysis, but now with clustering on all
## protein groups
## change the number of clusters to 7
set.seed(1)
grpProtein <- kmeans(food[,-1], centers=7, nstart=10)
o=order(grpProtein$cluster)
data.frame(food$Country[o],grpProtein$cluster[o])

food.Country.o. grpProtein.cluster.o.
1         Portugal                      1
2            Spain                      1
3          Denmark                      2
4          Finland                      2
5           Norway                      2
6           Sweden                      2
7          Austria                      3
8        E Germany                      3
9      Netherlands                      3
10       W Germany                      3
11  Czechoslovakia                      4
12         Hungary                      4
13          Poland                      4
14         Belgium                      5
15          France                      5
16         Ireland                      5
17     Switzerland                      5
```

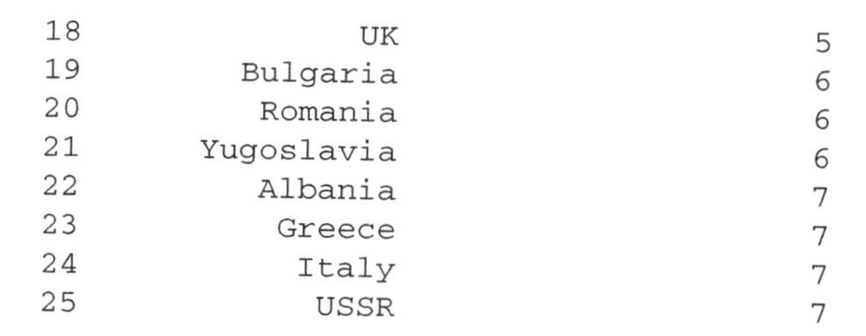

18	UK	5
19	Bulgaria	6
20	Romania	6
21	Yugoslavia	6
22	Albania	7
23	Greece	7
24	Italy	7
25	USSR	7

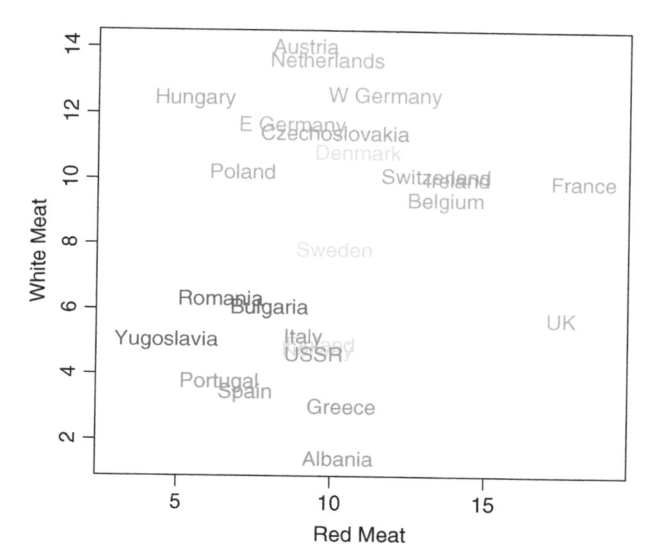

EXAMPLE 15.2 MONTHLY US UNEMPLOYMENT RATES

Our second example analyzes monthly seasonally adjusted unemployment rates covering the period January 1976 through August 2010 for the 50 US states ($n = 50$). In the figure shown later, we overlay the time series plots of three states; Iowa (green), New York (red), and California (black). The objective is to cluster states into groups that are alike. Here each state is characterized by a feature vector of very large dimension ($p = 416$), with its components representing the 416 monthly observations. Assume, for illustration, that New York and California form a cluster. Calculate the 416 monthly averages (of two observations each); this vector of averages is the centroid for that cluster. The sum of the squared distances from the centroid of this cluster (for New York and California there are $2(416) = 832$ such distances) expresses the within-cluster sum of squares.

```
## read the data; series are stored column-wise with labels in first
## row
raw <- read.csv("C:/DataMining/Data/unempstates.csv")
raw[1:3,]
```

```
  AL  AK   AZ  AR  CA  CO  CT  DE   FL  GA  HI  ID  IL  IN  IA  KS  KY  LA  ME
1 6.4 7.1 10.5 7.3 9.3 5.8 9.4 7.7 10.0 8.3 9.9 5.5 6.4 6.9 4.2 4.3 5.7 6.2 8.8
2 6.3 7.0 10.3 7.2 9.1 5.7 9.3 7.8  9.8 8.2 9.8 5.4 6.4 6.6 4.2 4.2 5.6 6.2 8.6
3 6.1 7.0 10.0 7.1 9.0 5.6 9.2 7.9  9.5 8.1 9.6 5.4 6.4 6.4 4.1 4.2 5.5 6.2 8.5

  MD   MA   MI  MN  MS  MO  MT  NE  NV  NH   NJ  NM   NY  NC  ND  OH  OK   OR
1 6.9 11.1 10.0 6.2 7.0 5.8 5.8 3.6 9.8 7.2 10.5 8.9 10.2 6.7 3.2 8.3 6.4 10.1
2 6.7 10.9  9.9 6.0 6.8 5.8 5.7 3.5 9.5 7.1 10.4 8.8 10.2 6.5 3.3 8.2 6.3  9.8
3 6.6 10.6  9.8 5.8 6.6 5.8 5.7 3.3 9.3 7.0 10.4 8.7 10.1 6.3 3.3 8.0 6.1  9.5

  PA  RI  SC  SD  TN  TX  UT  VT  VA  WA  WV  WI  WY
1 8.1 7.8 7.6 3.6 5.9 5.9 6.1 8.8 6.2 8.7 8.3 5.9 4.2
2 8.1 7.8 7.4 3.5 5.9 5.9 5.9 8.7 6.1 8.7 8.1 5.7 4.1
3 8.0 7.9 7.2 3.4 5.9 5.8 5.7 8.6 5.9 8.7 7.9 5.6 4.0

## time sequence plots of three series
plot(raw[,5],type="l",ylim=c(0,12),xlab="month",
+    ylab="unemployment rate") ## CA
points(raw[,32],type="l", cex = .5, col = "dark red")     ## New York
points(raw[,15],type="l", cex = .5, col = "dark green")   ## Iowa
```

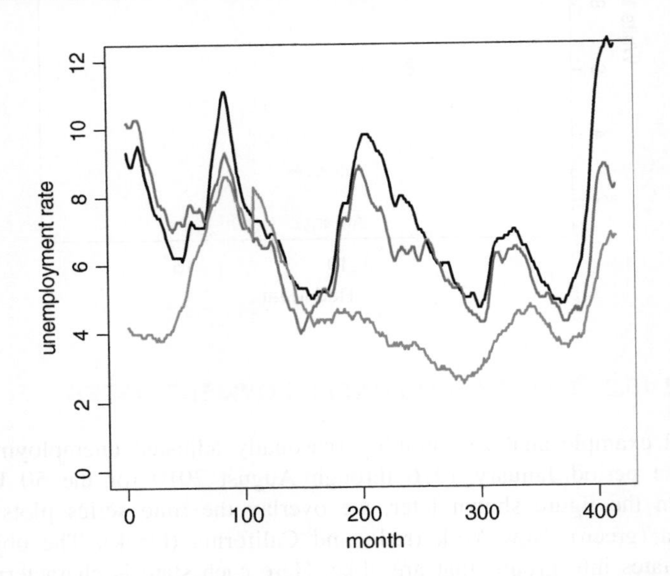

The outcome of k-means clustering for different number of clusters is shown as follows:

```
## transpose the data
## then we have 50 rows (states) and 416 columns (time periods)
rawt=matrix(nrow=50,ncol=416)
rawt=t(raw)
rawt[1:3,]

   [,1] [,2] [,3] [,4] [,5] [,6] [,7] [,8] [,9] [,10] [,11] [,12] [,13] [,14]
AL  6.4  6.3  6.1  6.0  6.0  6.0  6.2  6.3  6.4   6.5   6.6   6.7   6.9   7.0
AK  7.1  7.0  7.0  7.0  7.0  7.1  7.4  7.7  8.0   8.3   8.5   8.7   8.9   9.1
AZ 10.5 10.3 10.0  9.8  9.6  9.5  9.5  9.5  9.6   9.6   9.6   9.5   9.4   9.3
```

```
  [,15] [,16] [,17] [,18] [,19] [,20] [,21] [,22] [,23] [,24] [,25] [,26]
AL  7.1   7.2   7.2   7.1   6.9   6.7   6.5   6.3   6.1   6.0   5.9   5.8
AK  9.3   9.4   9.6   9.7   9.8   9.8   9.9  10.0  10.2  10.4  10.7  10.8
AZ  9.1   8.9   8.6   8.4   8.3   8.2   8.1   7.9   7.6   7.3   7.0   6.7
```

```
## k-means clustering in 416 dimensions
set.seed(1)
grpunemp2 <- kmeans(rawt, centers=2, nstart=10)
sort(grpunemp2$cluster)
```

```
CO CT DE GA HI IA KS ME MD MA MN MT NE NH NC ND OK SD UT VT VA WI WY AL AK AZ
 1  1  1  1  1  1  1  1  1  1  1  1  1  1  1  1  1  1  1  1  1  1  1  2  2  2
AR CA FL ID IL IN KY LA MI MS MO NV NJ NM NY OH OR PA RI SC TN TX WA WV
 2  2  2  2  2  2  2  2  2  2  2  2  2  2  2  2  2  2  2  2  2  2  2  2
```

```
grpunemp3 <- kmeans(rawt, centers=3, nstart=10)
sort(grpunemp3$cluster)
```

```
AL AK AR IL KY LA MI MS NM OH OR TN WA WV CO HI IA KS MD MN NE NH ND OK SD UT
 1  1  1  1  1  1  1  1  1  1  1  1  1  1  2  2  2  2  2  2  2  2  2  2  2  2
VT VA WY AZ CA CT DE FL GA ID IN ME MA MO MT NV NJ NY NC PA RI SC TX WI
 2  2  2  3  3  3  3  3  3  3  3  3  3  3  3  3  3  3  3  3  3  3  3  3
```

```
grpunemp4 <- kmeans(rawt, centers=4, nstart=10)
sort(grpunemp4$cluster)
```

```
AK LA MI MS WV CO HI IA KS MN NE NH ND OK SD UT VT VA WY AZ CT DE FL GA ME MD
 1  1  1  1  1  2  2  2  2  2  2  2  2  2  2  2  2  2  2  3  3  3  3  3  3  3
MA NJ NY NC RI AL AR CA ID IL IN KY MO MT NV NM OH OR PA SC TN TX WA WI
 3  3  3  3  3  4  4  4  4  4  4  4  4  4  4  4  4  4  4  4  4  4  4  4
```

```
grpunemp5 <- kmeans(rawt, centers=5, nstart=10)
sort(grpunemp5$cluster)
```

```
CO ID IA MN MO MT OK TX UT WI WY AZ CT DE FL GA ME MD MA NJ NY NC RI AK LA MI
 1  1  1  1  1  1  1  1  1  1  1  2  2  2  2  2  2  2  2  2  2  2  2  3  3  3
MS WV AL AR CA IL IN KY NV NM OH OR PA SC TN WA HI KS NE NH ND SD VT VA
 3  3  4  4  4  4  4  4  4  4  4  4  4  4  4  4  5  5  5  5  5  5  5  5
```

Other ways of setting up the clustering are possible. For example, we can calculate summary statistics for each state, such as the average and the standard deviation of the unemployment rates, and then use these two calculated features of the monthly unemployment rates as the attributes for clustering. The data file *unemp.csv* includes the average and the standard deviation for each state. The results for three clusters are indicated on the scatter plot of the standard deviations against the means. In general, a state's standard deviation in unemployment increases with its level. We see groups of states with low unemployment and low variability, and states with high unemployment and high variability. Note that this approach to clustering does not incorporate differences or similarities in the state-specific time-patterns of the unemployment rates.

```
## data set unemp.csv with means and standard deviations for
## each state
## k-means clustering on 2 dimensions (mean, stddev)
```

```
unemp <- read.csv("C:/DataMining/Data/unemp.csv")
unemp[1:3,]

  state     mean   stddev
1    AL 6.644952 2.527530
2    AK 8.033173 1.464966
3    AZ 6.120673 1.743672

set.seed(1)
grpunemp <- kmeans(unemp[,c("mean","stddev")], centers=3,
+     nstart=10)
## list of cluster assignments
o=order(grpunemp$cluster)
data.frame(unemp$state[o],grpunemp$cluster[o])
plot(unemp$mean,unemp$stddev,type="n",xlab="mean",
+     ylab="stddev")
text(x=unemp$mean,y=unemp$stddev,labels=unemp$state,
+     col=grpunemp$cluster+1)
```

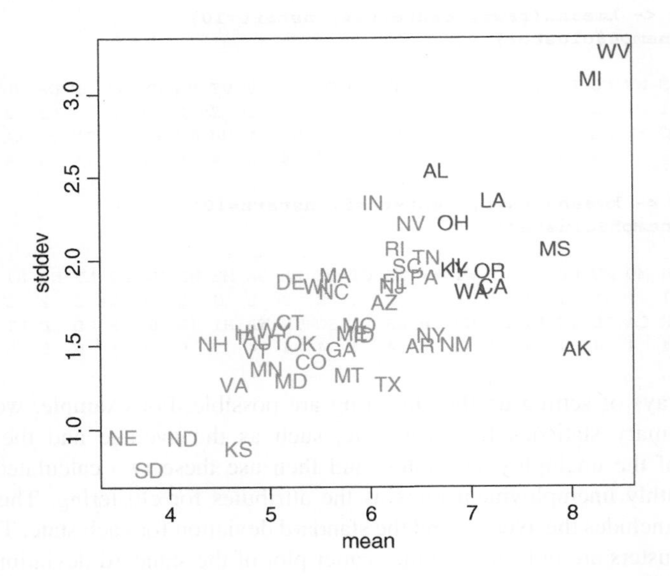

15.2 ANOTHER WAY TO LOOK AT CLUSTERING: APPLYING THE EXPECTATION-MAXIMIZATION (EM) ALGORITHM TO MIXTURES OF NORMAL DISTRIBUTIONS

The *k*-means algorithm is a variant of the generalized expectation-maximization algorithm. The "assignment" step in the *k*-means clustering algorithm represents the expectation step, while the "update" step represents the maximization step. Each unit is assigned a variable that determines its allocation into a certain cluster. The assignment variable can be thought of as a latent variable z. The k-means clustering algorithm makes a unique "hard" choice for this latent variable; one unit may be assigned to cluster 2, while another unit to cluster 5.

We can improve on the approach that makes a hard choice on each unit and obtains cluster centroids by averaging over the units that are associated with a particular value of z. Instead, we can determine for each unit probabilities for the various cluster associations (i.e., probabilities for the various outcomes of z) and then use these probabilities to estimate the group centroids from a weighted average over the entire set of units. The resulting algorithm, referred to as a *soft* clustering approach, is the type of algorithm normally associated with expectation maximization (EM). The probabilities used to compute the weighted averages across the units are called the *soft counts* (probability weights) as compared to the *hard counts* (0/1) that are used in the k-means algorithm. The probabilities computed for z are posterior probabilities and they are computed in the E-step. The M-step uses these probabilities to update the model parameters.

An EM algorithm that works on mixtures of normal distributions is described next. It makes a "soft" choice for the latent variable, and furthermore, allows for correlation among the features. Note that k-means clustering uses a spherical distance measure, and implicitly assumes equal importance of the features and independence among the features. The features can be scaled such that they have equal standard deviations, and the k-means algorithm can be applied to the standardized data. While this gets around problems of scale and avoids giving too much influence to features with large variability it will not adjust the analysis for possible correlation among the features.

Let $x = (x_1, x_2, \ldots, x_n)$ be a sample of independent observations from a mixture of *two* multivariate normal distributions of dimension p. The extension to 3 or more groups is fairly straightforward and is omitted here. Let $z = (z_1, z_2, \ldots, z_n)$ be the latent variables, with each z_i taking on two outcomes (1 and 2) that determine the distribution from which the observations are drawn. We assume p-variate normal distributions

$$x_i|(z_i = 1) \sim N_p(\mu_1, \Sigma_1) \text{ and } x_i|(z_i = 2) \sim N_p(\mu_2, \Sigma_2),$$

and probabilities

$$P(z_i = 1) = \tau_1 \text{ and } P(z_i = 2) = \tau_2 = 1 - \tau_1.$$

The aim is to estimate the unknown parameters: the "mixing" probability $\tau_1 = \tau$ for the two normal distributions, and the means (μ_1, μ_2) and covariance matrices (Σ_1, Σ_2) of the two normal distributions; that is,

$$\theta = (\tau, \mu_1, \Sigma_1, \mu_2, \Sigma_2).$$

The likelihood function is

$$L(\theta; x, z) = P(x, z|\theta) = \prod_{i=1}^{n}\sum_{j=1}^{2} I(z_i = j)\tau_j f(x_i; \mu_j, \Sigma_j),$$

where I is an indicator function and f is the probability density function of the multivariate normal distribution. The likelihood function can be re-written in exponential family form,

$$L(\theta; x, z) = \exp\left\{\sum_{i=1}^{n}\sum_{j=1}^{2} I\left(z_i = j\right)\left[\log \tau_j - 0.5\log|\Sigma_j|\right.\right.$$
$$\left.\left.-0.5(x_i - \mu_j)'\Sigma_j^{-1}(x_i - \mu_j) - \left(\frac{n}{2}\right)\log(2\pi)\right]\right\}.$$

Since the likelihood function includes the unknown latent variables $z = (z_1, z_2, \ldots, z_n)$, we use the iterative EM algorithm to estimate the parameters $\theta = (\tau, \mu_1, \Sigma_1, \mu_2, \Sigma_2)$.

15.2.1 E-Step

Given a current estimate of the parameters $\theta^{(t)}$, the conditional distribution of z_i is determined by Bayes theorem,

$$T_{j,i}^{(t)} = P(z_i = j|x_i; \theta^{(t)}) = \frac{\tau_j^{(t)} f(x_i; \mu_j^{(t)}, \Sigma_j^{(t)})}{\tau_1^{(t)} f(x_i; \mu_1^{(t)}, \Sigma_1^{(t)}) + \tau_2^{(t)} f(x_i; \mu_2^{(t)}, \Sigma_2^{(t)})}; \quad j = 1, 2$$

This discrete conditional distribution is used to evaluate the expected value of the log-likelihood function,

$$Q(\theta|\theta^{(t)}) = E[\log L(\theta; x, z)] = \left\{\sum_{i=1}^{n}\sum_{j=1}^{2} T_{j,i}^{(t)}\left[\log \tau_j - 0.5\log|\Sigma_j|\right.\right.$$
$$\left.\left.-\,0.5(x_i - \mu_j)'\Sigma_j^{-1}(x_i - \mu_j) - \left(\frac{n}{2}\right)\log(2\pi)\right]\right\}.$$

15.2.2 M-Step

The determination of the maximizing values of θ in $Q(\theta|\theta^{(t)})$ is relatively straightforward. Note that τ, (μ_1, Σ_1) and (μ_2, Σ_2) can be maximized independently of each other since they appear in separate linear terms.

First, consider the estimation of $\tau = \tau_1$, which satisfies the constraint $\tau_1 + \tau_2 = 1$:

$$\tau^{(t+1)} = \arg\max_{\tau} Q(\theta|\theta^{(t)}) = \arg\max_{\tau} \left\{ \left[\sum_{i=1}^{n} T_{1,i}^{(t)} \right] \log \tau_1 + \left[\sum_{i=1}^{n} T_{2,i}^{(t)} \right] \log \tau_2 \right\}.$$

This has the same form as the maximum likelihood estimate (MLE) of the parameter in the binomial distribution. Hence,

$$\tau_j^{(t+1)} = \frac{\sum_{i=1}^{n} T_{j,i}^{(t)}}{\sum_{i=1}^{n} (T_{1,i}^{(t)} + T_{2,i}^{(t)})} = \frac{1}{n} \sum_{i=1}^{n} T_{j,i}^{(t)} \quad \text{for } j = 1, 2.$$

The estimates of (μ_1, Σ_1) are found by maximizing

$$\begin{aligned}(\mu_1^{(t+1)}, \Sigma_1^{(t+1)}) &= \arg\max_{\mu_1, \Sigma_1} Q(\theta|\theta^{(t)}) \\ &= \arg\max_{\mu_1, \Sigma_1} \sum_{i=1}^{n} T_{1,i}^{(t)} \{-0.5 \log|\Sigma_1| - 0.5(x_i - \mu_1)' \Sigma_1^{-1} (x_i - \mu_1)\}.\end{aligned}$$

This maximization has the same form as a weighted MLE of the parameters of a normal distribution. Hence,

$$\mu_1^{(t+1)} = \frac{\sum_{i=1}^{n} T_{1,i}^{(t)} x_i}{\sum_{i=1}^{n} T_{1,i}^{(t)}} \quad \text{and} \quad \Sigma_1^{(t+1)} = \frac{\sum_{i=1}^{n} T_{1,i}^{(t)} (x_i - \mu_1^{(t+1)})(x_i - \mu_1^{(t+1)})'}{\sum_{i=1}^{n} T_{1,i}^{(t)}}$$

and by symmetry,

$$\mu_2^{(t+1)} = \frac{\sum_{i=1}^{n} T_{2,i}^{(t)} x_i}{\sum_{i=1}^{n} T_{2,i}^{(t)}} \quad \text{and} \quad \Sigma_2^{(t+1)} = \frac{\sum_{i=1}^{n} T_{2,i}^{(t)} (x_i - \mu_2^{(t+1)})(x_i - \mu_2^{(t+1)})'}{\sum_{i=1}^{n} T_{2,i}^{(t)}}.$$

We iterate on these two steps until we reach convergence. The final estimates are used to compute the posterior distribution of $z = (z_1, z_2, \ldots, z_n)$ (given in the first equation of the section on the E-step), and units are assigned to the group with the largest posterior probability. Software is available for the EM estimation and the calculation of the posterior probabilities. Here we use the function *vnormalmixEM* from the R package **mixtools**. Alternatively, one could use the **bayesm** package for a full Bayesian analysis using Markov Chain Monte Carlo (MCMC) methods.

EXAMPLE 15.3 EUROPEAN PROTEIN CONSUMPTION REVISITED

For illustration we use the data set we had used for k-means clustering in Example 15.1. We have data on 25 European countries ($n = 25$) that describe the protein intake (in percent) from nine major food sources ($p = 9$).

We consider the first two features, protein from red meat and protein from white meat, and fit a mixture of two normal distributions. We allow the covariance matrices to be different across the two groups. The clustering result is virtually identical to the one given by the *k*-means algorithm.

```
library(mixtools)

## for a brief description of mvnormalmixEM
## mvnormalmixEM(x, lambda = NULL, mu = NULL, sigma = NULL, k = 2,
##              arbmean = TRUE, arbvar = TRUE, epsilon = 1e-08,
##              maxit = 10000, verb = FALSE)
## arbvar=FALSE   same cov matrices
## arbvar=TRUE (default) different cov matrices
## arbmean=TRUE (default) different means
## k number of groups

food <- read.csv("C:/DataMining/Data/protein.csv")
## Consider just Red and White meat clusters
food[1:3,]

Country RedMeat WhiteMeat Eggs Milk Fish Cereals Starch Nuts Fr.Veg
1 Albania   10.1       1.4  0.5  8.9  0.2    42.3    0.6  5.5    1.7
2 Austria    8.9      14.0  4.3 19.9  2.1    28.0    3.6  1.3    4.3
3 Belgium   13.5       9.3  4.1 17.5  4.5    26.6    5.7  2.1    4.0

X=cbind(food[,2],food[,3])
X[1:3,]

     [,1] [,2]
[1,] 10.1  1.4
[2,]  8.9 14.0
[3,] 13.5  9.3

set.seed(1)
## here we use an iterative procedure and the results in repeated
## runs may not be exactly the same
## set.seed(1) is used to obtain reproducible results

## mixtures of two normal distributions on the first 2 features
## we consider different variances
out2<-mvnormalmixEM(X,arbvar=TRUE,k=2,epsilon=1e-02)
out2

$lambda
[1] 0.4418574 0.5581426

$mu
$mu[[1]]
[1] 8.117431 4.388409

$mu[[2]]
[1] 11.18218 10.67281

$sigma
$sigma[[1]]
          [,1]      [,2]
```

```
[1,]  3.3012533 -0.9631818
[2,] -0.9631818  1.8514426

$sigma[[2]]
          [,1]      [,2]
[1,] 12.505241 -4.493872
[2,] -4.493872  4.555115

$loglik
[1] -122.0898

$posterior
            comp.1       comp.2
 [1,] 9.999989e-01 1.090445e-06
 [2,] 1.438527e-13 1.000000e+00
 [3,] 3.599562e-08 1.000000e+00
 [4,] 9.978098e-01 2.190198e-03
 [5,] 1.991626e-08 1.000000e+00
 [6,] 5.149435e-08 9.999999e-01
 [7,] 8.655365e-08 9.999999e-01
 [8,] 9.990689e-01 9.311203e-04
 [9,] 7.324676e-16 1.000000e+00
[10,] 9.999738e-01 2.616119e-05
[11,] 1.373398e-07 9.999999e-01
[12,] 4.869934e-10 1.000000e+00
[13,] 9.992147e-01 7.853391e-04
[14,] 3.769919e-13 1.000000e+00
[15,] 9.995283e-01 4.717496e-04
[16,] 4.250826e-04 9.995749e-01
[17,] 9.999983e-01 1.679236e-06
[18,] 9.989033e-01 1.096650e-03
[19,] 9.999983e-01 1.702685e-06
[20,] 5.184376e-02 9.481562e-01
[21,] 2.905501e-09 1.000000e+00
[22,] 3.507084e-07 9.999996e-01
[23,] 9.996851e-01 3.149250e-04
[24,] 1.712852e-12 1.000000e+00
[25,] 9.999864e-01 1.355387e-05
```

```
prob1=round(out2$posterior[,1],digits=3)
prob2=round(out2$posterior[,2],digits=3)
prob=round(out2$posterior[,1])
o=order(prob)
data.frame(food$Country[o],prob1[o],prob2[o],prob[o])
```

```
   food.Country.o. prob1.o. prob2.o. prob.o.
1          Austria    0.000    1.000       0
2          Belgium    0.000    1.000       0
3   Czechoslovakia    0.000    1.000       0
4          Denmark    0.000    1.000       0
5        E Germany    0.000    1.000       0
6           France    0.000    1.000       0
7          Hungary    0.000    1.000       0
8          Ireland    0.000    1.000       0
9      Netherlands    0.000    1.000       0
10          Poland    0.000    1.000       0
```

```
11         Sweden   0.052   0.948   0
12    Switzerland   0.000   1.000   0
13             UK   0.000   1.000   0
14      W Germany   0.000   1.000   0
15        Albania   1.000   0.000   1
16       Bulgaria   0.998   0.002   1
17        Finland   0.999   0.001   1
18         Greece   1.000   0.000   1
19          Italy   0.999   0.001   1
20         Norway   1.000   0.000   1
21       Portugal   1.000   0.000   1
22        Romania   0.999   0.001   1
23          Spain   1.000   0.000   1
24           USSR   1.000   0.000   1
25     Yugoslavia   1.000   0.000   1
```

```
plot(food$Red, food$White, type="n",xlab="Red Meat",
+     ylab="White Meat")
text(x=food$Red,y=food$White,labels=food$Country,col=prob+1)
```

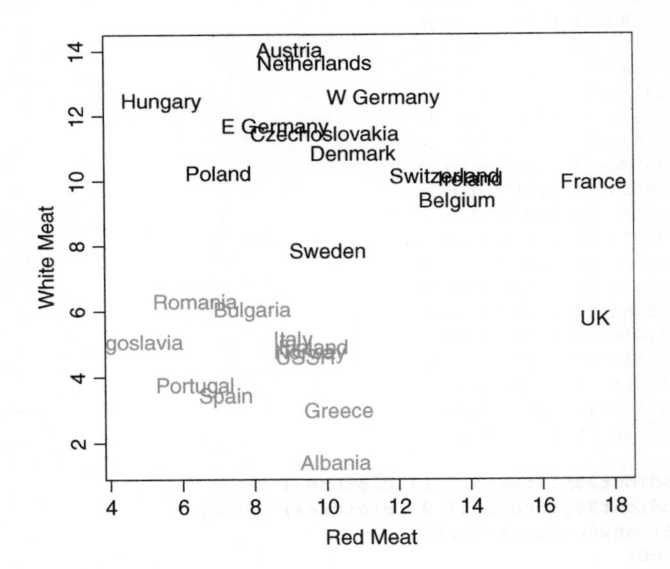

Next, we consider all nine features and fit a mixture of two normal distributions. We constrain the covariance matrices to be equal across the two groups. Even in this restricted situation one has to estimate 45 parameters in the common 9×9 covariance matrix; this is not easy considering the fact that we have observations from only 25 countries.

```
## mixtures of two normal distributions on all 9 features
## we consider equal variances
X1=cbind(food[,2],food[,3],food[,4],food[,5],food[,6],
food[,7],food[,8],food[,9],food[,10])
X1[1:3,]
```

```
      [,1] [,2] [,3] [,4] [,5] [,6] [,7] [,8] [,9]
[1,] 10.1  1.4  0.5  8.9  0.2 42.3  0.6  5.5  1.7
[2,]  8.9 14.0  4.3 19.9  2.1 28.0  3.6  1.3  4.3
[3,] 13.5  9.3  4.1 17.5  4.5 26.6  5.7  2.1  4.0
```

```
set.seed(1)
out2all<-mvnormalmixEM(X1,arbvar=FALSE,k=2,epsilon=1e-02)
out2all
prob1=round(out2all$posterior[,1],digits=3)
prob2=round(out2all$posterior[,2],digits=3)
prob=round(out2all$posterior[,1])
data.frame(food$Country,prob1,prob2,prob)
```

```
     food.Country prob1 prob2 prob
1         Albania     0     1    0
2         Austria     1     0    1
3         Belgium     1     0    1
4        Bulgaria     0     1    0
5  Czechoslovakia     1     0    1
6         Denmark     1     0    1
7       E Germany     1     0    1
8         Finland     1     0    1
9          France     1     0    1
10         Greece     0     1    0
11        Hungary     0     1    0
12        Ireland     1     0    1
13          Italy     1     0    1
14    Netherlands     1     0    1
15         Norway     1     0    1
16         Poland     1     0    1
17       Portugal     1     0    1
18        Romania     0     1    0
19          Spain     1     0    1
20         Sweden     1     0    1
21    Switzerland     1     0    1
22             UK     1     0    1
23           USSR     0     1    0
24      W Germany     1     0    1
25     Yugoslavia     0     1    0
```

```
o=order(prob)
data.frame(food$Country[o],prob[o])
```

```
   food.Country.o. prob.o.
1          Albania       0
2         Bulgaria       0
3           Greece       0
4          Hungary       0
```

```
5           Romania    0
6              USSR    0
7        Yugoslavia    0
8           Austria    1
9           Belgium    1
10   Czechoslovakia    1
11          Denmark    1
12        E Germany    1
13          Finland    1
14           France    1
15          Ireland    1
16            Italy    1
17      Netherlands    1
18           Norway    1
19           Poland    1
20         Portugal    1
21            Spain    1
22           Sweden    1
23      Switzerland    1
24               UK    1
25        W Germany    1
```

15.3 HIERARCHICAL CLUSTERING PROCEDURES

Hierarchical clustering is yet another approach of clustering n units (or objects), each described by p features, into a smaller number of groups. Hierarchical clustering creates a hierarchy of clusters which can be represented in a treelike diagram, called a *dendrogram*. In the dendrogram, units in the same cluster are joined by a horizontal line, with the scale on the y-axis of the dendrogram reflecting a measure of the distances of the units within the cluster. The leaves at the bottom of the dendrogram represent the individual units; leaves are combined to form small branches, small branches are combined into larger branches, until one reaches the trunk or root of the tree that represents a single cluster containing all units. Dendrograms are quite useful as they give us a visual representation of the clusters.

Algorithms for hierarchical clustering are either *agglomerative*, in which we start at the individual leaves and successively merge clusters together, or *divisive*, in which we start at the root and recursively split the clusters. Agglomerative procedures represent a "bottom-up" approach, where each unit starts in its own cluster and pairs of clusters are merged as we move up the hierarchy. Divisive procedures represent a "top down" approach, where all units start in one cluster and splits are performed recursively as we move down the hierarchy. We concentrate our discussion on the commonly used agglomerative procedures.

Hierarchical clustering procedures require both a distance measure and a linkage criterion. A *distance measure* between two objects with feature vectors $x_i = (x_{i1}, x_{i2}, \ldots, x_{ip})$ and $x_j = (x_{j1}, x_{j2}, \ldots, x_{jp})$ is non-negative and symmetric (i.e., $d(x_i, x_j) = d(x_j, x_i) \geq 0$, with $d(x_i, x_j) = 0$ implying $x_i = x_j$) and satisfies the triangle inequality (i.e., $d(x_i, x_j) \leq d(x_i, x_k) + d(x_j, x_k)$). The Euclidian (L2) norm $d(x_i, x_j) = \sqrt{\sum_{r=1}^{p} (x_{ir} - x_{jr})^2}$ and the L1 norm $d(x_i, x_j) = \sum_{r=1}^{p} |x_{ir} - x_{jr}|$ are commonly used. For categorical variables, with features representing the absence or presence of certain characteristics, we can define a measure of distance from the number of matches and mismatches. Sometimes, distance is expressed in terms of a similarity measure, which is the opposite (inverse) of the distance measure.

The choice of the clusters, consisting of one or more units, to be merged (or split in divisive clustering) is determined by the *linkage criterion*, which is a function of all pairwise distances among the units in the two different clusters being considered for distance evaluation.

The following illustrative example considers five units, $\{a\}$ $\{b\}$ $\{c\}$ $\{d\}$ and $\{e\}$. Our agglomerative bottom-up method will build the hierarchy from the individual units by progressively merging clusters. The first step is to determine which two units to merge into the first cluster. We will merge the two closest units, as determined by the distance measure that we have selected. Suppose that we have merged the two closest units c and e, giving us the following clusters: $\{a\}$ $\{b\}$, $\{d\}$, and $\{c\ e\}$. For further merging, we need to determine the distance between $\{a\}$ and $\{c\ e\}$, for example. We need to define what we mean by the distance between two clusters. The distance between two clusters A and B, each of them consisting of one or more units (elements), can be defined in several different ways:

As the *maximum* pairwise distance between units of different clusters, $\max\{d(x, y) : x \in A, y \in B\}$. This is referred to as *complete-linkage clustering.*

As the *minimum* pairwise distance between units of different clusters, $\min\{d(x, y) : x \in A, y \in B\}$. This is referred to as *single-linkage clustering.*

As the *average* pairwise distance between units of different clusters, $1/nu(A)nu(B) \left(\sum_{x \in A} \sum_{y \in B} d\,(x, y) \right)$. This is referred to as *average-linkage clustering.*

Several other linkage functions have been considered in the literature, but are not discussed in this introduction.

Here is another way to think about the agglomerative clustering process. We start with an $n \times n$ *distance matrix* where the number in the i-th row and j-th column is the distance between the i-th and j-th units. The distance matrix is symmetric with zeros in the diagonal, so only the lower (or upper) triangular region needs to be filled in. Then, as the clustering progresses, rows and columns are combined as clusters are merged and the distances between the merged rows and columns are updated. This is how agglomerative clustering is implemented on computers.

We describe this approach for our illustration with the five units $\{a\}$ $\{b\}$ $\{c\}$ $\{d\}$ and $\{e\}$. Assume that the pairwise distances are given in the 5×5 distance matrix as shown in Table 15.1. The distance between units $\{c\}$ and $\{e\}$ is smallest; hence our first cluster becomes $\{c\ \ e\}$; its units are two units apart. A new distance matrix, now a 4×4 matrix, is constructed and the distances in this matrix need to be updated. Assume that we use single-linkage clustering with linkage function $\min\{d(x, y) :$

Table 15.1 Example of Single-Linkage Clustering

Original 5 × 5 distance matrix, with subsequent single-linkage clustering

$$\begin{matrix} a \\ b \\ c \\ d \\ e \end{matrix} \begin{bmatrix} 0 & & & & \\ 9 & 0 & & & \\ 3 & 7 & 0 & & \\ 6 & 5 & 9 & 0 & \\ 11 & 10 & \mathbf{2} & 8 & 0 \end{bmatrix}$$

$$\begin{matrix} (ce) \\ a \\ b \\ d \end{matrix} \begin{bmatrix} 0 & & & \\ \mathbf{3} & 0 & & \\ 7 & 9 & 0 & \\ 8 & 6 & 5 & 0 \end{bmatrix}$$

$$\begin{matrix} (ace) \\ b \\ d \end{matrix} \begin{bmatrix} 0 & & \\ 7 & 0 & \\ 6 & \mathbf{5} & 0 \end{bmatrix}$$

$$\begin{matrix} (ace) \\ (bd) \end{matrix} \begin{bmatrix} 0 & \\ \mathbf{6} & 0 \end{bmatrix}$$

The numbers in bold face refer to the minimum distances.

$x \in A, y \in B\}$. The distance between the cluster $\{c\ \ e\}$ and unit $\{a\}$ is the minimum of the distance between $\{c\}$ and $\{a\}$, which is 3, and the distance between $\{e\}$ and $\{a\}$, which is 11. The minimum of these two numbers, 3, is entered in the updated 4×4 distance matrix as the distance between $\{c\ \ e\}$ and $\{a\}$. Next, we calculate the distance between $\{c\ \ e\}$ and $\{b\}$. It is the minimum of $d(\{c\},\ \{b\}) = 7$ and $d(\{e\},\ \{b\}) = 10$. The minimum is 7, and it is entered. The distance between $\{c\ \ e\}$ and $\{d\}$ is min $\{d(\{c\},\ \{d\}),\ d(\{e\},\ \{d\})\} = \min\ \{9,\ 8\} = 8$. The other distances in the 4×4 matrix are distances among individual units and can be copied from the initial distance matrix.

The smallest distance in the 4×4 matrix is between the cluster $\{c\ e\}$ and $\{a\}$, with cluster distance 3, leading to the formation of a new cluster of three elements $\{a\ \ c\ \ e\}$. We need to update the distance matrix to form a 3×3 matrix. The distance between $\{a\ \ c\ \ e\}$ and $\{b\}$, under single linkage, is $\min\ \{d\ (\{a\},\ \{b\}),\ d(\{c\},\ \{b\}),\ d(\{e\},\ \{b\})\} = \min\ \{9,\ 7,\ 10\} = 7$ using the pairwise distances among units that are given in the first 5×5 distance matrix. Even simpler, we could have calculated the distance between $\{a\ c\ e\}$ and $\{b\}$ from $\min\ \{d(\{a\},\{b\}),\ d(\{c\ e\},\{b\})\} = \min\{9,\ 7\} = 7$, using the updated 4×4 distance matrix. The distance between $\{a\ \ c\ \ e\}$ and $\{d\}$ is min $\{d(\{a\},\{d\}),\ d(\{c\ e\},\{d\})\} = \min\{6,\ 8\} = 6$, and the distance between $\{b\}$ and $\{d\}$ is 5. These become the entries in the 3×3 distance matrix. The smallest distance is now between $\{b\}$ and $\{d\}$. This becomes our next cluster and the updated 2×2 distance matrix is shown in Table 15.1. The distance between the cluster $\{a\ \ c\ \ e\}$ and the new one $\{b\ \ d\}$, under single linkage, is min $\{d(\{a\ c\ e\},\{b\}),\ d(\{a\ \ c\ \ e\},\{d\})\} = \min\{7,\ 6\} = 6$. The two clusters $\{a\ \ c\ \ e\}$ and $\{b\ d\}$ are linked at distance 6.

Table 15.2 Example of Complete-Linkage Clustering

Original 5 × 5 distance matrix, with subsequent complete-linkage clustering

$$\begin{array}{c} a \\ b \\ c \\ d \\ e \end{array} \begin{bmatrix} 0 & & & & \\ 9 & 0 & & & \\ 3 & 7 & 0 & & \\ 6 & 5 & 9 & 0 & \\ 11 & 10 & \mathbf{2} & 8 & 0 \end{bmatrix}$$

$$\begin{array}{c} (ce) \\ a \\ b \\ d \end{array} \begin{bmatrix} 0 & & & \\ 11 & 0 & & \\ 10 & 9 & 0 & \\ 9 & 6 & \mathbf{5} & 0 \end{bmatrix}$$

$$\begin{array}{c} (ce) \\ (bd) \\ a \end{array} \begin{bmatrix} 0 & & \\ 10 & 0 & \\ 11 & \mathbf{9} & 0 \end{bmatrix}$$

$$\begin{array}{c} (ce) \\ (abd) \end{array} \begin{bmatrix} 0 & \\ \mathbf{11} & 0 \end{bmatrix}$$

The numbers in bold face refer to the minimum distances.

Each agglomeration occurs at a greater distance between clusters than the previous agglomeration. We can decide to stop clustering either when the clusters are too far apart to be merged (distance criterion), or when there is a sufficiently small number of clusters (number criterion).

How would this aggregation work under complete linkage, with linkage function $\max\{d(x,y) : x \in A, y \in B\}$? The steps are shown in Table 15.2. With our first cluster $\{c\ \ e\}$, we update the entries in the 4×4 matrix as follows. The distance between the cluster $\{c\ \ e\}$ and unit $\{a\}$ is now the maximum of the distance between $\{c\}$ and $\{a\}$, which is 3, and the distance between $\{e\}$ and $\{a\}$, which is 11. The maximum of these two numbers, 11, is entered in the new distance matrix as the distance between $\{c\ \ e\}$ and $\{a\}$. Next, we calculate the distance between $\{c\ \ e\}$ and $\{b\}$. It is the maximum of $d(\{c\},\{b\}) = 7$ and $d(\{e\},\{b\}) = 10$. The maximum 10 is entered. The distance between $\{c\ \ e\}$ and $\{d\}$ is $\max\ \{d(\{c\},\{d\}),\ d(\{e\},\{d\})\} = \max\{9,\ 8\} = 9$. The other distances in the 4×4 matrix are distances among individual units and can be copied from the initial distance matrix.

The smallest distance in the 4×4 matrix is between the units $\{b\}$ and $\{d\}$, leading to the new cluster $\{b\ \ d\}$ with distance measure 5 and the updated distances in a 3×3 distance matrix. The distance between $\{c\ \ e\}$ and $\{b\ \ d\}$, under complete linkage, is $\max\ \{d(\{c\ \ e\},\{b\}),\ d(\{c\ \ e\},\{d\})\} = \max\{10,\ 9\} = 10$, using the entries in the 4×4 distance matrix. The distance between the new cluster $\{b\ \ d\}$ and $\{a\}$, under complete linkage, is $\max\ \{d(\{a\},\ \{b\}),\ d(\{a\},\{d\})\} = \max\{9,\ 6\} = 9$. The distance between $\{c\ \ e\}$ and $\{a\}$, 11, was already part of the 4×4 distance matrix calculated earlier.

The smallest distance in the 3×3 distance matrix is now between $\{b\ \ d\}$ and $\{a\}$, and hence $\{a\ \ b\ \ d\}$ becomes our next cluster. This cluster forms at distance 9. The updated 2×2 distance matrix is shown next. The distance between the cluster $\{c\ \ e\}$ and the newly formed cluster $\{a\ \ b\ \ d\}$ is the maximum $\{d(\{c\ \ e\}, \{b\ \ d\}), d(\{c\ \ e\}, \{a\})\} = \max\{10,\ \ 11\} = 11$. The clusters $\{c\ \ e\}$ and $\{a\ \ b\ \ d\}$ get linked at distance 11.

Dendrograms for single and complete linkage are shown in Figure 15.1. Objects in the same cluster are joined by a horizontal line whose height on the dendrogram

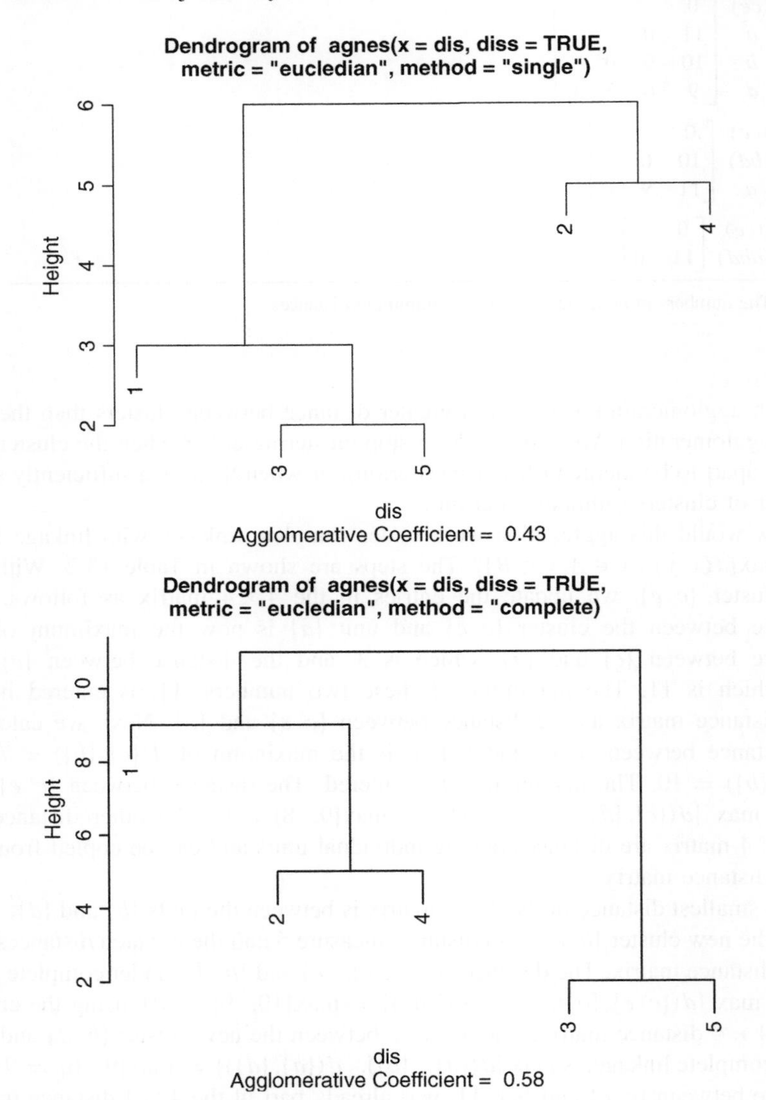

Figure 15.1 Dendrograms for single and complete linkage.

(y-axis) reflects the distance of the units within that cluster. The clustering and the scale on the dendrogram depends on the particular linkage criterion that has been adopted. By looking at the height on the y-scale of the dendrogram we can assess the price (expressed as distance among the units in the cluster) that is paid by clustering objects into larger and larger groups.

We could repeat the clustering and create a dendrogram under average linkage. Average linkage uses the average pairwise distance between units of different clusters, $1/nu(A)nu(B)\left(\sum_{x \in A} \sum_{y \in B} d\,(x, y)\right)$. The calculations for this linkage criterion are a bit more involved as we need to keep track of the distances between individual units when updating the distance measures. For complete and simple linkage the update is easier as we can update the new distance matrix from the one that comes immediately before. However, calculations are easily performed with computer programs, and these programs also display the dendrograms.

For a particular problem at hand it is always a good idea to try several clustering methods, using different distance measures and different linkage criteria. We need to explore whether the different methods are more or less consistent, and whether the results make sense and can be interpreted. Clustering methods are known to be sensitive to small perturbations and outliers, and also to differences in scale. If features involve measurements with very different scales it is always a good idea to first standardize the measurements so that they have similar means and variances. Also, note that there is no provision in hierarchical clustering for reallocating objects that have been linked together, perhaps incorrectly, at an earlier stage. Once a unit is included in a cluster, it stays there.

For agglomerative clustering we use the function *agnes* in the R package **cluster**. Alternatively, we can use the function *hclust* in the **stats** package. The help documentation in R will tell you about the various modeling and output options.

EXAMPLE 15.4 EUROPEAN PROTEIN CONSUMPTION REVISITED

```
library(cluster)
food <- read.csv("C:/DataMining/Data/protein.csv")
food[1:3,]
```

```
Country RedMeat WhiteMeat Eggs Milk Fish Cereals Starch Nuts Fr.Veg
1 Albania    10.1       1.4  0.5  8.9  0.2    42.3    0.6  5.5    1.7
2 Austria     8.9      14.0  4.3 19.9  2.1    28.0    3.6  1.3    4.3
3 Belgium    13.5       9.3  4.1 17.5  4.5    26.6    5.7  2.1    4.0
```

```
## we use the program agnes in the package cluster
## argument diss=FALSE indicates that we use the dissimilarity
## matrix that is being calculated from raw data.
## argument metric="euclidian" indicates that we use Euclidian
## distance
## no standardization is used as the default
## the default is "average" linkage
## Using data on all nine variables (features)
## Euclidean distance and average linkage
```

The dendrogram, resulting from agglomerative clustering with Euclidean distance and average linkage is shown below. A horizontal line at height 16 results in seven clusters. The countries in these clusters are quite similar (but not identical) to the clusters we obtained with the k-means clustering method in Section 15.1. Five clusters remain if you draw the line at height 19, and two clusters remain if you draw it at 25. Single and complete linkage will lead to somewhat different clusters and conclusions.

```
foodagg=agnes(food,diss=FALSE,metric="euclidian")
plot(foodagg) ## dendrogram
```

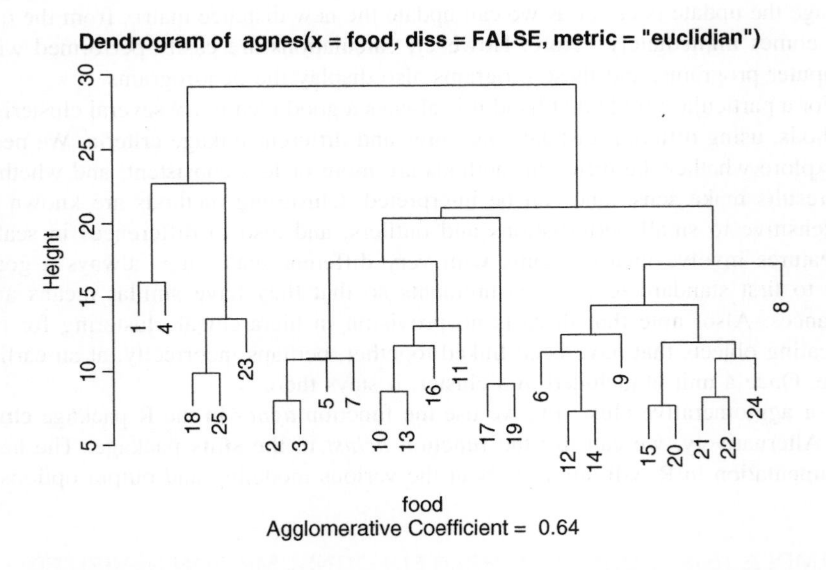

R programs for agglomerative clustering of the European protein data (output shown here) and the US unemployment data (output not shown) can be found on the web page that accompanies this book.

EXAMPLE 15.5 MONTHLY US UNEMPLOYMENT RATES REVISITED

We revisit the data on US unemployment that we discussed in Example 15.2. Here we use a 50×50 distance matrix which is constructed as follows: The (i,j)th element of the distance matrix is defined as 1 minus the correlation coefficient of the first temporal differences (monthly changes) in states i and j. It turns out that all pairwise correlations of the differenced time series are positive. For a correlation of one, the distance is zero and the two states are closely (in fact, perfectly) related. For a correlation close to zero, the distance is 1 and the states are different. The R program, when clustering on that particular distance matrix is listed on the web page that accompanies this book. We leave the interpretation of the results as an exercise. A second version of the analysis considers the correlations of the unemployment rates (the levels) between states i and j.

We may want to adjust a few outliers in the unemployment data set. States AZ, LA, and MS have outliers. An adjusted data set is also available. LA and MS data were adjusted to smooth out the impacts of Hurricane Katrina in the fall of 2005. The gap in AZ (we really do not know the reason for the sudden drop) was also smoothed out. The adjusted data are in the file *adj3unempstates.csv*. You may want to re-run the clustering with the adjusted data set using the program that is listed on the web page that accompanies this book.

REFERENCES

Hand, D.J., Daly, F., Lunn, A.D., McConway, K.J., and Ostrowski, E.: *A Handbook of Small Data Sets*. London: Chapman & Hall, 1994.

Hartigan, J.A.: *Clustering Algorithms*. New York: John Wiley & Sons, Inc., 1975.

Hartigan, J.A. and Wong, M.A.: A k-means clustering algorithm. *Applied Statistics*, Vol. 28 (1979), 100–108.

Weber, A.: *Agrarpolitik im Spannungsfeld der Internationalen Ernaehrungspolitik*. Kiel: Institut fuer Agrarpolitik und Marktlehre, 1973.

CHAPTER 16

Market Basket Analysis: Association Rules and Lift

Market basket analysis looks at purchase coincidence. It investigates whether two products are being purchased together, and whether the purchase of one product increases the likelihood of purchasing the other.

The data typically encountered in these applications can be arranged in a large matrix of rows and columns. Rows represent the different shoppers (or shopping trips) and columns represent the different products. The entries in the data matrix are the incidences (1 or 0) indicating whether or not the item in column j of the matrix (say Goose Island beer) is purchased by the shopper in row i (say Ledolter). The dimensions of the data matrix are usually quite large, with the information on incidences coming from many shoppers (rows) and many different products (columns).

Other common applications involve subscribers to internet radio sites making selections to listen to certain artists (bands). Internet radio (also known as *streaming radio* or *web radio*) has a huge following. While in the past services used to be free, owing to copyright considerations, sites such as Last.fm now charge the users a small monthly fee for gaining access to their play list. Data on the listening preferences of subscribers is being collected continuously. Again, dimensions of the data matrices involved tend to be huge, with many users (100,000s of users) and many different artists (10,000s of artists). The entries in the data matrix are incidences (1/0) indicating whether or not a certain user listens to a certain artist. Market basket analysis is also used by Netflix, the American provider of on-demand internet streaming media with information on movie viewing incidences on more than 25 million customers, and Amazon, the world's largest online retailer, with information on purchase incidences on thousands of products across millions of customers.

Knowing which supermarket products tend to get purchased together and knowing which pair of artists or pair of movies have high co-incidences brings important benefits. This knowledge allows for *targeted* (i.e., "smart") marketing, where customers who have bought one product or viewed one type of movie are being targeted (or "recommended") for advertisements that push related products that

Data Mining and Business Analytics with R, First Edition. Johannes Ledolter.

have high chances of being purchased. Recommender systems have become very common. Noting that a customer views certain products on Amazon.com, the store will recommend additional items. Netflix offers suggestions for movies that a user might like to watch. Many of these recommendations are based on incidence (and co-incidence) information that these companies have obtained from their customers.

Recall from elementary probability calculus the following results about probabilities:

$P(A)$: Probability that product A is being purchased (event A)—The proportion of times event A occurred is also referred to as the *support of A*. It is the relative frequency of 1s in column A of the incidence matrix.

$P(B)$: Probability that product B is being purchased (event B)—The proportion of times event B occurred is referred to as the *support of B*. It is the relative frequency of 1s in column B of the incidence matrix.

$P(A$ and $B)$: Probability that products A and B are being purchased at the same time—The proportion of times events A and B occurred together is referred to as the *support of A and B*. It is the relative frequency of having 1s (i.e., of co-incidence) in both columns A and B of the incidence matrix.

We look for association rules such as: $A =$ LHS (buy chips) $\rightarrow B =$ RHS (buy beer). The left-hand side is referred to as "antecedent"; the right-hand side is referred to as "consequent"; and the arrow expresses "is related to."

$P(B|A)$ is the conditional probability of B given A. It expresses the probability of event B (the RHS: buying beer), knowing that event A (the LHS: buying chips) has occurred. Recall from your study of probability that $P(B|A) = P(A \text{ and } B)/P(A)$.

The conditional probability of B (RHS) given A (LHS) is referred to as the *confidence of B*. It expresses our confidence that product B gets bought if A has been purchased. If this is a small number, the relationship between antecedent A and consequent B is not very relevant as in this case B is unlikely to occur. The confidence is calculated as the ratio, supp($A =$ LHS and $B =$ RHS)/supp($A =$ LHS), with the supports of these two events being obtained from the incidence matrix.

The *lift of A on B* is defined as the ratio

$$\text{lift}(A \rightarrow B) = \frac{P(B|A)}{P(B)} = \frac{P(A \text{ and } B)}{P(A)P(B)}.$$

It compares $P(B|A)$ with $P(B)$. If this ratio is larger than 1, we say that A (LHS) results in an upward lift on B (RHS). Knowing that A (the antecedent) has occurred, increases the chance that B (the consequent) occurs. The lift of A (LHS) on B (RHS) is calculated as the ratio supp(LHS and RHS)/[supp(LHS)supp(RHS)]. Note that this is the same as the lift of B on A, $P(A|B)/P(A) = P(A \text{ and } B)/P(A)P(B)$.

It is important to find antecedents that result in big lifts. But, big lifts are practically relevant only if the consequent has a reasonable chance of occurring. Hence, one screens for combinations that result in good lift as well as high confidence.

Some authors introduce yet another measure and define the *leverage of* $A =$ LHS on $B =$ RHS as

$$\begin{aligned}\text{leverage}(A \to B) &= P(A \text{ and } B) - P(A)P(B),\\ &= \text{lift}(A \to B)P(A)P(B) - P(A)P(B),\\ &= [\text{lift}(A \to B) - 1]P(A)P(B).\end{aligned}$$

The leverage is 0 if there is no association, and it is ≥ 0 if A has a lift on B. The multiplication with $P(A)P(B)$ incorporates the importance of both antecedent and consequent. If they are not likely, then the lift does not translate into leverage. A lower lift, but with a higher likelihood of both antecedent and consequent, may still be of practical interest.

Comment: Confidence and lift can all be obtained from the incidence matrix. However, with the usually very large dimensions of this matrix, supports of many events need to be computed. For 1000 columns (products), there is the need to form half a million pairs and calculate one million confidences and lifts, before the screening for the important events can even start. Considering that the incidence matrix may have 100,000 rows (customers), this certainly amounts to a lot of data crunching. Fortunately, efficient software is available to carry out the calculations and screen for good lift and high confidence, and an efficient R package for doing this will be described in the examples.

Comment: Furthermore, why restrict oneself to just one column or item when defining the conditioning (the antecedent, or LHS) variable? Why not consider pairs or triples of columns, and study the relationship between a consequent and an antecedent that is being described by the joint occurrence of several items (columns)? If this is done, then we also need to calculate the support of triples and quadruples of columns. This involves even more computations and searches, making it even more critical that the algorithms are efficient.

16.1 EXAMPLE 1: ONLINE RADIO

Online radio keeps track of everything you play. It uses this information for recommending music you are likely to enjoy and supports focused marketing that sends you advertisements for music you are likely to buy. Why waste scarce advertising dollars on items that customers are unlikely to purchase.

Suppose you were given data from a music community site. For each user you may have a log of every artist he/she had downloaded to their computer. You may even have demographic information on the user (such as age, sex, location, occupation, and interests). Your objective is to build a system that recommends new music to users in this community. From the available information, it is usually quite easy to determine the support for (i.e., the frequencies of listening to) various individual artists, as well as the joint support for pairs (or larger groupings) of artists. All you have to do is count the incidences (0/1) across all members of your network and divide those frequencies by the number of your members. From the support we can calculate the confidence and the lift.

For illustration we use a large data set with close to 300,000 records of song (artist) selections made by 15,000 users. Even larger data sets are available on the web (see, e.g., Celma (2010), and the data sets on his web page http://ocelma.net/MusicRecommendationDataset). Each row of our data set contains the name of the artist the user has listened to. Our first user, a woman from Germany, has listened to 16 artists, resulting in the first 16 rows of the data matrix. The two demographic variables listed here (gender and country) are not used in our analysis. However, it would be straightforward to stratify the following market basket analysis on gender and country of origin, and investigate whether findings change (we recommend that you do this as an exercise).

We use the R package **arules**, a very convenient and efficient package for mining association rules and for identifying frequent item sets (Hahsler et al., 2005). The first thing we need to accomplish is to transform the data as given here into an incidence matrix where each listener represents a row, with 0 and 1s across the columns indicating whether or not he or she has played a certain artist. The R template shown in the following illustrates how this can be done. The incidence matrix is stored in the R object “playlist.” The support for each of the 1004 artists is calculated, and the support is displayed for all artists with support larger than 0.08 (this means that artists shown on the graph are played by more than 8% of the users).

The last step in the program involves the construction of the association rules (using the function *apriori* in the R package **arules**). We look for artists (or groups of artists) who have support larger than 0.01 (1%) and who give confidence to another artist that is larger than 0.50 (50%). These requirements rule out rare artists. Observe that the program also calculates and lists antecedents (LHS) that involve more than one artist. For example, listening both to “Muse” and “The Beatles” has support larger than 0.01, and the confidence for “Radiohead,” given that someone listens to both “Muse” and “The Beatles,” is 0.507. These two numbers exceed the two requirements we had put on the screening. Antecedents that involve three artists do not come up in the list as they do not meet both requirements. You can see that many computations and logical queries need to be carried out before such a list is obtained.

The list is further narrowed down by also requiring that the lift is larger than 5, and the resulting list is ordered according to decreasing confidence. Listening to both “Led Zeppelin” and “The Doors” is quite predictive of listening to “Pink Floyd” (the confidence is about 50%). Knowing that someone listens to both “Led Zeppelin” and “The Doors” increases the chance of listening to “Pink Floyd” more than 5 – fold (lift = 5.69). Listening to “Judas Priest” lifts the chance of listening to the “Iron Maiden” by a factor of 8.56. If we know that a user listens to “Judas Priest,” we should definitely recommend that he also listens to “Iron Maiden.”

```
### *** Play counts *** ###

lastfm <- read.csv("C:/DataMining/Data/lastfm.csv")
```

```
lastfm[1:19,]

   user                  artist sex       country
1     1     red hot chili peppers   f       Germany
2     1  the black dahlia murder   f       Germany
3     1               goldfrapp   f       Germany
4     1        dropkick murphys   f       Germany
5     1                le tigre   f       Germany
6     1              schandmaul   f       Germany
7     1                   edguy   f       Germany
8     1            jack johnson   f       Germany
9     1               eluveitie   f       Germany
10    1             the killers   f       Germany
11    1            judas priest   f       Germany
12    1              rob zombie   f       Germany
13    1              john mayer   f       Germany
14    1                 the who   f       Germany
15    1             guano apes   f       Germany
16    1      the rolling stones   f       Germany
17    3        devendra banhart   m United States
18    3        boards of canada   m United States
19    3              cocorosie   m United States

length(lastfm$user)    ## 289,955 records in the file
lastfm$user <- factor(lastfm$user)
levels(lastfm$user)    ## 15,000 users
levels(lastfm$artist) ##  1,004 artists

library(arules) ## a-rules package for association rules
## Computational environment for mining association rules and
## frequent item sets

## we need to manipulate the data a bit before using arules
## we split the data in the vector x into groups defined in vector f
## in supermarket terminology, think of users as shoppers and artists
## as items bought
playlist <- split(x=lastfm[,"artist"],f=lastfm$user)
## split into a list of users
playlist[1:2]
## the first two listeners (1 and 3) listen to the following bands

$`1`
 [1] red hot chili peppers   the black dahlia murder goldfrapp
 [4] dropkick murphys        le tigre                schandmaul
 [7] edguy                   jack johnson            eluveitie
[10] the killers             judas priest            rob zombie
[13] john mayer              the who                 guano apes
[16] the rolling stones
1004 Levels: …and you will know us by the trail of dead [unknown] … zero 7

$`3`
 [1] devendra banhart     boards of canada     cocorosie
 [4] aphex twin           animal collective    atmosphere
 [7] joanna newsom        air                  portishead
[10] massive attack       broken social scene arcade fire
[13] plaid                prefuse 73           m83
```

```
[16] the flashbulb        pavement             goldfrapp
[19] amon tobin           sage francis         four tet
[22] max richter          autechre             radiohead
[25] neutral milk hotel   beastie boys         aesop rock
[28] mf doom              the books
1004 Levels: …and you will know us by the trail of dead [unknown] … zero 7

## an artist may be mentioned by the same user more than once
## it is important to remove artist duplicates before creating
## the incidence matrix

playlist <- lapply(playlist,unique)      ## remove artist duplicates

playlist <- as(playlist,"transactions")
## view this as a list of "transactions"
## transactions is a data class defined in arules

itemFrequency(playlist)
## lists the support of the 1,004 bands
## number of times band is listed to on the playlist of 15,000 users
## computes relative frequency of artist mentioned by the 15,000 users

itemFrequencyPlot(playlist,support=.08,cex.names=1.5)
## plots the item frequencies (only bands with > % support)
```

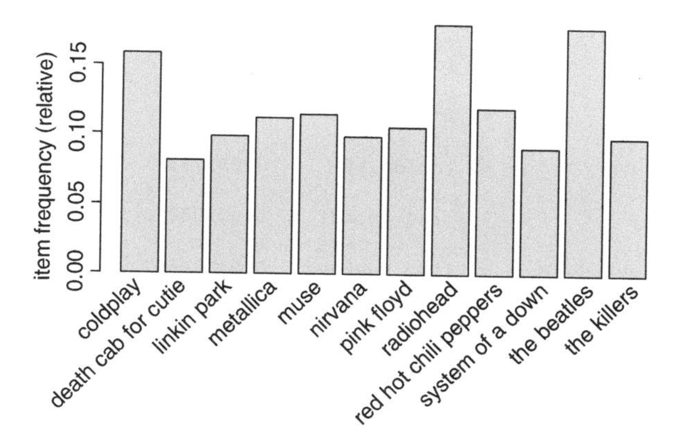


```
## Finally, we build the association rules
## only associations with support > 0.01 and confidence > .50
## this rules out rare bands

musicrules <- apriori(playlist,parameter=list(support=.01,
+    confidence=.5))

inspect(musicrules)

  lhs                        rhs              support    confidence  lift
1 {t.i.}                  => {kanye west}   0.01040000 0.5672727  8.854413
2 {the pussycat dolls}    => {rihanna}      0.01040000 0.5777778 13.415893
3 {the fray}              => {coldplay}     0.01126667 0.5168196  3.260006
4 {sonata arctica}        => {nightwish}    0.01346667 0.5101010  8.236292
5 {judas priest}          => {iron maiden}  0.01353333 0.5075000  8.562992
```

```
6  {the kinks}               => {the beatles}  0.01360000  0.5298701  2.979030
7  {travis}                  => {coldplay}     0.01373333  0.5628415  3.550304
8  {the flaming lips}        => {radiohead}    0.01306667  0.5297297  2.938589
9  {megadeth}                => {metallica}    0.01626667  0.5281385  4.743759
10 {simon & garfunkel}       => {the beatles}  0.01540000  0.5238095  2.944956
11 {broken social scene}     => {radiohead}    0.01506667  0.5472155  3.035589
12 {blur}                    => {radiohead}    0.01753333  0.5228628  2.900496
13 {keane}                   => {coldplay}     0.02226667  0.6374046  4.020634
14 {snow patrol}             => {coldplay}     0.02646667  0.5251323  3.312441
15 {beck}                    => {radiohead}    0.02926667  0.5092807  2.825152
16 {snow patrol,
    the killers}             => {coldplay}     0.01040000  0.5954198  3.755802
17 {radiohead,
    snow patrol}             => {coldplay}     0.01006667  0.6344538  4.002021
18 {death cab for cutie,
    the shins}               => {radiohead}    0.01006667  0.5033333  2.792160
19 {the beatles,
    the shins}               => {radiohead}    0.01066667  0.5673759  3.147425
20 {led zeppelin,
    the doors}               => {pink floyd}   0.01066667  0.5970149  5.689469
21 {pink floyd,
    the doors}               => {led zeppelin} 0.01066667  0.5387205  6.802027
22 {pink floyd,
    the doors}               => {the beatles}  0.01000000  0.5050505  2.839489
23 {the beatles,
    the strokes}             => {radiohead}    0.01046667  0.5607143  3.110471
24 {oasis,
    the killers}             => {coldplay}     0.01113333  0.6626984  4.180183
25 {oasis,
    the beatles}             => {coldplay}     0.01060000  0.5196078  3.277594
26 {oasis,
    radiohead}               => {coldplay}     0.01273333  0.5876923  3.707058
27 {beck,
    the beatles}             => {radiohead}    0.01300000  0.5909091  3.277972
28 {bob dylan,
    the rolling stones}      => {the beatles}  0.01146667  0.5910653  3.323081
29 {david bowie,
    the rolling stones}      => {the beatles}  0.01000000  0.5703422  3.206572
30 {led zeppelin,
    the rolling stones}      => {the beatles}  0.01066667  0.5776173  3.247474
31 {radiohead,
    the rolling stones}      => {the beatles}  0.01060000  0.5638298  3.169958
32 {coldplay,
    the smashing pumpkins}   => {radiohead}    0.01093333  0.6283525  3.485683
33 {the beatles,
    the smashing pumpkins}   => {radiohead}    0.01146667  0.6209386  3.444556
34 {radiohead,
    u2}                      => {coldplay}     0.01140000  0.5213415  3.288529
35 {coldplay,
    sigur rÃ3s}              => {radiohead}    0.01206667  0.5801282  3.218167
36 {sigur rÃ3s,
    the beatles}             => {radiohead}    0.01046667  0.6434426  3.569393
37 {bob dylan,
    pink floyd}              => {the beatles}  0.01033333  0.6150794  3.458092
38 {bob dylan,
    radiohead}               => {the beatles}  0.01386667  0.5730028  3.221530
39 {bloc party,
    the killers}             => {coldplay}     0.01106667  0.5236593  3.303150
```

```
40 {david bowie,
    pink floyd}               => {the beatles} 0.01006667 0.5741445 3.227949
41 {david bowie,
    radiohead}                => {the beatles} 0.01393333 0.5225000 2.937594
42 {placebo,
    radiohead}                => {muse}        0.01366667 0.5137845 4.504247
43 {led zeppelin,
    radiohead}                => {the beatles} 0.01306667 0.5283019 2.970213
44 {death cab for cutie,
    the killers}              => {coldplay}    0.01086667 0.5884477 3.711823
45 {death cab for cutie,
    the beatles}              => {radiohead}   0.01246667 0.5013405 2.781105
46 {muse,
    the killers}              => {coldplay}    0.01513333 0.5089686 3.210483
47 {red hot chili peppers,
    the killers}              => {coldplay}    0.01086667 0.5093750 3.213047
48 {the beatles,
    the killers}              => {coldplay}    0.01253333 0.5340909 3.368950
49 {radiohead,
    the killers}              => {coldplay}    0.01506667 0.5243619 3.307582
50 {muse,
    the beatles}              => {radiohead}   0.01380000 0.5073529 2.814458

## let's filter by lift > 5.
## Among those associations with support > 0.01 and confidence > .50,
## only show those with lift > 5

inspect(subset(musicrules, subset=lift > 5))

  lhs                        rhs               support confidence      lift
1 {t.i.}                  => {kanye west}   0.01040000  0.5672727  8.854413
2 {the pussycat dolls}    => {rihanna}      0.01040000  0.5777778 13.415893
3 {sonata arctica}        => {nightwish}    0.01346667  0.5101010  8.236292
4 {judas priest}          => {iron maiden}  0.01353333  0.5075000  8.562992
5 {led zeppelin,
   the doors}             => {pink floyd}   0.01066667  0.5970149  5.689469
6 {pink floyd,
   the doors}             => {led zeppelin} 0.01066667  0.5387205  6.802027

## lastly, order by confidence to make it easier to understand
inspect(sort(subset(musicrules, subset=lift > 5), by="confidence"))

  lhs                        rhs               support confidence      lift
1 {led zeppelin,
   the doors}             => {pink floyd}   0.01066667  0.5970149  5.689469
2 {the pussycat dolls}    => {rihanna}      0.01040000  0.5777778 13.415893
3 {t.i.}                  => {kanye west}   0.01040000  0.5672727  8.854413
4 {pink floyd,
   the doors}             => {led zeppelin} 0.01066667  0.5387205  6.802027
5 {sonata arctica}        => {nightwish}    0.01346667  0.5101010  8.236292
6 {judas priest}          => {iron maiden}  0.01353333  0.5075000  8.562992
```

16.2 EXAMPLE 2: PREDICTING INCOME

As second example we use the Adult data set from the UCI machine learning repository; this data set was analyzed by Hahsler et al. (2005) in their illustration of the R package **arules**. The data, taken from the US Census Bureau database, contains 48,842 individuals with data on income and several possible predictors of income such as age, work class, education, and so on. The first part of the R program in their paper (copied below) transforms the original variables into 115 binary indicator variables. Income, for example is coded as "small" (US$50,000 or less) and "large" (US$50,000 or more). We skip these details and focus our attention on the resulting transaction (incidence) matrix *Adult* with its 48,842 rows and 115 columns. Since the dimensions of the matrix are so large, the matrix is stored in a special, sparse, transaction matrix format. The item (column) with the largest number of 1s is "capital – loss = none" (46,560 out of 48,842). The number of 1s on a subject (also referred to as the *length*) varies between 9 and 13; this narrow band is not surprising as indicator variables were constructed from categorical variables, and we know that summing over indicators of each categorical variable must give one for every subject. But, it is the way how these 0s and 1s coincide that matters. We get a better understanding of how the strings of zeros and ones look like by considering the incidence matrix (which we create from the transaction matrix). The indicator for small income (0/1) is in the penultimate column; the indicator for large income is in the very last column.

Next, we apply the R function *apriori*. This function calculates and screens the support, the confidence and the lift of items. Consequents in *apriori* are single items (columns), but antecedents can be either single items or groups of items (item sets). Here we use the default specification for two of apriori's parameters, and specify minlen (an integer value for the minimal number of items per item set) as 1 and maxlen (an integer value for the maximal number of items per item set) as 10. This means that apriori searches over all item sets for which the sum of the number of items in the LHS item set and the (single) item on the RHS is between 1 and 10. This involves quite a lot of computations and queries. In this particular application we consider only those LHS and RHS item sets for which the support is at least 0.01 and the confidence is at least 0.60.

What happens if we had specified minlen = 3 and maxlen = 3? Then apriori searches over all combinations of item sets with two items in the antecedent group and one item as consequent; for 115 items, it calculates statistics for $(115)(114)(113)/2 = 740,715$ LHS and RHS arrangements.

Finally, we search over a subset of all rules (with support greater than 0.01 and confidence greater than 0.60, as specified previously) that have small income on the RHS and achieve a lift larger than 1.2. The combinations that satisfy these conditions are the ones that are able to predict small income earners from the coded explanatory information. Similarly, we search and list all rules that identify high income earners. We find that workers in the private sector working part-time tend to have a small income, while persons with high capital gain who are born in the United States tend to have a large income.

```
library(arules)
data(AdultUCI)
dim(AdultUCI)
AdultUCI[1:3,]
```

```
  age        workclass fnlwgt education education-num     marital-status
1  39        State-gov  77516 Bachelors            13      Never-married
2  50 Self-emp-not-inc  83311 Bachelors            13 Married-civ-spouse
3  38          Private 215646   HS-grad             9           Divorced
         occupation  relationship  race  sex capital-gain capital-loss
1      Adm-clerical Not-in-family White Male         2174            0
2   Exec-managerial       Husband White Male            0            0
3 Handlers-cleaners Not-in-family White Male            0            0
  hours-per-week native-country income
1             40  United-States  small
2             13  United-States  small
3             40  United-States  small
```

```
AdultUCI[["fnlwgt"]] <- NULL
AdultUCI[["education-num"]] <- NULL
AdultUCI[["age"]] <- ordered(cut(AdultUCI[["age"]], c(15, 25, 45, 65,
+    100)), labels = c("Young", "Middle-aged", "Senior", "Old"))
AdultUCI[["hours-per-week"]] <- ordered(cut(AdultUCI[["hours-per-
+    week"]], c(0, 25, 40, 60, 168)), labels = c("Part-time",
+    "Full-time", "Over-time", "Workaholic"))
AdultUCI[["capital-gain"]] <- ordered(cut(AdultUCI[["capital-gain"]],
+    c(-Inf, 0, median(AdultUCI[["capital-gain"]][AdultUCI
+    [["capital-gain"]] > 0]), Inf)),
+    labels = c("None", "Low", "High"))
AdultUCI[["capital-loss"]] <- ordered(cut(AdultUCI[["capital-loss"]],
+    c(-Inf, 0, median(AdultUCI[["capital-loss"]][AdultUCI
+    [["capital-loss"]] > 0]), Inf)), labels = c("none", "low", "high"))

Adult <- as(AdultUCI, "transactions")
Adult
```

```
transactions in sparse format with
 48842 transactions (rows) and
 115 items (columns)
```

```
summary(Adult)
```

```
transactions as itemMatrix in sparse format with
 48842 rows (elements/itemsets/transactions) and
 115 columns (items) and a density of 0.1089939

most frequent items:
           capital-loss=none            capital-gain=None
                       46560                        44807
native-country=United-States                   race=White
                       43832                        41762
           workclass=Private                      (Other)
                       33906                       401333

element (itemset/transaction) length distribution:
sizes
    9    10    11    12    13
   19   971  2067 15623 30162

   Min. 1st Qu.  Median    Mean 3rd Qu.    Max.
   9.00   12.00   13.00   12.53   13.00   13.00

includes extended item information - examples:
          labels variables       levels
```

```
1       age=Young        age        Young
2 age=Middle-aged        age Middle-aged
3      age=Senior        age       Senior

includes extended transaction information - examples:
  transactionID
1             1
2             2
3             3
```

aa=as(Adult,"matrix")
transforms transaction matrix into an incidence matrix
aa[1:2,]   # print the first two rows of the incidence matrix

```
  age=Young age=Middle-aged age=Senior age=Old workclass=Federal-gov
1         0               1          0       0                   0
2         0               0          1       0                   0
  workclass=Local-gov workclass=Never-worked workclass=Private
1                   0                      0                 0
2                   0                      0                 0
  workclass=Self-emp-inc workclass=Self-emp-not-inc workclass=State-gov
1                      0                          0                   1
2                      0                          1                   0
  workclass=Without-pay education=Preschool education=1st-4th education=5th-6th
1                     0                   0                 0                 0
2                     0                   0                 0                 0
  education=7th-8th education=9th education=10th education=11th education=12th
1                 0             0              0              0              0
2                 0             0              0              0              0
  education=HS-grad education=Prof-school education=Assoc-acdm
1                 0                     0                    0
2                 0                     0                    0
  education=Assoc-voc education=Some-college education=Bachelors
1                   0                      0                   1
2                   0                      0                   1
  education=Masters education=Doctorate marital-status=Divorced
1                 0                   0                       0
2                 0                   0                       0
  marital-status=Married-AF-spouse marital-status=Married-civ-spouse
1                                0                                 0
2                                0                                 1
  marital-status=Married-spouse-absent marital-status=Never-married
1                                    0                            1
2                                    0                            0
  marital-status=Separated marital-status=Widowed occupation=Adm-clerical
1                        0                      0                       1
2                        0                      0                       0
  occupation=Armed-Forces occupation=Craft-repair occupation=Exec-managerial
1                       0                       0                          0
2                       0                       0                          1
  occupation=Farming-fishing occupation=Handlers-cleaners
1                          0                            0
2                          0                            0
  occupation=Machine-op-inspct occupation=Other-service
1                            0                        0
2                            0                        0
  occupation=Priv-house-serv occupation=Prof-specialty
1                          0                         0
2                          0                         0
  occupation=Protective-serv occupation=Sales occupation=Tech-support
```

```
1                          0                0                       0
2                          0                0                       0
  occupation=Transport-moving relationship=Husband relationship=Not-in-family
1                          0                    0                          1
2                          0                    1                          0
  relationship=Other-relative relationship=Own-child relationship=Unmarried
1                          0                      0                      0
2                          0                      0                      0
  relationship=Wife race=Amer-Indian-Eskimo race=Asian-Pac-Islander race=Black
1                 0                       0                       0          0
2                 0                       0                       0          0
  race=Other race=White sex=Female sex=Male capital-gain=None capital-gain=Low
1          0          1          0        1                 0                1
2          0          1          0        1                 1                0
  capital-gain=High capital-loss=none capital-loss=low capital-loss=high
1                 0                 1                0                 0
2                 0                 1                0                 0
  hours-per-week=Part-time hours-per-week=Full-time hours-per-week=Over-time
1                        0                        1                        0
2                        1                        0                        0
  hours-per-week=Workaholic native-country=Cambodia native-country=Canada
1                         0                       0                     0
2                         0                       0                     0
  native-country=China native-country=Columbia native-country=Cuba
1                    0                       0                   0
2                    0                       0                   0
  native-country=Dominican-Republic native-country=Ecuador
1                                 0                      0
2                                 0                      0
  native-country=El-Salvador native-country=England native-country=France
1                          0                      0                     0
2                          0                      0                     0
  native-country=Germany native-country=Greece native-country=Guatemala
1                      0                     0                        0
2                      0                     0                        0
  native-country=Haiti native-country=Holand-Netherlands
1                    0                                 0
2                    0                                 0
  native-country=Honduras native-country=Hong native-country=Hungary
1                       0                   0                      0
2                       0                   0                      0
  native-country=India native-country=Iran native-country=Ireland
1                    0                   0                      0
2                    0                   0                      0
  native-country=Italy native-country=Jamaica native-country=Japan
1                    0                      0                    0
2                    0                      0                    0
  native-country=Laos native-country=Mexico native-country=Nicaragua
1                   0                     0                        0
2                   0                     0                        0
  native-country=Outlying-US(Guam-USVI-etc) native-country=Peru
1                                         0                   0
2                                         0                   0
  native-country=Philippines native-country=Poland native-country=Portugal
1                          0                     0                       0
2                          0                     0                       0
  native-country=Puerto-Rico native-country=Scotland native-country=South
1                          0                       0                    0
2                          0                       0                    0
```

```
  native-country=Taiwan native-country=Thailand native-country=Trinadad& Tobago
1                     0                       0                               0
2                     0                       0                               0
  native-country=United-States native-country=Vietnam native-country=Yugoslavia
1                            1                      0                         0
2                            1                      0                         0
  income=small income=large
1            1            0
2            1            0
```

```
itemFrequencyPlot(Adult[, itemFrequency(Adult) > 0.2], cex.names = 1)
```

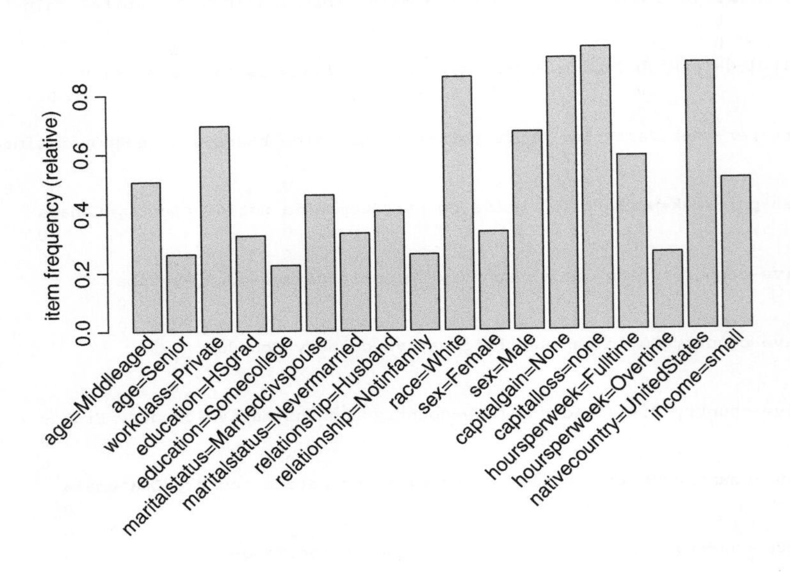


```
rules <- apriori(Adult, parameter = list(support = 0.01,
+    confidence = 0.6))

parameter specification:
 confidence minval smax arem  aval originalSupport support minlen maxlen target
        0.6    0.1    1 none FALSE            TRUE    0.01      1     10  rules
   ext
 FALSE

algorithmic control:
 filter tree heap memopt load sort verbose
    0.1 TRUE TRUE  FALSE TRUE    2    TRUE

apriori - find association rules with the apriori algorithm
version 4.21 (2004.05.09)        (c) 1996-2004   Christian Borgelt
set item appearances …[0 item(s)] done [0.00s].
set transactions …[115 item(s), 48842 transaction(s)] done [0.04s].
sorting and recoding items … [67 item(s)] done [0.01s].
creating transaction tree … done [0.05s].
checking subsets of size 1 2 3 4 5 6 7 8 9 10 done [1.14s].
writing … [276443 rule(s)] done [0.09s].
creating S4 object  … done [0.17s].
```

```
rules

set of 276443 rules

summary(rules)

set of 276443 rules

rule length distribution (lhs + rhs):sizes
    1     2     3     4     5     6     7     8     9    10
    6   432  4981 22127 52669 75104 67198 38094 13244  2588

   Min. 1st Qu.  Median    Mean 3rd Qu.    Max.
  1.000   5.000   6.000   6.289   7.000  10.000

summary of quality measures:
    support           confidence          lift
 Min.   :0.01001   Min.   :0.6000   Min.   : 0.7171
 1st Qu.:0.01253   1st Qu.:0.7691   1st Qu.: 1.0100
 Median :0.01701   Median :0.9051   Median : 1.0554
 Mean   :0.02679   Mean   :0.8600   Mean   : 1.3109
 3rd Qu.:0.02741   3rd Qu.:0.9542   3rd Qu.: 1.2980
 Max.   :0.95328   Max.   :1.0000   Max.   :20.6826

mining info:
  data ntransactions support confidence
 Adult         48842    0.01        0.6

rulesIncomeSmall <- subset(rules, subset = rhs %in% "income=small"
+     & lift > 1.2)
inspect(sort(rulesIncomeSmall, by = "confidence")[1:3])

  lhs                                    rhs              support    confidence    lift
1 {workclass=Private,
   marital-status=Never-married,
   relationship=Own-child,
   sex=Male,
   hours-per-week=Part-time,
   native-country=United-States} => {income=small} 0.01074895  0.7104195 1.403653
2 {workclass=Private,
   marital-status=Never-married,
   relationship=Own-child,
   sex=Male,
   hours-per-week=Part-time}      => {income=small} 0.01144507  0.7102922 1.403402
3 {workclass=Private,
   marital-status=Never-married,
   relationship=Own-child,
   sex=Male,
   capital-gain=None,
   hours-per-week=Part-time,
   native-country=United-States} => {income=small} 0.01046231  0.7097222 1.402276

rulesIncomeLarge <- subset(rules, subset = rhs %in% "income=large"
+     & lift > 1.2)
inspect(sort(rulesIncomeLarge, by = "confidence")[1:3])

  lhs                                    rhs           support    confidence    lift
1 {marital-status=Married-civ-spouse,
   capital-gain=High,
```

```
   native-country=United-States}  => {income=large} 0.01562180 0.6849192 4.266398
2 {marital-status=Married-civ-spouse,
   capital-gain=High,
   capital-loss=none,
   native-country=United-States}  => {income=large} 0.01562180 0.6849192 4.266398
3 {relationship=Husband,
   race=White,
   capital-gain=High,
   native-country=United-States}  => {income=large} 0.01302158 0.6846071 4.264454
```

REFERENCES

Celma, O.: *Music Recommendation and Discovery*. New York: Springer, 2010.

Hahsler, M., Gruen, B., and Hornik, K.: arules–A computational environment for mining association rules and frequent item sets. *Journal of Statistical Software*, Vol. 14 (2005), 1–25.

CHAPTER 17

Dimension Reduction: Factor Models and Principal Components

Scarcity of data is not a problem in data mining applications. Usually, the opposite is true: too much data. Assume that we study n units (objects or cases). A unit may be a person making a buying decision, a physical object such as a car, or the economy at a certain period of time. We are interested in the response a buyer makes (such as whether he buys and how much he buys), interested in whether the car experiences problems before the warranty period is over, and interested in next month's unemployment rate. Economists will immediately think of hundreds of variables that (may) have an effect on next month's unemployment rate: interest rates, money supply, employment, consumer confidence, inventory levels, exchange rates, wage rates, inflation, and so on. Monthly information on all these variables is available for the past couple of months, which gives us a very large number of possible predictor variables. The performance of a physical object (whether a car will fail or whether a high quality piece of steel will break at normal operating conditions) depends on the many factors that go into its production. Again, it is not very difficult to come up with a large number of variables that relate to this outcome. Similarly, the decision whether to buy or not to buy depends on numerous factors: the price of the product, the prices of competing products, their advertisement, the general economic climate, the personal finances of the buyer, whether the buyer already has similar products, whether other buyers have bought the product, and so forth.

Arrange the data on the many, many explanatory variables into a data matrix X with n rows (the number of units or cases, such as the number of buyers, the number of studied cars, and the number of months) and p columns (the number of variables). The number of columns will be quite large, perhaps larger than the number of rows (units). But the columns of the X matrix are usually highly correlated; while we may have 100 measures on the state of the economy for a certain month, these indicators measure the same thing, namely, the state of the economy at a certain period of time. The state of the economy can probably be represented reasonably well with a smaller number of "state" vectors or factors

Data Mining and Business Analytics with R, First Edition. Johannes Ledolter.

that combine the information from the 100 variables in a certain "smart" way. The objective of this section is to explore ways of transforming the columns of the X matrix such that its dimension is reduced. It may be possible to construct three or four combinations of the original variables, let us call them our new *factors*, which then explain most of the information that is present in the matrix X. This dimension-reduction step certainly makes any subsequent modeling of the response easier as now one relates the response to a much smaller number of constructed factors. In addition, note that a brute force regression modeling of the response in terms of hundreds of explanatory variables will not work in situations where the number of columns in the X matrix exceeds the number of its rows.

Even if there is no response to model, it makes good sense to reduce the dimensionality of a data matrix X. Just the fact that we have data on many variables does not mean that we know what is going on. Looking at a fewer number of transformed variables may give us more useful information. We attempt to reduce the very high dimensional information down to a few important constructed factors (much fewer than p). We do this by building a simple linear model for X that represents X in a lower dimensional space.

An obvious first step is to inspect all pairwise correlations among the p columns of the X matrix. The entries in the resulting $p \times p$ correlation matrix (a very large symmetric matrix, with 1's in the diagonal) can be shaded or colored according to the size of the correlations. Or, one can only show those correlations that exceed certain thresholds (such as correlations larger than 0.9 or smaller than -0.9). However, note that this initial step only looks at pairwise relationships. Approaches that look at all variables together are needed, and they are described next.

A *factor model* is a regression on a multivariate $n \times p$ matrix $X = \{x_{ij}\}$ such that

$$E[x_{ij}] = \phi_{j1} v_{i1} + \phi_{j2} v_{i2} + \cdots + \phi_{jk} v_{ik}, \tag{17.1}$$

where the $v_{i1}, v_{i2}, \ldots, v_{ik}$ are the realizations of $k \le p$ new *factors* (here defined for unit i) that explain, or are being shared by, the data matrix of the original p variables. The coefficients $\phi_{j1}, \phi_{j2}, \ldots, \phi_{jk}$ are the *loadings* of the original variable x_j $(j = 1, 2, \ldots, p)$ onto these k new factors.

Here is another way to think about this model. Consider the ith row of the matrix X (we now fix the unit; such as the buyer, the car, or the month) and consider the p-dimensional vector of its characteristics $x_{i1}, x_{i2}, \ldots, x_{ip}$. Treating $\widetilde{x}_i = (x_{i1}, x_{i2}, \ldots, x_{ip})'$ as a $p \times 1$ column vector, we can write the factor model as

$$E[\widetilde{x}_i] = \phi_1 v_{i1} + \phi_2 v_{i2} + \cdots + \phi_k v_{ik}. \tag{17.2}$$

This shows that the p original variables can be written as linear combinations of k factors. The coefficients $\phi_1, \phi_2, \ldots, \phi_k$ are now column vectors with p rows. The first vector, for example, consists of elements $\phi_{11}, \phi_{21}, \ldots, \phi_{p1}$. The last one consists of elements $\phi_{1k}, \phi_{2k}, \ldots, \phi_{pk}$. If $k < p$, we achieve a reduction in dimensionality.

It would be easy to get the loadings if the factors were known. Then the loadings could be estimated by regressing each of the p variables x_j on the factors $v_1, v_2, \ldots, v_k$; here each variable and all factors are vectors of length n. Many regressions (p of them) would have to be carried out. The regression of the first column of the X matrix on the k factors would give us the coefficients $\phi_{11}, \phi_{12}, \ldots, \phi_{1k}$; the regression of the second column of the X matrix on the k factors would give us the coefficients $\phi_{21}, \phi_{22}, \ldots, \phi_{2k}$; and so on. But the factors are not known, so this method cannot be used. Instead, one can use the method of principal components to obtain both the factors and the loadings at the same time.

Principal components analysis (PCA) is a mathematical procedure that uses an orthogonal transformation to convert a set of observations of possibly (and most likely) correlated variables $x_1, x_2, \ldots, x_p$ (i.e., the columns of the X matrix) into a set of values of uncorrelated variables called the *principal components*. The number of principal components, k, is less than or equal to the number of original variables p. The transformations are defined in such a way that the first principal component (a linear combination of the p original variables $x_1, x_2, \ldots, x_p$) has as high a variance as possible (in other words, it accounts for as much of the variability in the data as possible), and that each succeeding component, in turn, has the highest variance possible under the constraint that it is orthogonal to (uncorrelated with) the preceding components.

PCA is an important tool in exploratory data analysis and is used for preprocessing high dimensional explanatory information that can simplify any subsequent predictive modeling. PCA can be carried out by an eigenvalue decomposition of the data covariance matrix (a $p \times p$ matrix) or through a singular-value decomposition of the data matrix X, usually after mean-centering each column. We are skipping the mathematical details as they require a fair amount of matrix algebra. The output of a PCA gives us the weights that are used to project the original observations into the new directions; they are called the *loadings* and are identical to the vector loadings $\phi_1, \phi_2, \ldots, \phi_k$ in the factor model representation in Equation 17.2, with $k = p$ factors. The loadings are used to calculate the principal components scores (resulting in p values for each row or unit of the data matrix). These are the factor scores $v_{i1}, v_{i2}, \ldots, v_{ip}$ in the factor model representation with p factors.

PCA reveals the internal structure of the data in a way that best explains the variance in the data. The elements of the multivariate data set X can be visualized in a high dimensional space, with each of the p variables defining an axis. PCA supplies a lower dimensional picture of the data matrix, a projection or "shadow" of X when viewed from its most informative viewpoint. This projection is achieved by using the first few principal components so that the dimensionality of the transformed data is reduced.

Here is another and somewhat more basic explanation. Consider a data matrix consisting of just two columns, x_1 and x_2. A scatter plot of the two variables tells us about the direction in which the variability of the points is largest. Think about an elongated elliptic scatter around a line that runs through the data set.

This line describes the maximal variation in the data if we decide to reduce the dimensionality of the data from 2 dimensions to 1. A perpendicular projection of the data points onto this line gives us a set of one-dimensional values for which the variance will be maximal. These projected values are the scores of the first principal component. The coefficients of the linear projection transformation (here there are two coefficients, and they multiply x_{i1} and x_{i2} to obtain the projection, or the first principal component score v_{i1}) are the loadings of the first principal component. With just two columns of the X matrix and in two-dimensional space, the second principal component represents the line that is perpendicular to the first best fitting line. With more than two variables and columns in the X matrix, we first get the direction v_1 that has the largest variance. Then we search among all directions that are orthogonal to v_2 for the direction with the largest variance. This is the second principal component v_2. Then we look for the direction that is orthogonal to both v_1 and v_2 and that has the largest variance; this results in v_3; and so on.

A PCA partitions the total variance in the X matrix, $V(\text{Total}) = V(x_1) + V(x_2) + \cdots + V(x_p)$, into variances of p newly constructed principal component scores, $V(\text{Total}) = V(v_1) + V(v_2) + \cdots + V(v_p)$, that are now ordered from largest to smallest. The constructed variables are linear combinations of the original variables, and the weights are selected such that the linear combinations are orthogonal.

PCA can be applied to the raw data and covariance matrices or to standardized data and correlation matrices. Standardization is appropriate if the variables have different units (e.g., when one variable is measured in million dollars and the other in percentages). As PCA is sensitive to the relative scaling of the original variables, we recommend that all original variables $x_1, x_2, \ldots, x_p$ are scaled to have unit variance. Standardization is less useful if we analyze variables that have the same units (such as the unemployment rates of the 50 states).

The R program *prcomp* in the **stats** package is used for the subsequent analyses.

17.1 EXAMPLE 1: EUROPEAN PROTEIN CONSUMPTION

We considered this data set before when we used it to illustrate clustering in Chapter 15. The data set consists of 25 units (European countries; $n = 25$) and their protein intakes (in percentage) from nine major food sources ($p = 9$). Here we use the data set to illustrate PCA as a dimension-reduction technique. PCA finds the linear combination of the nine variables (the columns of the X matrix) that has the largest variability and thus explains most of the variation. After finding this linear combination (the first direction), PCA looks for a second linear combination (direction) that is orthogonal to the first one and that maximizes the variance in this second direction, and so on. The hope is that one needs only a few of these linear combinations (or directions) to describe the data. The scores (the linear combinations applied to the original variables of the n units) are called the *principal components*.

The output under "rotation" lists the matrix of loadings on the original variables (i.e., the weights in the linear combinations that transform the original variables into the principal components scores). These weights are also the factor loadings in our factor-model representation in Equation 17.2 (this is because the transformations are orthogonal, and because the inverse of an orthogonal matrix is equal to its transpose).

The first principal component (shown later in the output) puts weight −0.302 on RedMeat, −0.310 on WhiteMeat, and so on. These weights represent the entries in the $p \times 1$ vector of loadings ϕ_1. The second principal component puts weight −0.056 on RedMeat, −0.237 on WhiteMeat, and so on. These weights represent the entries in the vector of loadings ϕ_2. Until the last (ninth) principal component with weights 0.246 on RedMeat, 0.592 on WhiteMeat, and so on. These weights represent the entries in the vector of loadings ϕ_9.

Principal components scores result if the loadings of the variables (listed under rotation) are applied to the values on the original variables. We achieve this with the *predict* function in R. This results in nine $n \times 1$ vectors of principal components scores. These are equivalent to the factors $v_1, v_2, \ldots, v_p$ in our factor-model representation in Equation 17.1, with $p = 9$ factors.

The output under "standard deviations" lists the standard deviations of the nine constructed linear combinations. Note that they are ranked from large to small and note that their squares (the variances) add up to 9. This is because in this example we are working with standardized observations, and the sum of variances of the nine standardized variables equals 9. A graph of the variances shows that the first two or three linear combinations explain most of the total variability.

The factor model with $k = p = 9$ factors replicates the data matrix X exactly. The factor model with $k < p$ factors provides an approximation, but if the first few principal components explain much of the variability, the approximation will be excellent. The factor-model representation in two dimensions with $k = 2$ factors, for example, approximates the 9×1 vector $\widetilde{x}_i$ of row i through $E[\widetilde{x}_i] = \phi_1 v_{i1} + \phi_2 v_{i2}$, where ϕ_1 and ϕ_2 are the first two loadings (given under rotations in the R output) and where v_{i1} and v_{i2} are the ith scores on the first two principal components (given in the R output under predict).

A scatter plot of the first two principal components with labels for country shows that Portugal and Spain are quite high in the second dimension. The loadings for the second principal component describe a diet that is high on fish, starch (bread), and vegetables.

```
food <- read.csv("C:/DataMining/Data/protein.csv")
food
```

```
     Country RedMeat WhiteMeat Eggs Milk Fish Cereals Starch Nuts Fr.Veg
1    Albania    10.1       1.4  0.5  8.9  0.2    42.3    0.6  5.5    1.7
2    Austria     8.9      14.0  4.3 19.9  2.1    28.0    3.6  1.3    4.3
3    Belgium    13.5       9.3  4.1 17.5  4.5    26.6    5.7  2.1    4.0
4   Bulgaria     7.8       6.0  1.6  8.3  1.2    56.7    1.1  3.7    4.2
```

```
5  Czechoslovakia   9.7   11.4  2.8 12.5  2.0   34.3   5.0  1.1  4.0
6         Denmark  10.6   10.8  3.7 25.0  9.9   21.9   4.8  0.7  2.4
7       E Germany   8.4   11.6  3.7 11.1  5.4   24.6   6.5  0.8  3.6
8         Finland   9.5    4.9  2.7 33.7  5.8   26.3   5.1  1.0  1.4
9          France  18.0    9.9  3.3 19.5  5.7   28.1   4.8  2.4  6.5
10         Greece  10.2    3.0  2.8 17.6  5.9   41.7   2.2  7.8  6.5
11        Hungary   5.3   12.4  2.9  9.7  0.3   40.1   4.0  5.4  4.2
12        Ireland  13.9   10.0  4.7 25.8  2.2   24.0   6.2  1.6  2.9
13          Italy   9.0    5.1  2.9 13.7  3.4   36.8   2.1  4.3  6.7
14    Netherlands   9.5   13.6  3.6 23.4  2.5   22.4   4.2  1.8  3.7
15         Norway   9.4    4.7  2.7 23.3  9.7   23.0   4.6  1.6  2.7
16         Poland   6.9   10.2  2.7 19.3  3.0   36.1   5.9  2.0  6.6
17       Portugal   6.2    3.7  1.1  4.9 14.2   27.0   5.9  4.7  7.9
18        Romania   6.2    6.3  1.5 11.1  1.0   49.6   3.1  5.3  2.8
19          Spain   7.1    3.4  3.1  8.6  7.0   29.2   5.7  5.9  7.2
20         Sweden   9.9    7.8  3.5 24.7  7.5   19.5   3.7  1.4  2.0
21    Switzerland  13.1   10.1  3.1 23.8  2.3   25.6   2.8  2.4  4.9
22             UK  17.4    5.7  4.7 20.6  4.3   24.3   4.7  3.4  3.3
23           USSR   9.3    4.6  2.1 16.6  3.0   43.6   6.4  3.4  2.9
24      W Germany  11.4   12.5  4.1 18.8  3.4   18.6   5.2  1.5  3.8
25     Yugoslavia   4.4    5.0  1.2  9.5  0.6   55.9   3.0  5.7  3.2
```

```
## correlation matrix
cor(food[,-1])
```

```
               RedMeat  WhiteMeat        Eggs       Milk        Fish     Cereals
RedMeat     1.00000000  0.1530027  0.58560895  0.5029311  0.06095745 -0.49987746
WhiteMeat   0.15300271  1.0000000  0.62040916  0.2814839 -0.23400923 -0.41379691
Eggs        0.58560895  0.6204092  1.00000000  0.5755331  0.06557136 -0.71243682
Milk        0.50293110  0.2814839  0.57553312  1.0000000  0.13788370 -0.59273662
Fish        0.06095745 -0.2340092  0.06557136  0.1378837  1.00000000 -0.52423080
Cereals    -0.49987746 -0.4137969 -0.71243682 -0.5927366 -0.52423080  1.00000000
Starch      0.13542594  0.3137721  0.45223071  0.2224112  0.40385286 -0.53326231
Nuts       -0.34944855 -0.6349618 -0.55978097 -0.6210875 -0.14715294  0.65099727
Fr.Veg     -0.07422123 -0.0613167 -0.04551755 -0.4083641  0.26613865  0.04654808
                Starch       Nuts      Fr.Veg
RedMeat     0.13542594 -0.3494486 -0.07422123
WhiteMeat   0.31377205 -0.6349618 -0.06131670
Eggs        0.45223071 -0.5597810 -0.04551755
Milk        0.22241118 -0.6210875 -0.40836414
Fish        0.40385286 -0.1471529  0.26613865
Cereals    -0.53326231  0.6509973  0.04654808
Starch      1.00000000 -0.4743116  0.08440956
Nuts       -0.47431155  1.0000000  0.37496971
Fr.Veg      0.08440956  0.3749697  1.00000000
```

```
pcafood <- prcomp(food[,-1], scale=TRUE)
## we strip the first column (country labels) from the data set
## scale = TRUE: variables are first standardized. Default is FALSE
pcafood
```

```
Standard deviations:
[1] 2.0016087 1.2786710 1.0620355 0.9770691 0.6810568 0.5702026 0.5211586
[8] 0.3410160 0.3148204

Rotation:
                 PC1         PC2         PC3          PC4         PC5
RedMeat   -0.3026094 -0.05625165 -0.29757957 -0.646476536  0.32216008
WhiteMeat -0.3105562 -0.23685334  0.62389724  0.036992271 -0.30016494
```

```
Eggs        -0.4266785 -0.03533576  0.18152828 -0.313163873  0.07911048
Milk        -0.3777273 -0.18458877 -0.38565773  0.003318279 -0.20041361
Fish        -0.1356499  0.64681970 -0.32127431  0.215955001 -0.29003065
Cereals      0.4377434 -0.23348508  0.09591750  0.006204117  0.23816783
Starch      -0.2972477  0.35282564  0.24297503  0.336684733  0.73597332
Nuts         0.4203344  0.14331056 -0.05438778 -0.330287545  0.15053689
Fr.Veg       0.1104199  0.53619004  0.40755612 -0.462055746 -0.23351666
                   PC6         PC7         PC8        PC9
RedMeat    -0.45986989  0.15033385 -0.01985770  0.2459995
WhiteMeat  -0.12100707 -0.01966356 -0.02787648  0.5923966
Eggs        0.36124872 -0.44327151 -0.49120023 -0.3333861
Milk        0.61843780  0.46209500  0.08142193  0.1780841
Fish       -0.13679059 -0.10639350 -0.44873197  0.3128262
Cereals     0.08075842  0.40496408 -0.70299504  0.1522596
Starch      0.14766670  0.15275311  0.11453956  0.1218582
Nuts        0.44701001 -0.40726235  0.18379989  0.5182749
Fr.Veg      0.11854972  0.44997782  0.09196337 -0.2029503
```

```
foodpc <- predict(pcafood)
foodpc
```

```
             PC1         PC2         PC3        PC4         PC5         PC6
 [1,]  3.4853673 -1.63047985 -1.76123326 -0.2296580  0.02325397 -1.03426476
 [2,] -1.4226694 -1.04123130  1.33780391 -0.1680973 -0.93344658  0.21842810
 [3,] -1.6220323  0.15949557  0.21653445 -0.5207260  0.75509039 -0.28980510
 [4,]  3.1340813 -1.30106563  0.15128956 -0.2141894 -0.48474537 -0.69557793
 [5,] -0.3704646 -0.60266842  1.19594183  0.4639821  0.25682380 -0.82309047
 [6,] -2.3652688  0.28544582 -0.75226337  0.9673412 -0.75243310 -0.17032964
 [7,] -1.4222108  0.45030085  1.30254017  1.1359613  0.42294279 -0.64831247
 [8,] -1.5638563 -0.59600255 -2.04950734  1.4153084  0.03720310  0.83420035
 [9,] -1.4879824  0.78536517  0.00188261 -1.9574576  0.25045870 -0.89894837
[10,]  2.2397000  1.00105887 -0.88260339 -1.7943200 -0.40497731  1.14447671
[11,]  1.4574398 -0.81595115  1.91416751  0.2173883 -0.04139773  0.53910843
[12,] -2.6634775 -0.76370648 -0.01988068 -0.4347281  1.01438731  0.48232591
[13,]  1.5345653  0.39898708  0.12608962 -1.2224605 -0.80354036  0.21408555
[14,] -1.6414454 -0.91199089  0.76648819  0.1261517 -0.76127751  0.29752197
[15,] -0.9747029  0.82202867 -1.70407650  1.1376216 -0.41487370 -0.05645162
[16,] -0.1218695  0.53174194  1.47478926  0.4582224 -0.02321953  0.58830002
[17,]  1.7058540  4.28893399  0.04363280  0.8935596 -0.38528872 -0.69709651
[18,]  2.7568124 -1.11878536  0.07008085  0.6150113  0.31709607  0.13051679
[19,]  1.3118074  2.55352416  0.51528370 -0.3592043  0.51590218  0.66928818
[20,] -1.6337300 -0.20738445 -1.28037195  0.7341013 -0.81982482  0.04407662
[21,] -0.9123182 -0.75105865 -0.15425409 -1.1704447 -0.83095955 -0.09024236
[22,] -1.7353682 -0.09397944 -1.15268145 -1.7336921  1.08393948 -0.09656499
[23,]  0.7825965 -0.11077014 -0.36967910  0.9275729  1.66955744  0.18542992
[24,] -2.0938353 -0.29377901  0.80397944 -0.1087951 -0.06836077 -0.20099295
[25,]  3.6230077 -1.03802883  0.20604724  0.8215511  0.37768982  0.35391862
              PC7         PC8         PC9
 [1,] -0.47174197  0.76155126 -0.10325325
 [2,] -0.18115417 -0.25100249 -0.21744631
 [3,] -0.19559674 -0.20331176 -0.03317146
 [4,]  0.46478244 -0.80824466 -0.29986287
 [5,]  0.31494841  0.01229809 -0.14944825
 [6,] -0.22581590 -0.62102079  0.48027941
 [7,] -0.55478278 -0.16317720 -0.25990129
 [8,]  0.72623047  0.22591749 -0.13308980
 [9,]  0.94647514 -0.02222005  0.54360773
[10,] -0.14739068 -0.30583067  0.38796520
[11,] -0.76810179  0.14561849  0.53694446
```

```
[12,] -0.02866861  0.02299904 -0.07946587
[13,]  0.14999185 -0.08040607 -0.73235148
[14,] -0.06209574  0.45992565  0.26087199
[15,] -0.04278766 -0.10734586 -0.14732974
[16,]  1.26072252  0.19159633 -0.22099984
[17,]  0.04649963  0.20502225  0.26348804
[18,] -0.13307908 -0.02689368  0.33803173
[19,] -0.59721082  0.23532802 -0.47673575
[20,] -0.54116219 -0.07221780 -0.10839228
[21,]  0.51229089  0.52929748  0.06656513
[22,] -0.65096941 -0.23920906 -0.13193590
[23,]  0.57410168 -0.05202711  0.09168518
[24,] -0.45677673  0.35662909 -0.02527918
[25,]  0.06129122 -0.19327598  0.14922442

## how many principal components do we need?
plot(pcafood, main="")
mtext(side=1, "European Protein Principal Components",  line=1, font=2)
```

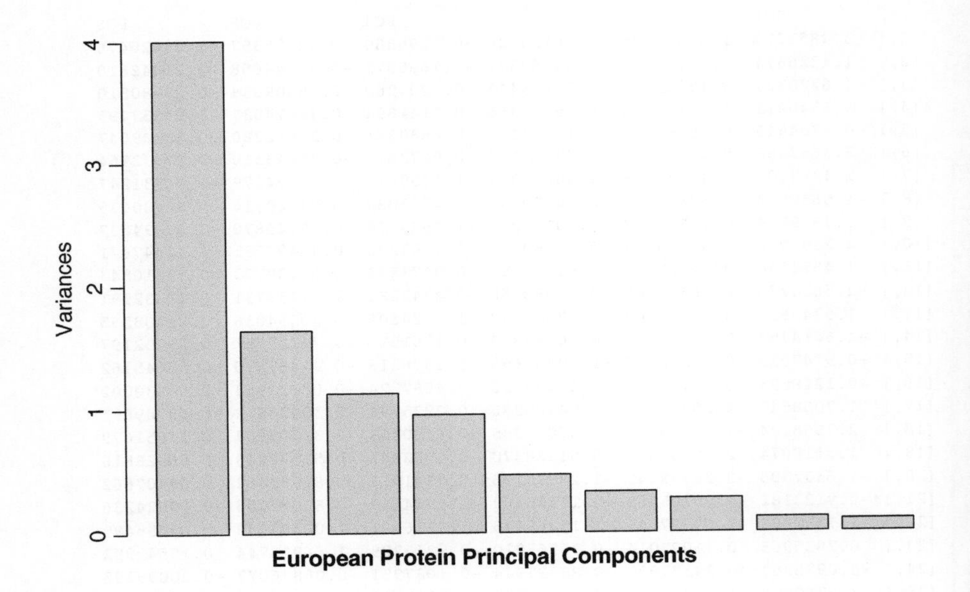


```
## how do the PCs look?
par(mfrow=c(1,2))
plot(foodpc[,1:2], type="n", xlim=c(-4,5))
text(x=foodpc[,1], y=foodpc[,2], labels=food$Country)
plot(foodpc[,3:4], type="n", xlim=c(-3,3))
text(x=foodpc[,3], y=foodpc[,4], labels=food$Country)
```

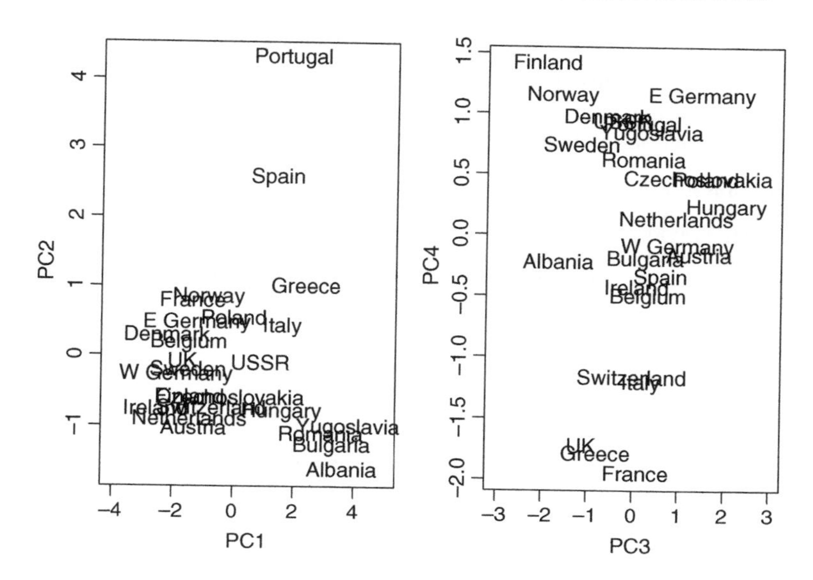


```
pcafood$rotation[,2]

    RedMeat   WhiteMeat        Eggs        Milk        Fish     Cereals
-0.05625165 -0.23685334 -0.03533576 -0.18458877  0.64681970 -0.23348508
     Starch        Nuts      Fr.Veg
 0.35282564  0.14331056  0.53619004
```

17.2 EXAMPLE 2: MONTHLY US UNEMPLOYMENT RATES

Here the data matrix X has $n = 50$ rows (representing the 50 states) and $p = 416$ columns (the monthly observations). Let us summarize the 416 monthly unemployment rates through a smaller set of constructed variables. We find that the first two or three principal components explain most of the variability. The first principal component represents the common trend component of the 50 time series; a time sequence plot of the average monthly unemployment rates and a graph of the 416 loadings of the variables on the first principal component (here loadings are multiplied by -1) are very similar (the absolute value of the correlation between the average monthly unemployment rates and the loadings on the first principal component is 0.83).

Principal components scores are obtained and the scatter plot of the first two principal components is shown on the following page. It is of interest to check whether an informal clustering of the 50 states on the basis of the first two principal component scores would be similar to the results of k-means clustering that uses the information on all 416 components. The cluster assignments from *k*-means clustering with three clusters are superimposed in color on the scatter plot of the first two principal components. The results show that much of the *k*-means clustering information is already contained in the first two principal components.

```
library(cluster) ## needed for cluster analysis
states=c("AL","AK","AZ","AR","CA","CO","CT","DE","FL","GA","HI","ID",
+    "IL","IN","IA","KS","KY","LA","ME","MD","MA","MI","MN","MS","MO",
+    "MT","NE","NV","NH","NJ","NM","NY","NC","ND","OH","OK","OR","PA",
+    "RI","SC","SD","TN","TX","UT","VT","VA","WA","WV","WI","WY")
states

[1] "AL" "AK" "AZ" "AR" "CA" "CO" "CT" "DE" "FL" "GA" "HI" "ID" "IL" "IN" "IA"
[16] "KS" "KY" "LA" "ME" "MD" "MA" "MI" "MN" "MS" "MO" "MT" "NE" "NV" "NH" "NJ"
[31] "NM" "NY" "NC" "ND" "OH" "OK" "OR" "PA" "RI" "SC" "SD" "TN" "TX" "UT" "VT"
[46] "VA" "WA" "WV" "WI" "WY"

raw <- read.csv("C:/DataMining/Data/unempstates.csv")
raw[1:3,]

  AL  AK   AZ  AR  CA  CO  CT  DE   FL  GA  HI  ID  IL  IN  IA  KS  KY  LA  ME
1 6.4 7.1 10.5 7.3 9.3 5.8 9.4 7.7 10.0 8.3 9.9 5.5 6.4 6.9 4.2 4.3 5.7 6.2 8.8
2 6.3 7.0 10.3 7.2 9.1 5.7 9.3 7.8  9.8 8.2 9.8 5.4 6.4 6.6 4.2 4.2 5.6 6.2 8.6
3 6.1 7.0 10.0 7.1 9.0 5.6 9.2 7.9  9.5 8.1 9.6 5.4 6.4 6.4 4.1 4.2 5.5 6.2 8.5
  MD   MA   MI  MN  MS  MO  MT  NE  NV  NH   NJ  NM   NY  NC  ND  OH  OK   OR
1 6.9 11.1 10.0 6.2 7.0 5.8 5.8 3.6 9.8 7.2 10.5 8.9 10.2 6.7 3.2 8.3 6.4 10.1
2 6.7 10.9  9.9 6.0 6.8 5.8 5.7 3.5 9.5 7.1 10.4 8.8 10.2 6.5 3.3 8.2 6.3  9.8
3 6.6 10.6  9.8 5.8 6.6 5.8 5.7 3.3 9.3 7.0 10.4 8.7 10.1 6.3 3.3 8.0 6.1  9.5
  PA  RI  SC  SD  TN  TX  UT  VT  VA  WA  WV  WI  WY
1 8.1 7.8 7.6 3.6 5.9 5.9 6.1 8.8 6.2 8.7 8.3 5.9 4.2
2 8.1 7.8 7.4 3.5 5.9 5.9 5.9 8.7 6.1 8.7 8.1 5.7 4.1
3 8.0 7.9 7.2 3.4 5.9 5.8 5.7 8.6 5.9 8.7 7.9 5.6 4.0

## transpose so that we have 50 rows (states) and 416 columns
rawt=matrix(nrow=50,ncol=416)
rawt=t(raw)
rawt[1:3,]

   [,1] [,2] [,3] [,4] [,5] [,6] [,7] [,8] [,9] [,10] [,11] [,12] [,13] [,14]
AL  6.4  6.3  6.1  6.0  6.0  6.0  6.2  6.3  6.4   6.5   6.6   6.7   6.9   7.0
AK  7.1  7.0  7.0  7.0  7.0  7.1  7.4  7.7  8.0   8.3   8.5   8.7   8.9   9.1
AZ 10.5 10.3 10.0  9.8  9.6  9.5  9.5  9.5  9.6   9.6   9.6   9.5   9.4   9.3
. . .

pcaunemp <- prcomp(rawt,scale=FALSE)
pcaunemp

plot(pcaunemp, main="")
mtext(side=1,"Unemployment: 50 states",line=1,font=2)
```

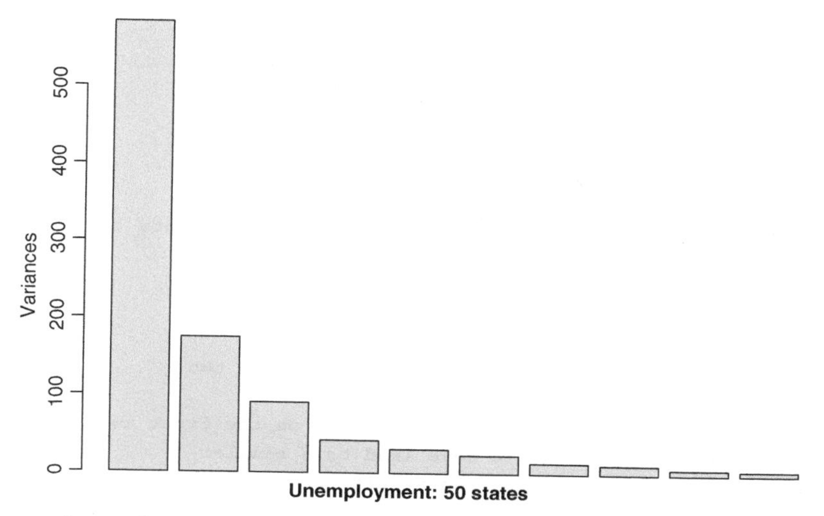


```
pcaunemp$rotation[,1]
pcaunemp$rotation[1:10,1]    ## just the first 10 values
```

```
[1] -0.03260897 -0.03208997 -0.03203986 -0.03153477 -0.03165013 -0.03200857
[7] -0.03212307 -0.03211913 -0.03232713 -0.03247573
```

```
ave=dim(416)
for (j in 1:416) {
ave[j]=mean(rawt[,j])
}

par(mfrow=c(1,2))
## plot negative loadings for first principal component
plot(-pcaunemp$rotation[,1])
## plot monthly averages of unemployment rates
plot(ave,type="l",ylim=c(3,10),xlab="month",ylab="ave
+     unemployment rate")
```

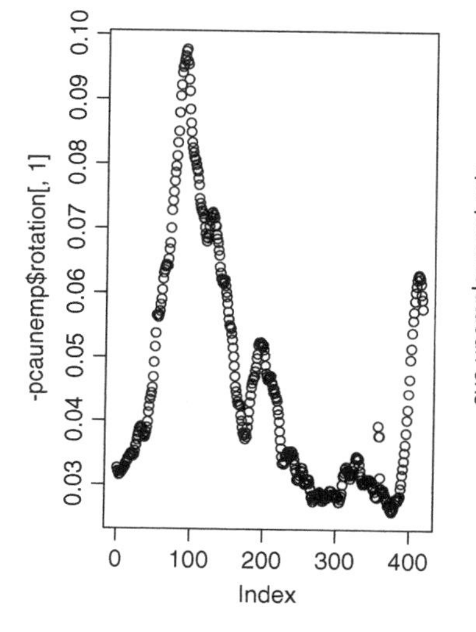

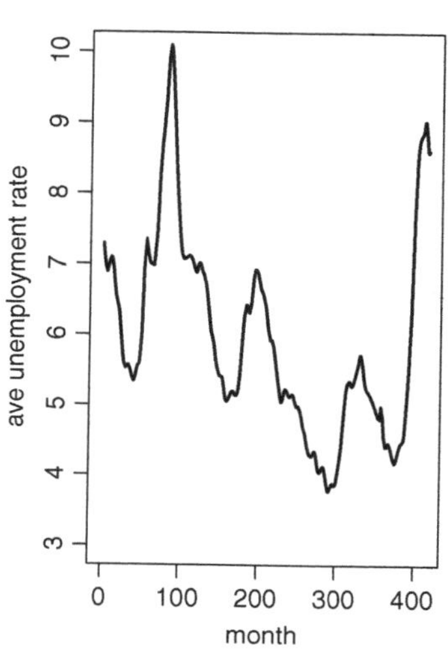

```
abs(cor(ave,pcaunemp$rotation[,1]))

[1] 0.8257265

pcaunemp$rotation[,2]
pcaunemp$rotation[,3]

## below we obtain the scores of the principal components
## the first 2-3 principal components do a good job
unemppc <- predict(pcaunemp)
unemppc

## below we construct a scatter plot of the first two
## principal components
## we assess whether an informal clustering on the first two
## principal components would have lead to a similar
## clustering than the clustering results of the k-means
## clustering approach applied on all 416 components
## the graph indicates that it does

set.seed(1)
grpunemp3 <- kmeans(rawt,centers=3,nstart=10)
par(mfrow=c(1,1))
plot(unemppc[,1:2],type="n")
text(x=unemppc[,1],y=unemppc[,2],labels=states,col=rainbow(7)
+     [grpunemp3$cluster])
```

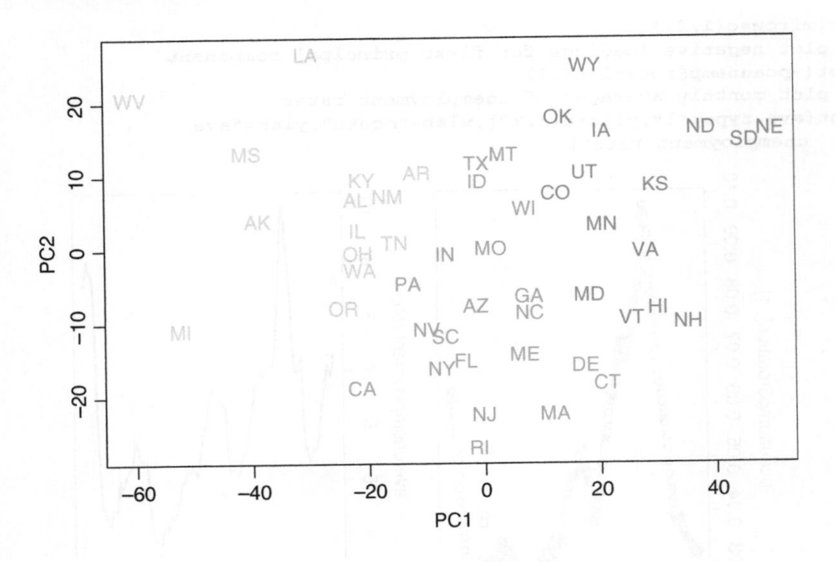

CHAPTER 18

Reducing the Dimension in Regressions with Multicollinear Inputs: Principal Components Regression and Partial Least Squares

In the previous chapter, we discussed how principal components analysis (PCA) can help reduce the dimensionality of a data matrix X consisting of n rows and p columns. In some applications of PCA, dimensionality reduction is the only objective, and the knowledge that certain linear combinations are able to explain most of the variability in the X matrix is all that is needed. But more often the goal is to predict a response y, and PCA is only a first step of the analysis. Often there are more predictors than the number of cases, which would make a brute-force regression impossible. PCA's role is to reduce the large number of correlated predictor variables down to a smaller and more manageable number of transformed predictors that represent most of the information in the original predictor matrix X. Instead of regressing the response y on the original predictor variables directly, *principal components regression* uses the principal components as predictors. Typically, one uses only a subset of the most important principal components in the regression. Regression on principal components is frequently used in econometrics where one encounters a large number of highly correlated predictor variables. Stock and Watson (2002a, 2002b) use PCA in their diffusion index approach to forecasting.

Note that in principal components regression, the response y plays no role in determining the transformed predictor variables. One wonders whether an approach that incorporates the response in the construction of the transformed predictors from the very beginning would work better. This is where *partial least squares* (PLS) comes into the picture. PLS, a regression technique for tough regression problems with many and usually highly-correlated covariates, constructs a set of linear combinations of the inputs in the matrix X for subsequent regression, but, unlike principal components regression, it uses y in the construction of these linear

Data Mining and Business Analytics with R, First Edition. Johannes Ledolter.

combinations. PLS, introduced by Wold (1975), is widely used in chemometrics, where ill-posed regression problems with many highly-correlated inputs seem to be the norm. For details on this method and how it relates to standard least squares, see Frank and Friedman (1993).

Like principal components regression, PLS is not scale invariant; so it is common to standardize each predictor column j so that the columns x_j (for $j = 1, 2, \ldots, p$) have all mean zero and variance one. PLS begins by computing, for each j, the inner product of x_j and the response vector y, $\widehat{\phi}_{1j} = x_j'y$. These projections, $\widehat{\phi}_{11}, \widehat{\phi}_{12}, \ldots, \widehat{\phi}_{1p}$, become the *loadings* when constructing the first (derived) input variable $z_1 = \sum_{j=1}^{p} \widehat{\phi}_{1j} x_j$; this is referred to as the *first PLS direction*. Different from principal components regression, the inputs x_j in the linear combination z_1 are weighted by their relationship with the response y. Next, we orthogonalize the input vectors $x_1, x_2, \ldots, x_p$ with respect to the first PLS direction z_1. We do this by regressing each x_j on z_1 and treating the residuals from this regression as the new $x_j^{(2)}$. With the new $x_j^{(2)}$ we determine a second set of loadings $\widehat{\phi}_{2j} = (x_j^{(2)})'y$ $(j = 1, 2, \ldots, p)$ and the second PLS direction $z_2 = \sum_{j=1}^{p} \widehat{\phi}_{2j} x_j^{(2)}$. Next, we orthogonalize the input vectors $x_1^{(2)}, x_2^{(2)}, \ldots, x_p^{(2)}$ with respect to the second PLS direction z_2, determine a third set of loadings and the third PLS direction, and so on. We continue this process, until $k \le p$ directions have been obtained. The regression of the response y on the first k PLS directions describes the PLS fit. If we construct all $k = p$ directions (assuming that this is actually possible and that the number of regressors $p \le n$), we get the usual least squares fit; using $k < p$ directions produces a reduced regression fit.

Instead of loading on the inner product of x_j and the response vector y, $\widehat{\phi}_{1j} = x_j'y$, one could also load on the correlation between the (standardized) x_j and the response vector y. Zero correlation implies that x_j is not part of z_1, while large correlations give x_j considerable input in the construction of the first PLS direction. Note that the change from the inner product to the correlation would not affect the first PLS direction; all it would do is multiply the direction vector with a scalar.

Algorithm 3.3 in the book by Hastie et al. (2008) gives a detailed description of the calculations behind PLS. One starts centering and standardizing the p columns of the covariate matrix X to obtain the standardized covariate matrix X^0 with columns $x_{ij}^0 = (x_{ij} - \overline{x}_j)/s_j$ (for $j = 1, 2, \ldots, p$). The initial $n \times 1$ vector of fitted values $\widehat{y}_0 = \overline{y}$ has the average response $\overline{y} = \sum_{i=1}^{n} y_i/n$ in all its rows. The K-direction PLS(K) fit is obtained through iterations $k = 1, \ldots, K$ as follows:

1. Calculate the $p \times 1$ vector of loadings $\phi^k = (X^{k-1})'y$ and the $n \times 1$ vector of kth directions $z_k = X^{k-1}\phi^k$.
2. Update the PLS(k) fitted values according to $\widehat{y}_k = \widehat{y}_{k-1} + [(z_k)'y/(z_k)'z_k]z_k$ and orthogonalize the predictor matrix with respect to direction z_k by transforming each column according to $x_j^k = x_j^{k-1} - [(z_k)'x_j^{k-1}/(z_k)'z_k]z_k$.

Algorithm 3.3 is simple to program, and an R program for it is given in Section 18.1.1. Alternatively, one can use the function *pls* in the R package **mixOmics**. Use the R help function to learn more about the parameters of this function.

18.1 THREE EXAMPLES

18.1.1 Example 1: Generated Data

We generate an X matrix of $n = 400$ rows and $p = 100$ columns. Each of the columns contains 400 independent N(0,1) realizations, and the 100 columns are generated independently. The column for the response contains 400 independent realizations from a normal distribution with mean 6 and standard deviation 2. The objective of this exercise is threefold: (i) to check that the results of our R program coincide with the results of the function *pls* in the R package **mixOmics**; (ii) to illustrate that in this example a PLS with just the first (or perhaps the first two) PLS directions comes very close to the regression that fits the response on all 100 columns. Admittedly, this data matrix X does not represent a very difficult case for ordinary regression as the columns of the X matrix are orthogonal. For orthogonal columns, the first PLS direction coincides with the fitted values from the regression on all 100 columns. So it is not surprising that the R-square from PLS with $K = 1$ is similar to the R-square in the regression on all 100 columns. (iii) This result also illustrates that by adding more and more columns, even if they are totally unrelated to the response, the R-square tends to reach a level that appears quite respectable (around 22% here). Keep in mind, however, that there is no relationship whatsoever between the response and the X matrix. All we do here is fitting noise.

```
## PLS algorithm, following algorithm 3.3 in Hastie et al.
## standardize X's. PLS depends on scale
## we can't have too many partial least squares directions (nc)
## otherwise problems
## here we simulate observations

set.seed(1)
nrow=400      ## row dimension of X
ncol=100      ## column dimension of X
nc=2          ## number of PLS directions
nc1=nc+1

Y1=dim(nrow)
X=matrix(nrow=nrow,ncol=ncol)
X1=matrix(nrow=nrow,ncol=ncol)
Y=matrix(nrow=nrow,ncol=nc1)
Z=matrix(nrow=nrow,ncol=nc)
F=matrix(nrow=nrow,ncol=ncol)
FN=matrix(nrow=nrow,ncol=ncol)
me=dim(ncol)
s=dim(ncol)
```

```
## enter data into matrix X1 and column Y1
## data simulation
for (jj in 1:ncol) {
X1[,jj]=rnorm(nrow)
}
Y1=rnorm(nrow)
Y1=2*Y1+6

## standardization
for (j in 1:ncol) {
me[j]=mean(X1[,j])
s[j]=sd(X1[,j])
}
for (j in 1:ncol) {
for (i in 1:nrow) {
X[i,j]=(X1[i,j]-me[j])/s[j]
}
}

## Algorithm 3.3 starts
y=Y1
F=X
Y[,1]=mean(y)*y/y

for (k in 1:nc) {
phi=t(F)%*%y
Z[,k]=F%*%phi
fra=(t(Z[,k])%*%y)/(t(Z[,k])%*%Z[,k])
Y[,k+1]=Y[,k]+fra*Z[,k]
for (j in 1:ncol) {
fru=(t(Z[,k])%*%F[,j])/(t(Z[,k])%*%Z[,k])
FN[,j]=F[,j]-fru*Z[,k]
}
F=FN
}
fp=Y[,nc+1]
## Algorithm 3.3 ends

cor(y,fp)**2

[1] 0.2108954 ## R-square of PLS solution for nc=2

## R-square of PLS solution for nc=1: 0.1815176 (not shown)

ZZ=data.frame(Z[,1:nc])
m1=lm(y~.,data=ZZ)
cor(y,m1$fitted)**2

[1] 0.2108954

## R-square of PLS solution calculated from regression on first
## nc=2 PLS directions: 0.2108954

XX=data.frame(X)
mall=lm(y~.,data=XX)
cor(y,mall$fitted)**2
```

```
[1] 0.2178761

## R-square on all 100 columns 0.2178761

## even with few PLS directions, R**2 of largest model is
## approached very quickly

## comparison with library(mixOmics)
library(mixOmics)
mpls=pls(X1,Y1,ncomp=2,mode="classic")
x1=mpls$variates$X[,1]
x2=mpls$variates$X[,2]
m3=lm(y~x1+x2)
cor(y,m3$fitted)**2

[1] 0.2108954

## R-square of PLS using pls(mixOmics) with nc=2: 0.2108954

fmpls=m3$fitted
## fmpls and fp (and m1$fitted are all the same)
```

18.1.2 Example 2: Predicting Next Month's Unemployment Rate of a Certain State From Past Unemployment Rates of All 50 States

We use the unemployment rates on all $n = 50$ US states over a period of 416 months (from January 1976 through August 2010). We have looked at this data set previously (in the contexts of clustering in Chapter 15 and PCA in Chapter 17). Here we treat this data set as a multivariate (50-variate) time series spanning 416 periods. Vector autoregressive (VAR) models are commonly used to predict vector time series. There one regresses the observations (unemployment rates) of a certain state (say the first state) on the previous (lagged) unemployment rates of that state, as well as those of all other states. The order of the VAR tells us about how many lags to consider. Order 4 implies that we consider the information from the previous 4 months. This order is probably sufficient for seasonally adjusted rates as we can reasonably expect that the memory of the system is shorter than four periods; in other words, we assume that the previous four unemployment rates summarize the information on the future trajectory. With order 4, the first row of the response vector y corresponds to month 5 as we are losing the first four periods because of lags. The first row of the data matrix X contains unemployment rates at periods 4, 3, 2, and 1. We have 412 $(= 416 - 4)$ rows for the response vector and the data matrix X. The matrix X contains $(50)(4) = 200$ columns; four columns of lags for each of the 50 states. If we had taken the previous 12 lags, the data matrix X would have $416-12 = 404$ rows and $(50)(12) = 600$ columns, resulting in more columns than rows. The response vector for the first state is regressed on the columns of the data matrix X. A brute force regression on so many columns (which furthermore are highly multicollinear because lagged unemployment rates of the same state are similar and because unemployment rates from different states are usually alike) would be difficult and impossible if the number of columns in the X matrix exceeds the number of rows (as in a model with 12 lags). A principal components regression where one regresses the response vector on the first several principal components, and a PLS regression on the first couple of PLS directions are two things to try. The analysis shown in the following considers the prediction for a single state (here the first one, Alaska). The calculations need to be repeated for the other states. Note that the response vector changes for different states, but the data matrix X stays the same.

A small error (or a large R-square) in the regression expresses the fact that the explanatory variables are very capable of predicting the unemployment rate for the next month. The PLS results for the first state (Alaska) shows that the regression on the first PLS direction already explains 89.1% of the variation. The regression on the first two PLS directions explains 93.3% of the variation. This is already very close to the R-square from the regression on all 200 columns of the X matrix that explains 99.98%. Note that this last R-square is identical to the R-square we get from the regression on all 200 PLS directions.

```
library(mixOmics)

nrow=412        ## row dimension of X
ncol=200        ## column dimension of X
nstates=50      ## number of states

X=matrix(nrow=nrow,ncol=ncol)
Y=matrix(nrow=nrow,ncol=nstates)

raw <- read.csv("C:/DataMining/Data/unempstates.csv")
raw[1:3,]

   AL  AK   AZ  AR  CA  CO  CT  DE   FL  GA  HI  ID  IL  IN  IA  KS  KY  LA  ME
1 6.4 7.1 10.5 7.3 9.3 5.8 9.4 7.7 10.0 8.3 9.9 5.5 6.4 6.9 4.2 4.3 5.7 6.2 8.8
2 6.3 7.0 10.3 7.2 9.1 5.7 9.3 7.8  9.8 8.2 9.8 5.4 6.4 6.6 4.2 4.2 5.6 6.2 8.6
3 6.1 7.0 10.0 7.1 9.0 5.6 9.2 7.9  9.5 8.1 9.6 5.4 6.4 6.4 4.1 4.2 5.5 6.2 8.5
   MD   MA   MI  MN  MS  MO  MT  NE  NV  NH   NJ  NM   NY  NC  ND  OH  OK   OR
1 6.9 11.1 10.0 6.2 7.0 5.8 5.8 3.6 9.8 7.2 10.5 8.9 10.2 6.7 3.2 8.3 6.4 10.1
2 6.7 10.9  9.9 6.0 6.8 5.8 5.7 3.5 9.5 7.1 10.4 8.8 10.2 6.5 3.3 8.2 6.3  9.8
3 6.6 10.6  9.8 5.8 6.6 5.8 5.7 3.3 9.3 7.0 10.4 8.7 10.1 6.3 3.3 8.0 6.1  9.5
   PA  RI  SC  SD  TN  TX  UT  VT  VA  WA  WV  WI  WY
1 8.1 7.8 7.6 3.6 5.9 5.9 6.1 8.8 6.2 8.7 8.3 5.9 4.2
2 8.1 7.8 7.4 3.5 5.9 5.9 5.9 8.7 6.1 8.7 8.1 5.7 4.1
3 8.0 7.9 7.2 3.4 5.9 5.8 5.7 8.6 5.9 8.7 7.9 5.6 4.0

X=matrix(nrow=412,ncol=200)
Y=matrix(nrow=412,ncol=50)

for (j in 1:50) {
for (i in 1:412) {
Y[i,j]=raw[i+4,j]
}
}

for (j in 1:50) {
for (i in 1:412) {
```

```
X[i,j]=raw[i+3,j]
X[i,j+50]=raw[i+2,j]
X[i,j+100]=raw[i+1,j]
X[i,j+150]=raw[i,j]
}
}

nc=1      ## number of PLS directions
## pls on nc components
mpls=pls(X,Y[,1],ncomp=nc,mode="classic")
m1=lm(Y[,1]~.,data.frame(mpls$variates$X))
summary(m1)
cor(Y[,1],m1$fitted)**2

[1] 0.8913555

nc=2      ## number of PLS directions
## pls on nc components
mpls=pls(X,Y[,1],ncomp=nc,mode="classic")
m2=lm(Y[,1]~.,data.frame(mpls$variates$X))
summary(m2)
cor(Y[,1],m2$fitted)**2

[1] 0.9326079

nc=3      ## number of PLS directions
## pls on nc components
mpls=pls(X,Y[,1],ncomp=nc,mode="classic")
m3=lm(Y[,1]~.,data.frame(mpls$variates$X))
summary(m3)
cor(Y[,1],m3$fitted)**2

[1] 0.9649273

## regression on all columns of X
mreg=lm(Y[,1]~.,data.frame(X))
mreg
cor(Y[,1],mreg$fitted)**2

[1] 0.9997558
```

18.1.3 Example 3: Predicting Next Month's Unemployment Rate. Comparing Several Methods in Terms of Their Out-of-Sample Prediction Performance

In the previous example, we assessed the model fit by looking at the R-square. It is not surprising that a model with many estimated coefficients fits well. But, assessing the out-of-sample prediction performance is different from assessing the in-sample fit, and this is what we do in the next example.

We investigate the performance of the following methods:

Univariate AR(4) model fit to individual series. In this method, the current unemployment rate is regressed on its previous four lags. This is done separately for each state. Univariate AR models are commonly used in forecasting, where one relates the current observation to observations from previous periods. For discussion, see Box et al. (1994) or Abraham and Ledolter (1983).

Multivariate VAR(4). In this method, the current unemployment rate of one state is regressed on its previous four lags and the four lags from all other states. This amounts to a regression on 200 predictors. In total, 50 regressions (one for each state) onto the same set of 200 predictors have to be carried out. Component-by-component estimation is equivalent to the joint least squares estimation, even for a general covariance matrix of the error terms; see Johnson and Wichern (1988, Chapter 7).

VAR(4) with LASSO constraints. A problem with fitting a regression on so many similar predictor variables is that many predictor variables (perhaps most) are not needed. Regression methods that constrain the magnitudes of the estimates may help overcome the problem of estimating too many unimportant coefficients. LASSO methods under several constraint assumptions are evaluated. See Chapter 6 for a discussion of LASSO.

PCA regression and PLS. The matrix of predictors is high-dimensional and its columns are highly multicollinear. This is because lagged unemployment rates of the same state are similar and unemployment rates from different states are alike. Principal components regression where we regress the response vector on the first 10 (200) principal components, and PLS regression on the first 10 (100) PLS directions are being considered.

Forecast evaluations. We evaluate the forecasts on 25 randomly selected time periods (the evaluation sample). The models are estimated on the remaining 387 observations (the training sample). Models are fit for all 50 states, resulting in (25)(50) one-step-ahead forecast errors for each studied method. Methods are evaluated on the root mean square error (RMSE) that is being calculated from the (25)(50) one-step-ahead forecast errors. The random selection of the evaluation and training samples is repeated 50 times, and box-plots of the root mean square forecast errors are prepared.

Discussion. The R program on the book's webpage is used for all calculations. The results are summarized in the box plots in Figure 18.1. The following conclusions can be made.

1. The univariate AR(4) model, the model that probably comes to mind first and a model that is rather easy to estimate (a regression on the previous four lags), performs quite well. Its average RMSE is about 0.13 percentage points. This says that if the true unemployment rate is 5%, the predictions from this model could be off by about $\pm(2)(0.13) = 0.26$ percentage points.
2. The VAR(4) model with its regression on 200 (quite a few!) predictors performs worse; its average RMSE is 0.15. However, one can see that LASSO restrictions will improve the forecast performance. LASSO with $s = 0.25$ (i.e., the L1 norm of the LASSO estimates is one-quarter of the L1 norm of the unrestricted least squares estimates) leads to an average RMSE of 0.11. This illustrates the power of a regularization estimation approach that "zeros out" many of the unwanted coefficients.

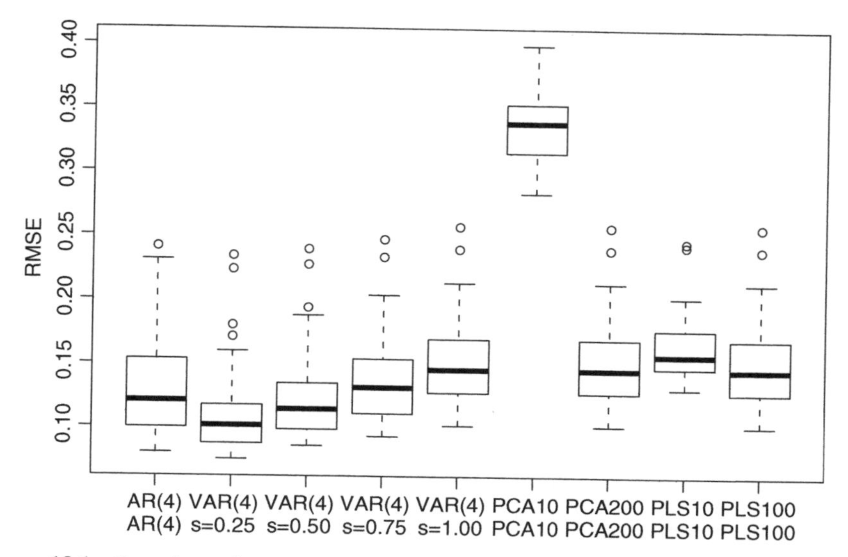

Figure 18.1 Box plots of root mean square forecast errors: Predicting the levels of unemployment rates.

3. The PCA regression on the first 10 principal components is rather bad; its average RMSE is 0.33. This implies that we would need a lot more principal components to predict the unemployment rate reliably. Using 200 principal components (i.e., all of them) gives us the same box-plot as the VAR(4) model without LASSO restrictions. This is comforting to see as we know that a regression on all principal components is identical to a regression on all predictors.
4. The forecast performance of PLS on the first 10 PLS directions (average $RMSE = 0.165$) is considerably better than that of the PCA regression on 10 components, but it is still worse than that of the simple univariate AR(4). PLS on 100 PLS directions has already approximated the VAR(4) results without any LASSO regularization.
5. We learn that neither PCA regression nor PLS can compete with the VAR(4) model that uses some LASSO regularization.
6. Why does PCA regression perform so poorly? Take the extreme case when regressing the unemployment rates on just the first principal component. For nonstationary data, as we have in this illustration, the loadings on the first principal component represent the average trend in the data; see the discussion in Section 17.2. A regression on the average trend will fit individual series poorly, certainly much poorer than an approach that relates the current value of a certain state to previous values for that state (as autoregressive models do). So it is not surprising that even with 10 principal components, the results of principal components regression are rather poor. The results should change when predicting successive differences (or changes), $y_t - y_{t-1}$,

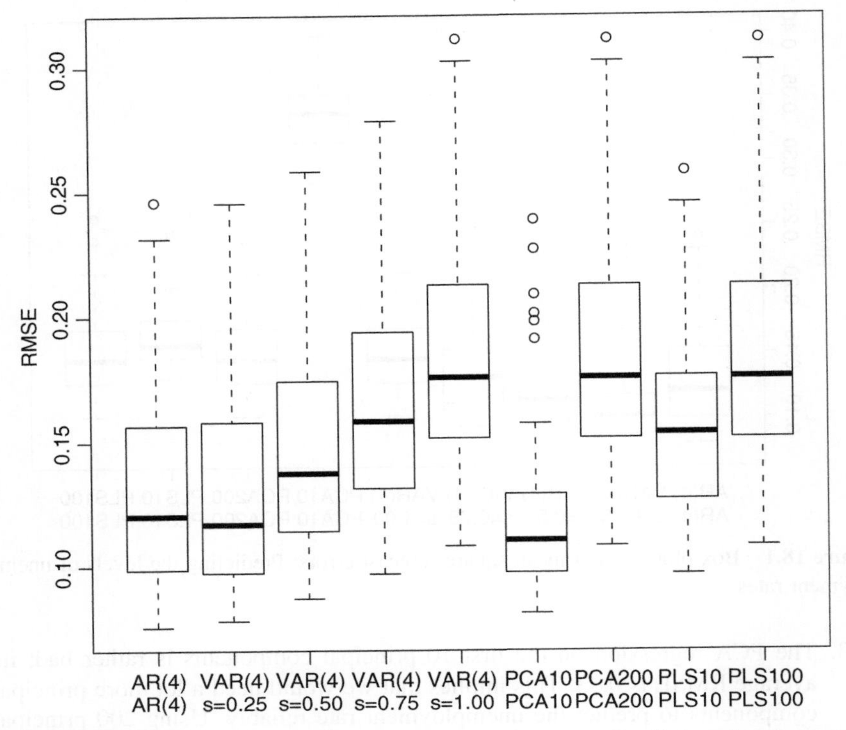

Figure 18.2 Box plots of root mean square forecast errors: Predicting the changes of unemployment rates.

as differencing removes the trend component and makes the differences stationary. Stock and Watson mention that for PCA regression to work well, the columns of the matrix X should be stationary.

7. We repeat the analysis for first successive differences. When fitting models on first successive differences and when predicting changes, principal components regression performs very well; see the box plots shown in Figure 18.2. Successive first differences of unemployment rates are virtually independent, and averages of first differences are about zero, for all states. The naïve predictor (the one that uses zero as prediction for future changes and does so for every state) performs quite well (average RMSE is 0.155). This is quite different from the analysis of levels. There the naïve predictor that averages the current levels of the 50 states to predict the future level of a state performs very poorly (average RMSE is 1.724). This is yet another explanation why the regression on the first few principal components fails if we try to predict nonstationary levels. For PCA regression to work, we need to assume that the series are stationary.

The R programs for predicting levels and changes of the US unemployment rates can be found on the web page that accompanies this book.

REFERENCES

Abraham, B. and Ledolter, J.: *Statistical Methods for Forecasting*. New York: John Wiley & Sons, Inc., 1983.

Box, G.E.P., Jenkins, G.M., and Reinsel, G.C.: *Time Series Analysis, Forecasting and Control*. Third edition. Englewood Cliffs: Prentice Hall, 1994.

Frank, I.E. and Friedman, J.H.: A statistical view of some chemometrics regression tools. *Technometrics*, Vol. 35 (1993), 109–135.

Hastie, T., Tibshirani, R., and Friedman, J.: *The Elements of Statistical Learning: Data Mining, Inference and Prediction*. Second edition. New York: Springer, 2009.

Johnson, R.A. and Wichern, D.W.: *Applied Multivariate Statistical Analysis*. Second edition. Englewood Cliffs, NJ: Prentice Hall, 1988.

Stock, J.H. and Watson, M.W.: Forecasting using principal components from a large number of predictors. *Journal of the American Statistical Association*, Vol. 97 (2002a), 1167–1179.

Stock, J.H. and Watson, M.W.: Macroeconomic forecasting using diffusion indexes, *Journal of Business & Economic Statistics*, Vol. 20 (2002b), 147–162.

Wold, H.: Soft modeling by latent variables: The nonlinear iterative partial least squares approach. In *Perspectives in Probability and Statistics, Papers in Honour of M.S. Bartlett*; Gani, J., Ed.;. Sheffield: Applied Probability Trust, 1975.

CHAPTER 19

Text as Data: Text Mining and Sentiment Analysis[1]

Text is a vast source of data for business. Examples are earnings announcements, communications to shareholders, press releases, restaurant reviews, political speeches in Congress, Supreme Court opinions, and so on.

Text data is extremely high dimensional. The analysis of phrase counts from text documents is the current state of the art. The "bag of words" representation of text assigns frequencies to words or combinations of words. However, considerable preprocessing of text is needed before one can obtain frequency information on words and before one can start the statistical analysis. Information retrieval and the appropriate "tokenization" of the information are very important.

The first step when faced with a raw text document is to *stem* the words. This means that one cuts words to their root: for example, "tax" from taxing, taxes, and taxation. The Porter stemming algorithm, named after Martin Porter, has become the English standard. It is an algorithm for removing common morphological and inflexional endings from English words.

The next step is to search the text documents for a list of *stop words* containing irrelevant words marked for removal. If, and, but, who, what, the, they, their, a, or, and so on are examples of stop words that need to be removed. But one needs to be careful because one person's stop word is another's key term.

Consider the following passage from Shakespeare's "As You Like It":

> All the world's a stage, and all the men and women merely players: they have their exits and their entrances; and one man in his time plays many parts, his acts being seven ages.

What the statistician sees are "cleaned up" words and their frequencies:

world	stage	men	women	play	exit	entrance	time	part	act	seven	age
1	1	2	1	2	1	1	1	1	1	1	1

[1]Chapter 19 on the analysis of text data draws heavily on the research of Professor Matt Taddy of Chicago's Booth School of Business. His contribution is most gratefully acknowledged.

Data Mining and Business Analytics with R, First Edition. Johannes Ledolter.

Also, one usually removes words that are extremely rare. Imagine one wants to compare all of Shakespeare's plays. A play is referred to as a *document*, and the set of all plays is referred to as the *corpus*. If a certain term occurs only once, it will not be useful for comparing documents. A reasonable rule removes words with relative frequencies below 0.5%.

More sophisticated removal rules have been considered in the literature. For example, in a corpus of n documents (e.g., all of Shakespeare's plays), one can base the screening of words on the tf-idf (term frequency/inverse document frequency) score,

$$\text{tf-idf} = f_{ij} \times \log\left(\frac{n}{d_j}\right),$$

where the term frequency, f_{ij}, is the (relative) frequency of word j in document i (its frequency count divided by the total number of words in that document), n is the number of documents, d_j is the number of documents containing word j, and $\log(n/d_j)$ is the inverse document frequency (idf). One omits all words with tf-idf below a certain threshold.

EXAMPLE 19.1 Consider a document containing 10,000 words wherein the word *donkey* appears 300 times. Following the earlier definition, the term frequency (tf) for *donkey* is $(300/10,000) = 0.03$. Now, assume we have 1,000 documents and *donkey* appears in 10 of these. Then, its inverse document frequency (idf) is calculated as log $(1000/10) = 2$. The tf-idf score is the product of these quantities: $0.03 \times 2 = 0.06$.

Several other preprocessing steps can be used to get a meaningful list of words and their counts (frequencies). Words can be single words, or bigrams of words. *Bigrams* are groups of two adjacent words, and such bigrams are commonly used as the basis for the statistical analysis of text. Bigrams can be extended to trigrams (three adjacent words) and, more general, *n-grams*, which are sequences of n adjacent words.

19.1 INVERSE MULTINOMIAL LOGISTIC REGRESSION

In the following example, we analyze text information that comes from restaurant reviews. The text from a corpus of 6166 restaurant reviews was reduced to 2640 bigrams (Taddy, 2012a, 2012b). We treat the 2640 bigrams as the $g = 2640$ possible outcome categories of a multinomial distribution. Now consider a certain review (i.e., review 1). Assume that this review has a total of n_1 occurrences of these 2640 bigrams. These $n_1 = n_{11} + n_{12} + \cdots + n_{1,g=2640}$ occurrences represent n_{11} occurrences for bigram 1, n_{12} occurrences for bigram 2, ... , $n_{1,g=2640}$ occurrences for bigram 2640. Fixing the row (document, here the review) sum of occurrences, the $g \times 1$ outcome vector of counts follows a multinomial distribution with parameter n_1 and $g = 2640$ probabilities that sum to one. Each of these 2640 probabilities is modeled as a function of covariates. In the restaurant review example, we have just one covariate x, namely the overall restaurant rating on a scale from 1 to 5. But there may be more covariates such as ratings on multiple categories that relate to food, service, and attractiveness of the restaurant, the price of a meal, the restaurant's location, and so on.

In multinomial logistic regression (see Chapter 11), we model the probabilities through

$$p_k = P[y = k] = \frac{\exp(\alpha_k + x\beta_k)}{\sum_{h=1}^{g} \exp(\alpha_h + x\beta_h)}, \quad \text{for } k = 1, 2, \ldots, g = 2640.$$

One pair (α_k, β_k) of the parameters can be set to zero without loss of generality as the multinomial probabilities must add to one. Note that the multinomial logistic regression representation leads to a very large number of parameters; in our case, there are $2(g - 1) = 5278$ coefficients that need to be estimated; and this is for a single covariate. It becomes clear that one needs penalty-based estimation/variable selection methods for the estimation of these parameters. The routine *mnlm* in the R library **textir** does just that. It may take the program a few minutes (sometimes also longer) to get the estimation done, especially if there is a large number of bigrams and many parameters that need to be estimated.

After obtaining the estimates, we use the fitting results for the *inverse prediction* of x. Taddy (2012a) explains how to obtain the inverse prediction of the covariate x for each review with given word count distribution. For the reviews in our estimation sample, this inverse prediction provides us with the fitted (in-sample prediction) rating. Imagine now that you are reading the text of a new review, but you do not know the rating. You would want a method that can take the text information and give you a prediction of the unknown rating. This is exactly what the inverse multinomial logistic regression can do.

There are many other applications. Consider one on political sentiment studied by Gentzkow and Shapiro (2010) in their paper "What drives media slant? Evidence from U.S. daily newspapers". Gentzkow and Shapiro analyzed text information in political speeches made by members of Congress. They summarized the speeches of 529 members of Congress in form of trigrams and their associated frequencies. Furthermore, they collected covariates on each speaker, among them the Republican vote percentage in the speaker's district from the last election. This is an important variable because it represents the political ideology (the conservative/liberal orientation) of the speaker. Imagine now that you are given another speech (or an article in a newspaper), but that you do not know its author and his/her political ideology. From the frequency distribution of the speech's trigrams, you want to infer the political sentiment of the speech. Again, as we will show in Section 19.3, this is can be achieved with the inverse multinomial logistic regression model.

We use the routine *mnlm* in the R package **textir** for the analysis. Details of the statistical methodology are discussed in the paper by Taddy (2012a).

19.2 EXAMPLE 1: RESTAURANT REVIEWS

We are given 6166 restaurant reviews, with counts on 2640 bigrams. Each review has an average of 90 words. Furthermore, each review is accompanied by several ratings on a five star scale: an overall rating, and separate ratings for food, service, value, and atmosphere. For the following discussion, we use only the overall rating. This data set is taken from the paper by Taddy (2012a), and the analysis that follows is described in his paper. The data are given in the object *we8there* in the R package **textir**.

Imagine that you are sent the text of a new review, but not its overall rating. How would you predict the rating from just the text?

Here are two examples from among the 6166 restaurant reviews. The first number in bold font at the end of the review represents the overall rating.

Excellent: Waffle House, Bossier City LA

I normally would not review a Waffle House, but this one deserves it. The workers, Amanda, Amy, Cherry, James, and J.D. were the most pleasant crew I have seen. Although it was only lunch, B.L.T., and chili, it was great. The best thing was the 1950's rock and roll music, not too loud not too soft. This is a rare exception to what we all think a Waffle House is. Keep up the good work. [**5**: 5555]

Terrible: Sartin's Seafood, Nassau Bay TX

Had a very rude waitress and the manager was not nice either. [**1**: 1115]

An R program for carrying out the text analysis is listed below. We have annotated the output and explain the various steps of the analysis. The variable predinv contains the inverse prediction of x. The ROC analysis (see Chapter 8 for details) tells us how to best use the inverse prediction when rating a restaurant.

```
library(textir)
data(we8there)          ## 6166 reviews and 2640 bigrams

dim(we8thereCounts)

[1] 6166 2640

dimnames(we8thereCounts)

$Docs
   [1] "1"    "2"    "5"    "11"   "12"   "13"   "14"   "15"   "17"   "18"
  [11] "19"   "20"   "21"   "22"   "23"   "24"   "25"   "26"   "27"   "28"
  . . .
$Terms
   [1] "veri good"        "go back"          "dine room"
   [4] "dine experi"      "great food"       "food great"
   [7] "realli good"      "ice cream"        "high recommend"
  [10] "great place"      "food servic"      "look like"
  . . .
```

```
dim(we8thereRatings)

[1] 6166    5

we8thereRatings[1:3,]
## ratings (restaurants ordered on overall rating from 5 to 1)

  Food Service Value Atmosphere Overall
1    5       5     5          5       5
2    5       5     5          5       5
5    5       5     4          4       5

as.matrix(we8thereCounts)
as.matrix(we8thereCounts)[12,400] ## count for bigram 400 in review 12

## get to know what's in the matrix

g1=min(as.matrix(we8thereCounts)[,]) ## min count over reviews/bigrams
g2=max(as.matrix(we8thereCounts)[,]) ## max count over reviews/bigrams
g1
g2
## a certain bigram in a certain review was mentioned 13 times

hh=as.matrix(we8thereCounts)[,1000]
hh
## here we look at the frequencies of the bigram in column 1000
## the data are extremely sparce

overall=as.matrix(we8thereRatings[,5])
## overall rating

## we determine frequencies of the 2640 different bigrams
## this will take some time
nn=2640
cowords=dim(nn)
for (i in 1:nn) {
cowords[i]=sum(as.matrix(we8thereCounts)[,i])
}
cowords
cowords[7]

[1] 251
    ## "realli good"   mentioned 251 times
    ## 10 times is the minimum

plot(sort(cowords,decreasing=TRUE))

## analysis per review
## we determine the frequencies of bigrams per review
## this will take some time
nn=6166
coreview=dim(nn)
for (i in 1:nn) {
coreview[i]=sum(as.matrix(we8thereCounts)[i,])
}
plot(sort(coreview,decreasing=TRUE))
```

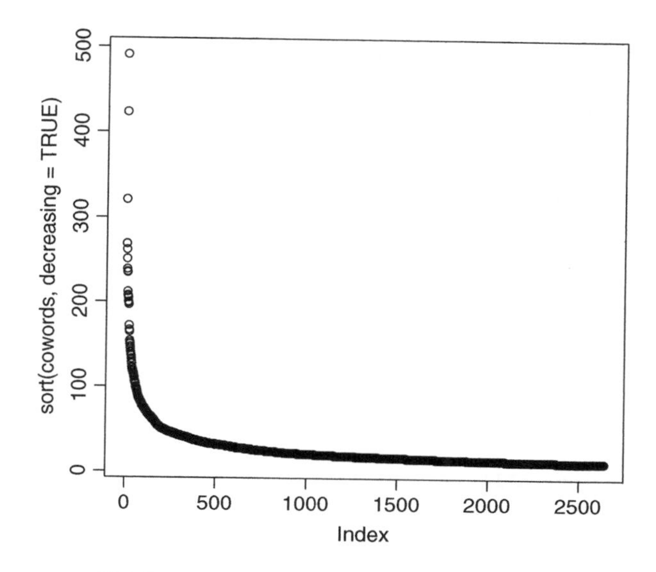

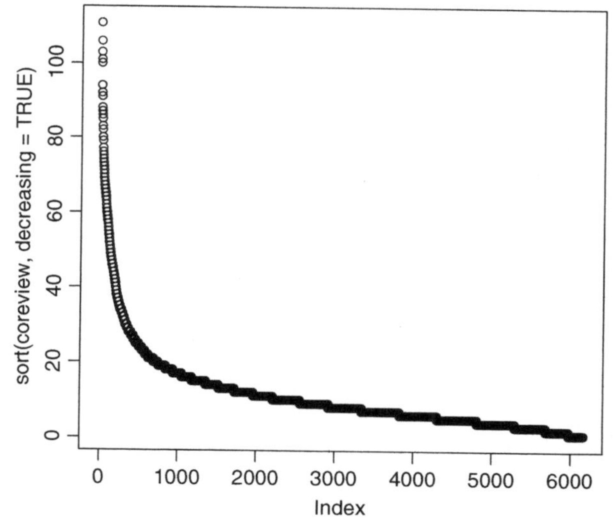


```
## Multinomial logistic regression and fitted reduction
we8mnlm=mnlm(we8thereCounts,overall,bins=5)
## bins: for faster inference if covariates are factors
## covariate is a factor with 5 levels
we8mnlm
we8mnlm$intercept   ## estimates of alphas
we8mnlm$loadings    ## estimates of betas
fitted(we8mnlm)
as.matrix(fitted(we8mnlm))[1,]
## fitted counts for first review
```

```
## following provides fitted multinomial probabilities
pred=predict(we8mnlm,overall,type="response")
pred[1,] ## predicted multinomial probs for review 1
sum(pred[1,]) ## must add to one

## following predicts inverse prediction (fitted reduction)
predinv=predict(we8mnlm,we8thereCounts,type="reduction")
predinv[1:10] ## prints predicted ratings for first 10 reviews
plot(predinv)
plot(predinv~overall)
corr(predinv,overall)

          [,1]
[1,] 0.706245

boxplot(predinv~overall)
## procedure works. Predicted ratings increase with actual
## ratings. Question of cutoff. Which cutoff to use for
## excellent review?
```

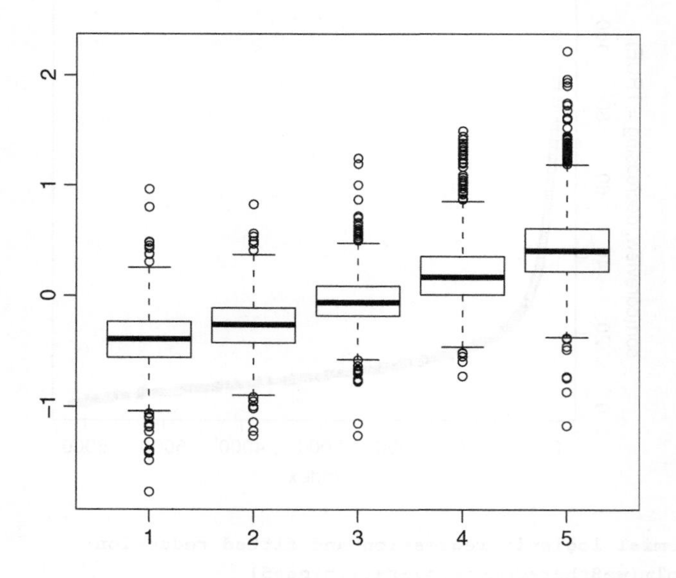


```
## ROC curve for classification of y with p
roc <- function(p,y){
  y <- factor(y)
  n <- length(p)
  p <- as.vector(p)
  Q <- p > matrix(rep(seq(0,1,length=500),n),ncol=500,
+    byrow=TRUE)
```

```
  fp <- colSums((y==levels(y)[1])*Q)/sum(y==levels(y)[1])
  tp <- colSums((y==levels(y)[2])*Q)/sum(y==levels(y)[2])
  plot(fp, tp, xlab="1-Specificity", ylab="Sensitivity")
  abline(a=0,b=1,lty=2,col=8)
}

c2=overall==4
c3=overall==5
c=c2+c3
min=min(predinv)
max=max(predinv)
pp=(predinv-min)/(max-min)

## plot of ROC curve
roc(p=pp, y=c)
```

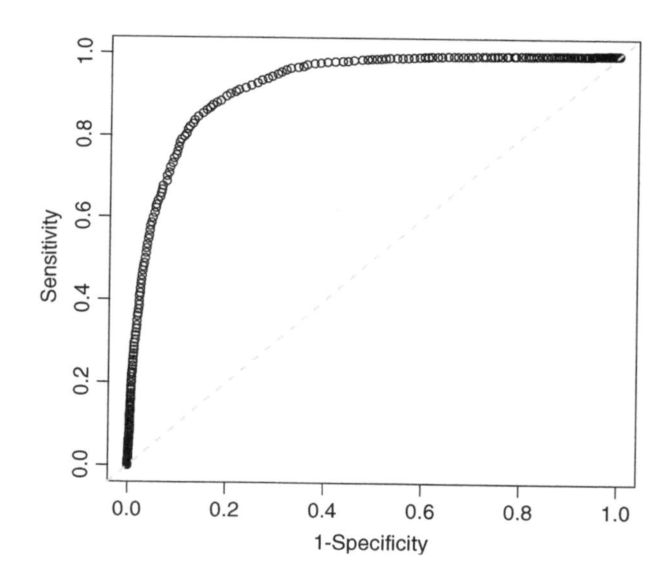


```
cut <- 0
truepos <- c==1 & predinv>=cut
trueneg <- c==0 & predinv<cut
## sensitivity (predict good review if review is good)
sum(truepos)/sum(c==1)

[1] 0.8938914

sum(trueneg)/sum(c==0)

[1] 0.8098511
```

```
## Zero may be a good cutoff.
## Sensitivity (true positive rate) of 0.89
## False positive rate of 1 - 0.81 = 0.19
## If inverse prediction > 0, conclude overall quality
## rating of 4 or 5.
```

The score on the inverse prediction helps us assess the overall quality rating of a review. The box plots of predinv against the overall rating shows this quite clearly. The ROC analysis (for details, see Chapter 8) tells us about a reasonable cutoff on predinv that separates good reviews (those with overall ratings 4 and 5) from negative reviews (3 or lower). A rule that classifies a review as positive if the inverse prediction is larger than 0 and as negative if the inverse prediction is smaller than 0 makes for an excellent decision rule. Eighty nine percent of positive reviews are identified as positive, and 81 percent of negative reviews are identified as negative. One can use this rule to classify new incoming reviews. Using its bigram counts, we determine its inverse prediction. If the inverse prediction is greater than 0, we classify the new review as good (rating 4 or higher).

19.3 EXAMPLE 2: POLITICAL SENTIMENT

We are given text information on political speeches made by members of Congress. Gentzkow and Shapiro (2010) summarized the text of 529 members of Congress through 1000 word trigrams and their associated frequencies. Furthermore, they collected the Republican vote percentage in the speaker's district from the last election; this is an important variable as it represents the political ideology (conservative/liberal leaning) of the speaker. The data are given in the object *congress109* in the R package **textir**. We have annotated the output and have explained the various steps of the analysis.

```
library(textir)
data(congress109)                    ## 529 speakers 1000 trigrams
dimnames(congress109Counts)
as.matrix(congress109Counts)[1,] ## Chris Cannon's counts
as.matrix(congress109Counts)[,1] ## "gifted.talented.student" counts
congress109Ideology
as.matrix(congress109Ideology)[,1]
repshare=as.matrix(congress109Ideology[,5])
repshare ## Republican vote share

## get to know what is in the matrix

g1=min(as.matrix(congress109Counts)[,])
g2=max(as.matrix(congress109Counts)[,])
g1
g2
## a certain trigram was mentioned by a certain speaker 631 times
```

```
hh=as.matrix(congress109Counts)[,1000]
hh
## here we look at the frequencies of bigram in column 1000

## Multinomial logistic regression and fitted reduction
congmnlm=mnlm(congress109Counts,repshare)
congmnlm
congmnlm$intercept ## estimates of alphas
congmnlm$loadings  ## estimates of betas
fitted(congmnlm)
as.matrix(fitted(congmnlm))[1,] ## fitted counts for first rep
maxf=max(as.matrix(fitted(congmnlm))[1,])
maxf
maxc=max(as.matrix(congress109Counts)[1,])
maxc

## following provides fitted multinomial probabilities
pred=predict(congmnlm,repshare,type="response")
pred[1,] ## predicted multinomial probs for first rep

## following predicts inverse prediction (fitted reduction)
predinv=predict(congmnlm,congress109Counts,type="reduction")
predinv[1:10] ## prints predicted ratings for first 10 reps

[1] 0.6347460 0.6592663 0.2897289 0.1356864 0.5949578 0.3614053 0.6640827
[8] 0.3546728 0.3819283 0.5861080

plot(predinv~repshare)
plot(repshare~predinv)
```

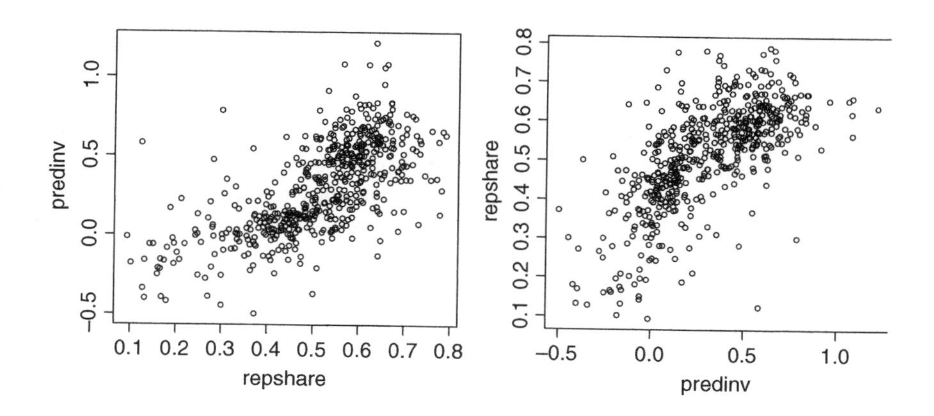


```
corr(predinv,repshare)

          [,1]
[1,] 0.6748002

model1=lm(repshare~predinv)
model1
```

```
Call:
lm(formula = repshare ~ predinv)

Coefficients:
(Intercept)      predinv
     0.4184       0.3120

plot(repshare~predinv)
abline(model1)
```

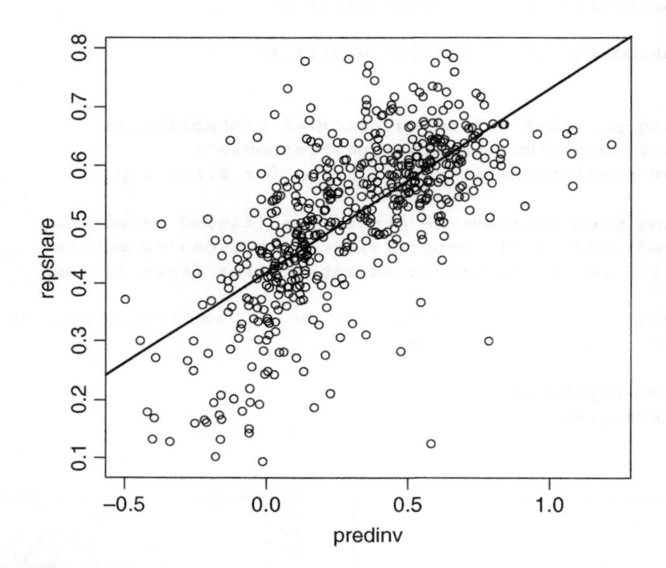

A scatter plot of the Republican vote share on the inverse prediction, with the least squares fit added, shows a linear relationship. Assume a new text leads to an inverse prediction of 0. This value implies that the unknown Republican vote share is about 0.4. We would conclude that this text comes from the liberal side. An inverse prediction of 0.6 implies a Republican vote share of about 0.6; this would tell us that the text comes from the conservative side. A threshold of about 0.3 on the inverse prediction represents a good cutoff for deciding whether a certain text has come from the conservative side.

APPENDIX 19.A RELATIONSHIP BETWEEN THE GENTZKOW/ SHAPIRO ESTIMATE OF "SLANT" AND PARTIAL LEAST SQUARES

The analysis conducted by Gentzkow and Shapiro is actually slightly different. They consider an ordinary regression (not a multinomial logistic regression) of the relative frequencies of trigrams on the vote-shares that they then use, in a second step, to determine the estimate of the political "slant" of each representative.

Their approach is as follows:

Consider the matrix of phrase frequencies $F = \{f_{ij}\}$, where $i = 1, 2, \ldots, n$ (number of rows/representatives) and $j = 1, 2, \ldots, p$ (number of columns/trigrams). The f_{ij} are relative frequencies, with counts normalized by the row sum. For each trigram regress the phrase frequencies (column j of the F matrix, which we denote here by f_j) onto the covariate x (here the Republican vote share), $f_j = a_j + b_j x + \varepsilon$, and obtain the intercept a_j and slope b_j (for $j = 1, 2, \ldots, p$). Then subtract the intercept a_j from each column of frequencies in F, and for each row i, regress the row of this new matrix, $f_{ij} - a_j$, on the b_j. The estimate of x_i in the regression $f_{ij} - a_j = x_i b_j + \varepsilon_j$ is the *slant* of row (representative) i; from elementary regression results, we know that the slants are given by

$$\widehat{x}_i = \frac{\sum_{j=1}^{p} b_j (f_{ij} - a_j)}{\sum_{j=1}^{p} (b_j)^2}, \quad \text{for } i = 1, 2, \ldots, n.$$

It is not difficult to show (Taddy, 2012a) that for *standardized* relative frequencies (with mean 0 and variance 1), the slants are the fitted values in a partial least squares regression of the vector x (the Republican vote share) on the matrix of phrase frequencies $F = \{f_{ij}\}$, using just the first partial least squares direction. See Chapter 18 for a discussion of partial least squares.

We have written an R program for estimating the slant as originally proposed by Gentzkow/Shapiro. We carry out two versions of the slant. One uses the unstandardized relative frequencies, while the other standardizes the relative frequencies to have mean 0 and variance 1. We illustrate that for relative frequencies that have been standardized, the approach by Gentzkow/Shapiro and the PLS regression on the first PLS direction leads to identical results. We use the *congress109* data for illustration.

```
library(textir)
data(congress109) ## data form Gentzkow/Shapiro

## Gentzkow/Shapiro slant (unstandardized relative
## frequencies)
a=dim(529)
b=dim(529)
d=dim(1000)
hh=as.matrix(freq(congress109Counts))
x=congress109Ideology$repshare
for (j in 1:1000) {
m1=lm(hh[,j]~x)
a[j]=m1$coef[1]
b[j]=m1$coef[2]
}
for (i in 1:529) {
d[i]=sum((hh[i,]-a)*b)
}
```

```
cor(d,x)**2

[1] 0.3723954

## Gentzkow/Shapiro slant (standardized relative frequencies)
hh=as.matrix(freq(congress109Counts))
for (j in 1:1000) {
hh[,j]=(hh[,j]-mean(hh[,j]))/sd(hh[,j])
}
x=congress109Ideology$repshare
for (j in 1:1000) {
m1=lm(hh[,j]~x)
a[j]=m1$coef[1]
b[j]=m1$coef[2]
}
for (i in 1:529) {
d[i]=sum((hh[i,]-a)*b)
}
cor(d,x)**2

[1] 0.5665803

## We get a higher correlation beween the slant and the
## covariate x (Republican vote share). Standardization
## helps!!

## Using PLS (textir) on first partial least squares direction
## scaling FALSE means unstandardized relative frequencies
## are used
library(textir)
fit=pls(freq(congress109Counts),
+   congress109Ideology$repshare,scale=FALSE,K=1)
cor(congress109Ideology$repshare,fit$fitted)**2

[1,] 0.5735825
## NOT the same as slant on un-standardized relative
## frequencies

## Using PLS (textir) on first partial least squares direction
## scaling TRUE means standardized relative frequencies
## mean zero and variance 1
library(textir)
fit=pls(freq(congress109Counts),congress109Ideology$repshare,
+   scale=TRUE,K=1)
cor(congress109Ideology$repshare,fit$fitted)**2

[1,] 0.5665803
## SAME as slant on standardized relative frequencies

## Using PLS (mixOmics) on first partial least squares
## direction
```

```
## standardized relative frequencies (mean zero and
## variance 1)
library(mixOmics)
mpls=pls(freq(congress109Counts),congress109Ideology$repshare,
+    ncomp=1,mode="classic",freqCut=0.000001,uniqueCut
+    =0.000001)
x1=mpls$variates$X[,1]
m1=lm(congress109Ideology$repshare~x1)
fmpls=m1$fitted
cor(x,m1$fitted)**2

[1] 0.5665803
## Same as with textir (with scaling TRUE)
## Same as slant on standardized relative frequencies
```

This shows that with standardized relative frequencies, the slants and the fitted values from PLS on the first partial least squares direction are identical.

Note that the equivalence holds for standardized relative word frequencies, but not for nonstandardized relative frequencies. The slants that use nonstandardized relative frequencies are not the same as the fitted values from PLS(1) on nonstandardized relative frequencies.

The PLS approach (which obtains an estimate of the slant, the fitted value of the covariate, from the results of an ordinary regression of phrase frequencies on the covariate) and the inverse multinomial logistic regression approach (which starts from the notion that language is caused by ideology and then inversely predicts the ideology) have much in common.

REFERENCES

Gentzkow, M. and Shapiro, J.: What drives media slant? Evidence from U.S. daily newspapers. *Econometrica*, Vol. 78 (2010), 35–71.

Taddy, M.: Multinomial inverse regression for text analysis. 2012a. Available at http://arxiv.org/abs/1012.2098. Accessed 2013 Jan 18. To appear in Journal of American Statistical Association, Vol. 108 (2013).

Taddy, M.: textir (R package), 2012b.

CHAPTER 20

Network Data

Network data consist of *nodes* (also called *vertices*; such as individuals in a social network, or companies in a trade network) and *edges* (the links between them). The links describe the presence or absence of connections among the nodes. Links can be either directed or undirected. A connection from node x to node y is called *directed* if y is a direct successor of x; in this case the edge from x to y includes an arrow that points to y. For example, student x in a social network may declare student y as his friend (a directed friendship connection pointing from x to y), while student y may not indicate a friendship with student x, implying the absence of an edge pointing from y to x. In a directed network, the graph between x and y is defined as an ordered pair. This differs from an *undirected* graph, where x and y are connected mutual friends (if I am your friend, then you are also my friend). In an undirected network, the graph between x and y is defined as an unordered pair, and the arrows on either side of the edge between x and y can be omitted. Covariate information on the nodes (i.e., characteristics on individuals in a social network, such as gender, age, and race) may affect the absence or presence of connections among the nodes, and such information needs to be incorporated into the analysis of network data. Finding out how a certain covariate affects the links between nodes may be an important aspect of the investigation.

Many good computer programs for visualizing networks are available, such as the **igraph** and **statnet** packages in R, as well as stand-alone packages such as **Gephi** and **Pajek**. Measures of network connectivity can be calculated to summarize important properties of the network, and formal statistical models, similar to but much more complicated than regression models, can be fitted to the data.

A network of nodes and their interconnections can be represented with an *adjacency matrix*. Below, we list the adjacency matrix A for a directed graph with three nodes; the relationships among the nodes are expressed with directed arrows. For a directed network, the adjacency matrix need not be symmetric. The first node in our illustration points to the second node and to the third node. The second node

Data Mining and Business Analytics with R, First Edition. Johannes Ledolter.

points to the first node, and the third node points to the second node.

$$A = \begin{bmatrix} 0 & 1 & 1 \\ 1 & 0 & 0 \\ 0 & 1 & 0 \end{bmatrix}$$

```
library(igraph)
m=matrix(nrow=3,ncol=3)
m[1,1]=0
m[1,2]=1
m[1,3]=1
m[2,1]=1
m[2,2]=0
m[2,3]=0
m[3,1]=0
m[3,2]=1
m[3,3]=0
m
```

```
     [,1] [,2] [,3]
[1,]    0    1    1
[2,]    1    0    0
[3,]    0    1    0
```

```
lab=c(1,2,3)
object <- graph.adjacency(m,mode="directed")
set.seed(1)
plot(object,vertex.label=lab)
```

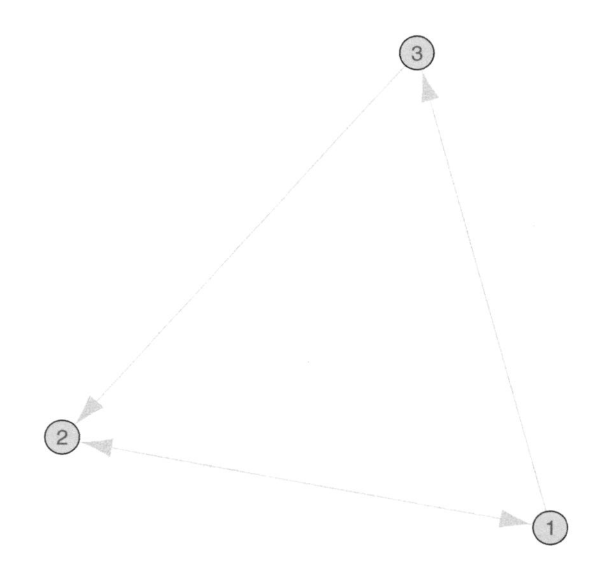

20.1 EXAMPLE 1: MARRIAGE AND POWER IN FIFTEENTH CENTURY FLORENCE

Early Renaissance Florence was ruled by an oligarchy of powerful families. By the fifteenth century, the Medicis emerged supreme, and the Medici Bank became the largest in Europe. How did the Medici win?

Political ties were established via marriage. The adjacency matrix for the 16 most powerful families in Florence shown in the following includes a one whenever two families were connected in marriage. This adjacency matrix represents an undirected graph. A graphical representation of the network, created with the R package **igraph**, is shown.

The data set analyzed here has been compiled from extensive data collected by John Padgett. The data includes information on families who were locked in a struggle for political control of the city of Florence around 1430. Two factions were dominant in this struggle: one revolved around the infamous Medicis, the other around the powerful Strozzis. The data can be downloaded from http://vlado.fmf.uni-lj.si/pub/networks/data/ucinet/ucidata.htm#padgett. The papers by Padgett (1994), Padgett and Ansel (1993), Breiger and Pattison (1986), and Wasserman and Faust (1994) provide a detailed discussion of this fascinating data set that has been constructed from historical documents. All we see here is a simple adjacency matrix of zeros and ones, but not the enormous research effort that must have gone into constructing this matrix.

```
library(igraph) ## load the package
## read the data
florence <- as.matrix(read.csv("C:/DataMining/Data/firenze.csv"))
florence
```

```
             Acciaiuoli Albizzi Barbadori Bischeri Castellani Ginori Guadagni
Lamberteschi Medici Pazzi Peruzzi Pucci Ridolfi Salviati Strozzi Tornabuoni
Acciaiuoli     0  0  0  0  0  0  0  0  1  0  0  0  0  0  0  0
Albizzi        0  0  0  0  0  1  1  0  1  0  0  0  0  0  0  0
Barbadori      0  0  0  0  1  0  0  0  1  0  0  0  0  0  0  0
Bischeri       0  0  0  0  0  0  1  0  0  0  1  0  0  0  1  0
Castellani     0  0  1  0  0  0  0  0  0  0  1  0  0  0  1  0
Ginori         0  1  0  0  0  0  0  0  0  0  0  0  0  0  0  0
Guadagni       0  1  0  1  0  0  0  1  0  0  0  0  0  0  0  1
Lamberteschi   0  0  0  0  0  0  1  0  0  0  0  0  0  0  0  0
Medici         1  1  1  0  0  0  0  0  0  0  0  0  1  1  0  1
Pazzi          0  0  0  0  0  0  0  0  0  0  0  0  0  1  0  0
Peruzzi        0  0  0  1  1  0  0  0  0  0  0  0  0  0  1  0
Pucci          0  0  0  0  0  0  0  0  0  0  0  0  0  0  0  0
Ridolfi        0  0  0  0  0  0  0  0  1  0  0  0  0  0  1  1
Salviati       0  0  0  0  0  0  0  0  1  1  0  0  0  0  0  0
Strozzi        0  0  0  1  1  0  0  0  0  0  1  0  1  0  0  0
Tornabuoni     0  0  0  0  0  0  1  0  1  0  0  0  1  0  0  0
```

```
marriage <- graph.adjacency(florence,mode="undirected", diag=FALSE)
## use the help function to understand the options for the graph
set.seed(1)
plot(marriage,layout=layout.fruchterman.reingold,
```

```
+     vertex.label=V(marriage)$name,vertex.color="red",
+     vertex.label.color="black",vertex.frame.color=0,
+     vertex.label.cex=1.5)
```

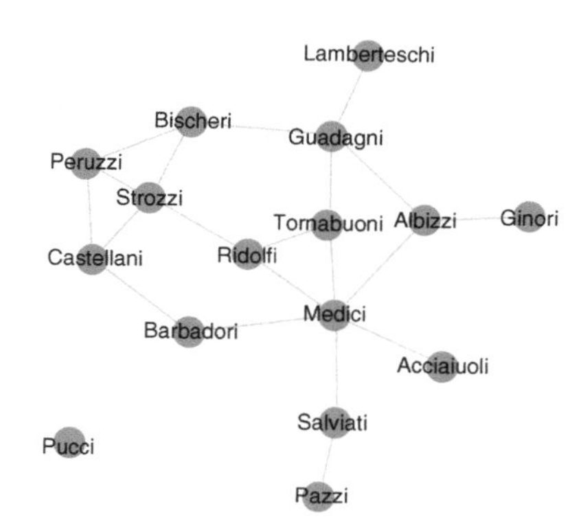

The presence and absence of network links can be used to measure network connectivity and "social importance." A node's *degree* in an undirected network is defined as its number of edges to other nodes. Medicis are connected. There are connections (edges) between the Medici and six other families, hence the degree is 6 for the Medici. The Lambertschi family, on the other hand, is connected to only one other family (the Guadagni); the Lambertschi family has degree one.

```
data.frame(V(marriage)$name,degree(marriage))
```

```
   V.marriage..name degree.marriage.
1        Acciaiuoli                1
2           Albizzi                3
3         Barbadori                2
4          Bischeri                3
5        Castellani                3
6            Ginori                1
7          Guadagni                4
8       Lamberteschi               1
9            Medici                6
10            Pazzi                1
11          Peruzzi                3
12            Pucci                0
13          Ridolfi                3
14         Salviati                2
15          Strozzi                4
16       Tornabuoni                3
```

A deeper measure of network structure is obtained through *betweenness*. Betweenness is a centrality measure of a node (or vertex) within a graph. Nodes that occur on many shortest paths between other nodes have higher betweenness than those that do not.

For an undirected graph with n nodes (vertices), the betweenness of node k is computed as follows:

1. For each unordered pair of nodes (i, j), compute all shortest paths between them. The shortest path is the path with the fewest steps between node i and node j.
2. For each unordered pair of nodes (i, j), determine the fraction of shortest paths that pass through the node in question (here, node k).
3. Sum this fraction over all unordered pairs of nodes (i, j).
4. Say $s_k(i,j)$ is the proportion of shortest paths between node i and node j containing node k. Then,

$$\text{betweenness}(k) = \sum_{(i<j):i\neq j, k\notin\{i,j\}} s_k(i,j).$$

As in degree, betweenness measures network connectivity of one node relative to others. Betweenness may be normalized, dividing the earlier expression by the number of unordered pairs of nodes not including k, which for an undirected graph is $(n-1)(n-2)/2$. For illustration, consider an undirected star graph. This is a graph with a center node that is connected to all other nodes, while none of the other nodes (the leaves in the star graph) are connected among themselves. Each shortest path between two star points goes through the center point; the betweenness of the center point is $(n-1)(n-2)/2$ (or 1, if normalized). The leaves in a star graph are not contained in shortest paths, and have betweenness 0.

Betweenness measures how much influence a node has over connections between other nodes. It measures total graph connectivity, rather than counting next door neighbors.

Comment: In directed graphs, the degree of a node is the number of links that point to or from the node. In directed graphs, $\text{betweenness}(k) = \sum_{(i,j):i\neq j, k\notin\{i,j\}} s_k(i,j)$, summing over all ordered pairs of nodes as the shortest path from node i to node j may be different from the shortest path from node j to node i. Normalization is achieved by dividing the sum by the number of ordered pairs of nodes not including k, which is $(n-1)(n-2)$.

Below, we illustrate betweenness for the nodes in the undirected network among the 16 families in Medevial Florence. Take Peruzzi, as example, with betweenness 2. It arises from the connections between Bisheri and Castellani (two shortest paths involving three nodes, with one going through Peruzzi), Bisheri and Barbadori (two shortest paths involving four nodes, with one going through Peruzzi), Guadagni and Castellani (two shortest paths involving four nodes, with one path going through Perruzi), and Lamberteschi and Castellani (two shortest paths involving five nodes with one going through Peruzzi). Hence betweenness for Peruzzi is $(1/2) + (1/2) + (1/2) + (1/2) = 2$.

```
## calculate and plot the shortest paths
V(marriage)$color <- 8
E(marriage)$color <- 8
PtoA <- get.shortest.paths(marriage, from="Peruzzi", to="Acciaiuoli")
E(marriage, path=PtoA[[1]])$color <- "magenta"
V(marriage)[PtoA[[1]] ]$color <- "magenta"
GtoS <- get.shortest.paths(marriage, from="Ginori", to="Strozzi")
E(marriage, path=GtoS[[1]])$color <- "green"
V(marriage)[ GtoS[[1]] ]$color <- "green"
V(marriage)[ "Medici" ]$color <- "cyan"

set.seed(1)
plot(marriage,  layout=layout.fruchterman.reingold,
+     vertex.label=V(marriage)$name,vertex.label.color="black",
+     vertex.frame.color=0, vertex.label.cex=1.5)
```

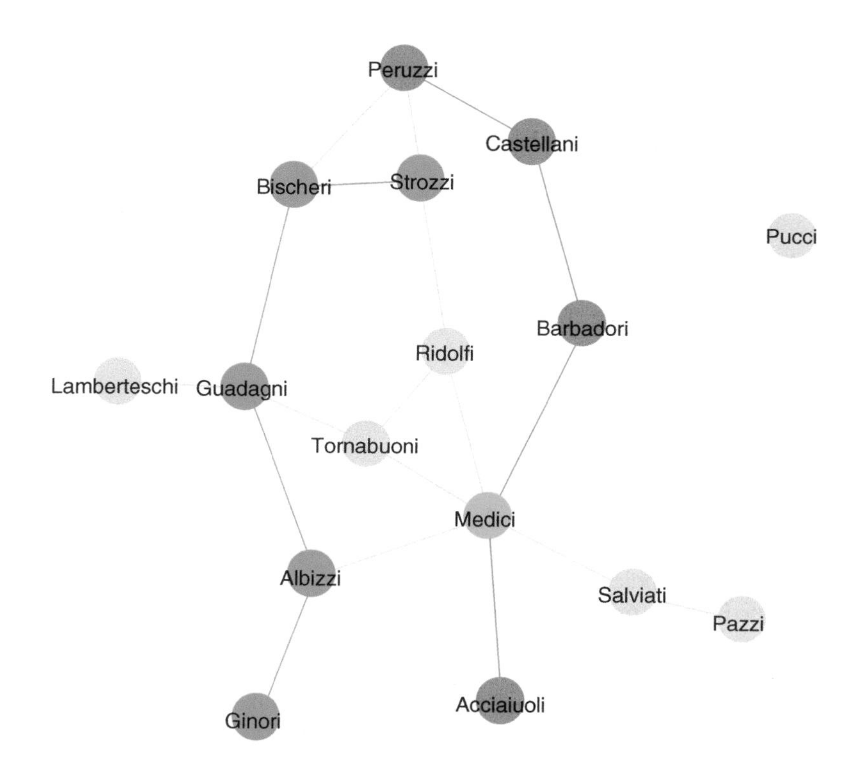


```
data.frame(V(marriage)$name, betweenness(marriage))

   V.marriage..name betweenness.marriage.
1        Acciaiuoli              0.000000
2           Albizzi             19.333333
3         Barbadori              8.500000
```

```
4           Bischeri         9.500000
5         Castellani         5.000000
6             Ginori         0.000000
7           Guadagni        23.166667
8       Lamberteschi         0.000000
9             Medici        47.500000
10             Pazzi         0.000000
11           Peruzzi         2.000000
12             Pucci         0.000000
13           Ridolfi        10.333333
14          Salviati        13.000000
15           Strozzi         9.333333
16        Tornabuoni         8.333333
```

Betweenness versus Degree. The Medici have the highest degree (largest number of edges), but only by a factor of 3/2 over the Strozzi's. But the Medici's betweenness (47.5) is five times higher than that of the Strozzi (9.33). Betweenness measures total graph connectivity, rather than counting the next door neighbors.

Burt's Structural Holes. Burt (2004) discusses social capital and how it can lead to brokerage opportunities. His ideas are similar than betweenness, but he puts extra weight on connections to isolated nodes. Isolated nodes, or "holes" can act like bottlenecks in companies, and they lead to unexpected employees having excess power and influence. But if you are the employee, it may be a fast track to promotion.

20.2 EXAMPLE 2: CONNECTIONS IN A FRIENDSHIP NETWORK

The second example represents a simulation of an in-school friendship network for a school community in the rural western United States, with a student body that is largely Hispanic and Native American. The network object *faux.mesa.high*, included in the R library **statnet**, has 205 vertices (students, in this case) and 203 undirected edges (mutual friendships). The vertex (node) attributes are Grade, Sex, and Race. The Grade attribute has values 7 through 12, indicating each student's grade in school. The Race attribute is based on the answers to two questions, one on Hispanic identity and one on race, and takes six possible values: White, Black, Hispanic, Asian, Native American, and Other. See Resnick et al. (1997).

The R package **statnet** (Handcock et al., 2008) contains much useful software for statistical network analysis. The package includes tools for the informative graphical display of networks (network visualization) and programs for model estimation, model evaluation, and model-based network simulation. There is a large and growing literature on statistical models for network data, and **statnet** implements recent advances on *e*xponential-family *r*andom *g*raph *m*odels (in short, **ergm**;

also an R package). We use **statnet** for the analysis of this example, but will not discuss the formal statistical models in this brief introduction.

```
library(statnet)

data(faux.mesa.high)              ## load the network object
summary(faux.mesa.high)           ## summarize the data set

Network attributes:
  vertices = 205
  directed = FALSE                undirected graph
  hyper = FALSE
  loops = FALSE
  multiple = FALSE
  bipartite = FALSE
  total edges = 203               in the undirected graph
  missing edges = 0
  non-missing edges = 203
  density = 0.009708274           203/(205*204/2)

Vertex attributes:

 Grade:
   numeric valued attribute
   attribute summary:
   Min. 1st Qu.  Median    Mean 3rd Qu.    Max.
  7.000   7.000   9.000   8.732  10.000  12.000

 Race:
   character valued attribute
   attribute summary:
Black  Hisp NatAm Other White
    6   109    68     4    18

 Sex:
   character valued attribute
   attribute summary:
  F   M
 99 106

No edge attributes

Network edgelist matrix:
       [,1] [,2]
  [1,]   25    1   node 25 and node 1 are connected
  [2,]   52    1   node 52 and node 1 are connected
  [3,]   58    1
. . .
. . .
```

```
[201,]  182  181
[202,]  190  183
[203,]  191  189   node 191 and node 189 are connected
```

There are 203 undirected connections among the 205 nodes. The edge list summarizes the unordered pairs of nodes that are connected. The 205×205 symmetric adjacency matrix has zeros in the diagonal (as a node cannot be connected to itself), and many zeros and (2)(213) ones as its off-diagonal elements. The network density is given by $(2)(203)/[(205)(204)] = 203/[(205)(204)/2] = 0.0097$. Only about 1% of all possible network connections are realized.

```
lab=network.vertex.names(faux.mesa.high)=c(1:205)
## assigns numbers to nodes
grd=faux.mesa.high%v%"Grade"
sx=faux.mesa.high%v%"Sex"
race=faux.mesa.high%v%"Race"
## we don't look at race in this example
vs=c(4,12)[match(sx,c("M","F"))]
## used for graph later on; boys by square (4 sides);
## girls by 12-sided
col=c(6,5,3,7,4,2)                  ## used for graph later on
as.sociomatrix(faux.mesa.high)      ## gives adjacency matrix
faux.mesa.high[1,]
faux.mesa.high[5,]
faux.mesa.high[,3]
m=faux.mesa.high[,]                 ## adjacency matrix
network.density(faux.mesa.high)
## density of network = NuEdges/[nodes*(nodes-1)/2]
```

```
[1] 0.009708274
```

The degree of a certain node in an undirected network is obtained by adding the number of connections that exist between the node and all other nodes of the network. Statnet, in its calculation of degree, counts an edge between nodes i and j twice as it considers both the direction from and the direction to the node. For an undirected network we divide the degrees that are obtained by statnet by 2, to make the results consistent with our earlier definition and the results from igraph. The same adjustment needs to be made for betweenness as statnet sums over all paths that move from one node to the other and vice versa.

```
deg=degree(faux.mesa.high)/2
## degree of network nodes (number of connections)
## Statnet double-counts the connections in an undirected network
## Edge between nodes i and j in an undirected network is counted
## twice
## We divide by 2 in order to make the results consistent with our
```

```
## discussion in the text and the output from igraph (in Example 1)
deg

  [1] 13  4  0  0  1  0  0  3  4  0  2  0  2  1  4  2  1  4  1  0  3  6  1  0  7
 [26]  0  2  0  2  4  2  1  1  2  0  3  0  1  0  1  0  0  3  3  0  0  9  0  0  0
 [51]  3  4  2  3  9  2  1  2  3  2  3  0  2  5  2  4  0  1  0  3  1  0  0  5  2
 [76]  2  2  1  5  0  1  1  2  0  0  1 10  2  2  1  1  4  1  0  0  7  1  1  3  5
[101]  1  4  2  5  2  0  0  3  2  5  1  1  0  3  3  0  1  0  0  0  1  1  7  3  1
[126]  0  5  1  3  0  2  2  1  3  0  5  4  3  6  5  1  3  0  1  0  1  1  1  3  4
[151]  2  0  3  0  1  1  4  4  0  8  3  0  0  3  4  1  1  0  0  1  0  0  4  1  0
[176]  1  0  3  4  1  1  2  3  0  5  1  4  0  7  4  2  2  1  3  3  2  0  1  1  1
[201]  2  1  0  3  1

betw=betweenness(faux.mesa.high)/2
## betweenness of network
## Statnet double-counts the betweenness in an undirected network
## We divide by 2 in order to make the results consistent with our
## discussion in the text and the output from igraph
betw

  [1] 3661.957143   46.033333    0.000000    0.000000    0.000000    0.000000
  [7]    0.000000    0.000000  235.000000    0.000000    0.000000    0.000000
 [13]   38.625000    0.000000   92.809524  118.000000    0.000000  235.500000
 [19]    0.000000    0.000000 1401.026190  189.080952    0.000000    0.000000
 [25]  463.500000    0.000000    1.000000    0.000000    0.000000  678.000000
 [31]    0.000000    0.000000    0.000000  118.000000    0.000000    3.000000
 [37]    0.000000    0.000000    0.000000    0.000000    0.000000    0.000000
 [43]    0.500000    2.000000    0.000000    0.000000 1858.955952    0.000000
 [49]    0.000000    0.000000    2.000000   54.428571    0.000000  570.000000
 [55]  580.067857  118.000000    0.000000 2037.973810  263.000000    0.000000
 [61]   80.591667    0.000000  118.000000  117.273810    0.000000  119.450000
 [67]    0.000000    0.000000    0.000000  888.000000    0.000000    0.000000
 [73]    0.000000  792.000000  348.000000    0.000000    0.000000    0.000000
 [79]  572.000000    0.000000    0.000000    0.000000    0.000000    0.000000
 [85]    0.000000    0.000000  405.816667    0.000000    2.000000    0.000000
 [91]    0.000000    3.333333    0.000000    0.000000    0.000000  490.466667
 [97]    0.000000    0.000000    0.000000  397.661905    0.000000 1388.270238
[103]    1.000000  288.000000    0.000000    0.000000    0.000000  118.000000
[109]  118.000000  118.416667    0.000000    0.000000    0.000000    1.000000
[115]  118.000000    0.000000    0.000000    0.000000    0.000000    0.000000
[121]    0.000000    0.000000 2032.525000 1606.026190    0.000000    0.000000
[127]  297.750000    0.000000  234.000000    0.000000    2.000000  678.000000
[133]    0.000000   42.633333    0.000000  350.000000  119.166667  118.000000
[139]  647.722619 1295.026190    0.000000  234.000000    0.000000    0.000000
[145]    0.000000    0.000000    0.000000    0.000000 2074.973810  785.500000
[151]    0.000000    0.000000  193.158333    0.000000    0.000000    0.000000
[157]   47.569048  505.500000    0.000000 1891.026190 1635.442857    0.000000
[163]    0.000000    0.000000 2442.973810    0.000000    0.000000    0.000000
[169]    0.000000    0.000000    0.000000    0.000000    2.000000    0.000000
[175]    0.000000    0.000000    0.000000 1930.973810   84.516667    0.000000
[181]    0.000000    1.000000   25.916667    0.000000  575.000000    0.000000
[187]  798.000000    0.000000 1058.600000  158.259524    0.000000    1.000000
[193]    0.000000  118.000000  168.000000    2.000000    0.000000    0.000000
[199]    0.000000    0.000000    0.000000    0.000000    0.000000  235.000000
[205]    0.000000

plot(deg)
plot(betw)
hist(deg,breaks=c(-0.5,0.5,1.5,2.5,3.5,4.5,5.5,6.5,7.5,
+    8.5,9.5,10.5,11.5,12.5,13.5))
```

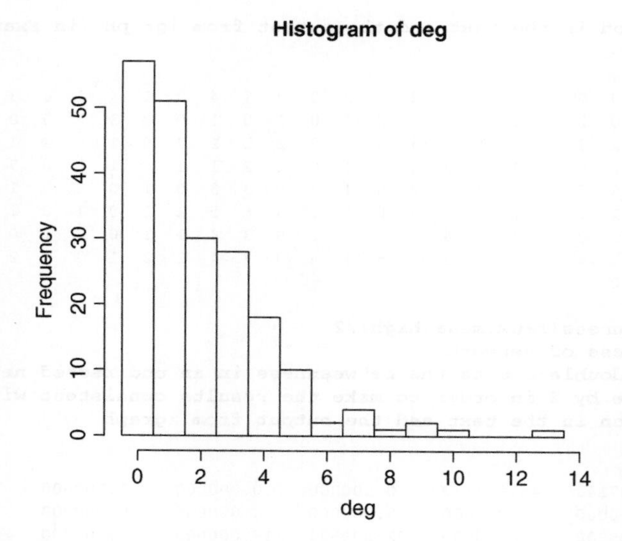

A couple of students (e.g., the first student on the list, a female seventh grader, with 13 connections) are well connected, but most students have relatively few connections. More than 25% of the students are "singletons" (i.e., have no connections to other nodes), and about 50% of the students have at most one connection.

```
plot(deg,betw)
```

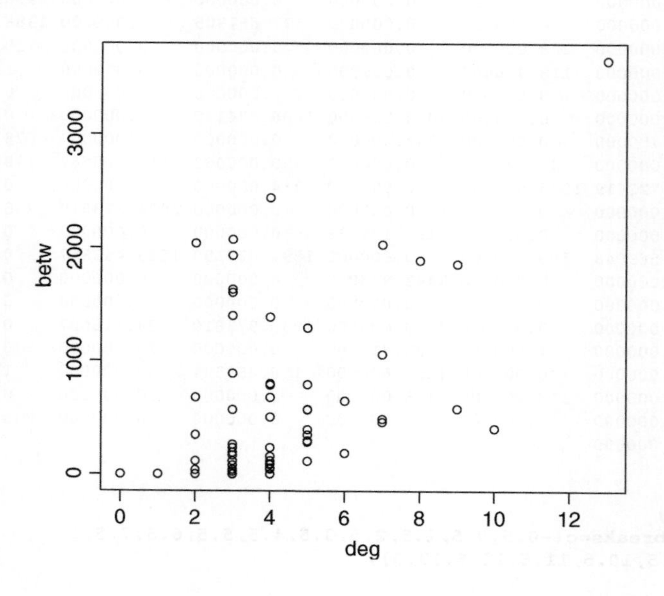

```
boxplot(deg~grd)
boxplot(deg~sx)
```

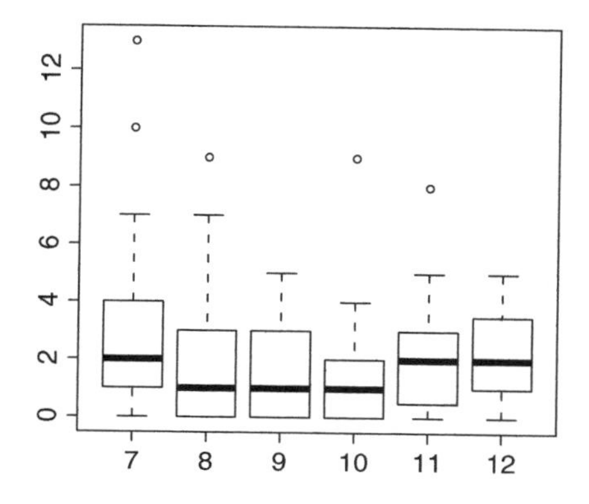

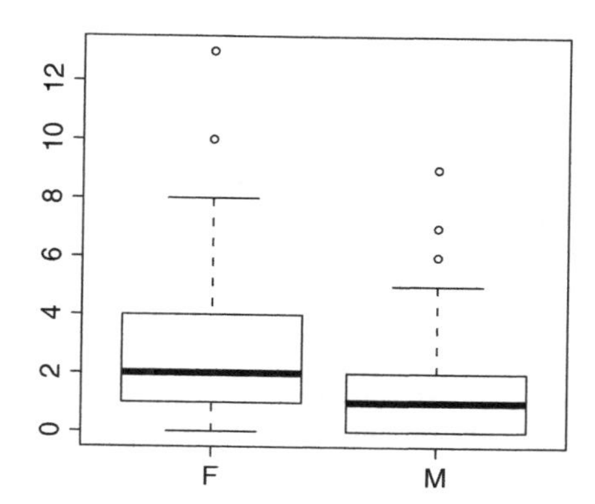

Females have more connections than males, and eighth to tenth graders have fewer connections than entering seventh graders and students who are close to graduation.

We have already shown how to obtain the adjacency matrix from a given statnet network object. Below we show how to export the edge list. We also show how to create a statnet network object from first principles when given the edge list (or adjacency matrix).

```
## faux.mesa.high is already a network object
## below we illustrate how to create an undirected network
## from the edge list
## first we obtain the edge list of a network object
attributes(faux.mesa.high)
vv=faux.mesa.high$mel
edge=matrix(nrow=203,ncol=2)
for (i in 1:203) {
vvv=vv[[203+i]]
edge[i,1]=vvv$inl
edge[i,2]=vvv$outl
}
edge
## edge contains the edge list
## in an undirected network, edge information is stored in the
## second half of faux.mesa.high$mel
faux1=network(edge,directed=FALSE,matrix.type="edgelist")
faux1
```

```
faux1[,]
deg=degree(faux1)/2
betw=betweenness(faux1)/2
plot(deg)
plot(betw)
plot(deg,betw)

## faux.mesa.high is already a network object
## below we illustrate how to create an undirected network
## from the adjacency matrix
## the adjacency matrix had been stored previously in m
faux2=network(m,directed=FALSE,matrix.type="adjacency")
faux2
faux2[,]
deg=degree(faux2)/2
betw=betweenness(faux2)/2
plot(deg)
plot(betw)
plot(deg,betw)
```

Visual displays of the network, with and without information on node attributes, are shown as follows:

```
## visual display of the network
set.seed(654)            ## to get reproducible graphs
plot(faux.mesa.high)
## generic graph without labels/covariates
set.seed(654)            ## to get reproducible graphs
plot(faux.mesa.high,label=lab) ## generic graph with labels
```

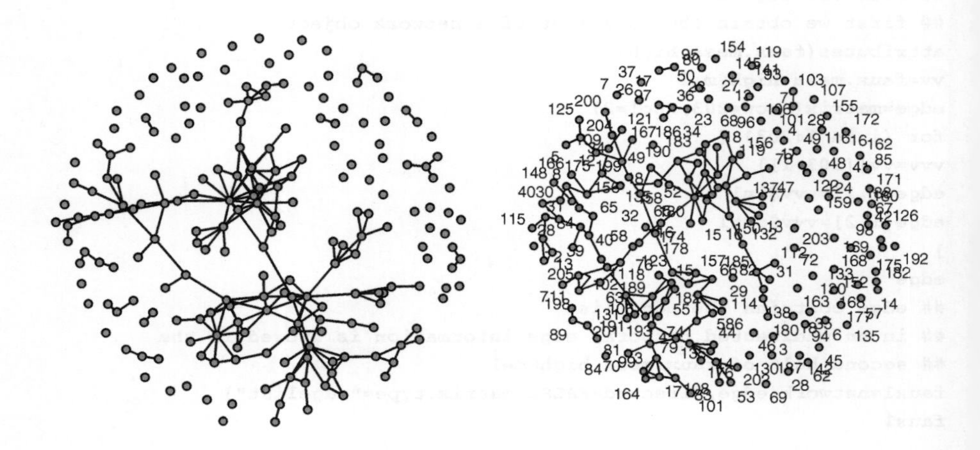

```
set.seed(654)          ## to get reproducible graphs
plot(faux.mesa.high,vertex.sides=vs,vertex.rot=45,
+    vertex.cex=2,vertex.col=col[grd-6],edge.lwd=2,
+    cex.main=3,displayisolates=FALSE)
legend("bottomright",legend=7:12,fill=col,cex=0.75)
## 45 rotates square
## isolates are not displayed
```

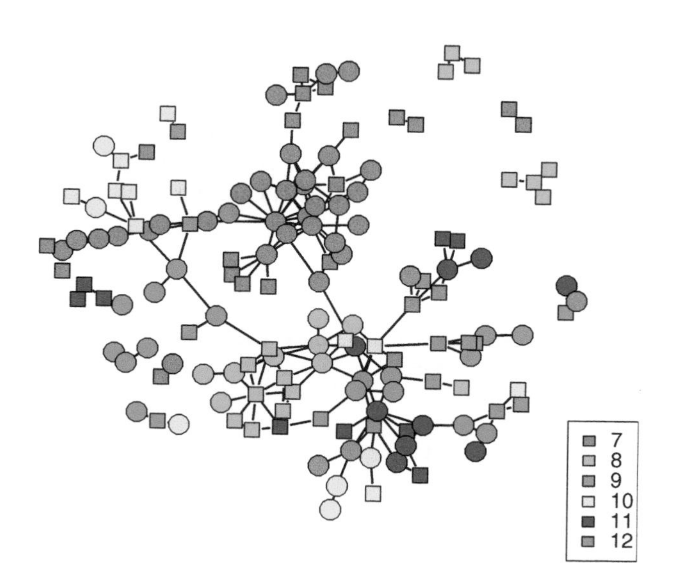

The last graph is quite informative. Females are represented by circles (12-sided objects) and males by squares; grade is indicated by color. Students, especially those in lower grades, interact mostly with students from the same grade; links with students from other grades are not that common. We notice a few isolated small groups, usually consisting of students from the same grade and gender, with very few ties to other students. A couple of students (e.g., student 1, a female seventh grader, with 13 connections) are well connected, but most students have relatively few connections. More than 25% of the students are isolated ("singletons"), and about 50% of the students have at most one connection. Note that "singletons" are not displayed on the last graph.

Let us look at the connections more closely and let us investigate whether connections are affected by the node characteristics, grade and gender. While we can use statistical models to estimate the impact of grade and gender (and the R package **statnet** can be used for this), the following descriptive analysis will be sufficient for our purpose.

Below, we use the adjacency matrix to calculate network densities for various subgroups. We measure the density of the

- interaction among students from a certain grade with students from the same grade;
- interaction among students from a certain grade with students from all grades;
- interaction among students from a certain gender group with students of the same gender;
- interaction among students from a certain gender group with students of either gender.

The network among females is more "dense" than the network among males (0.0169 vs 0.0089). Ties between female students and students of either gender are more prevalent than ties between male students and students of either gender (0.0116 vs 0.0079). The density among students within the same grade is largest for twelfth graders (0.0909). When considering connections with students from all grades, the densities for seventh graders (0.0121) and for twelfth graders (0.0114) are about the same.

```
## density of interaction among students from the
## same grade (ignoring gender)

m1=m[grd==7,grd==7]
sum(m1)/(nrow(m1)*(ncol(m1)-1))

[1] 0.03966155

m1=m[grd==8,grd==8]
sum(m1)/(nrow(m1)*(ncol(m1)-1))

[1] 0.04230769

m1=m[grd==9,grd==9]
sum(m1)/(nrow(m1)*(ncol(m1)-1))

[1] 0.02671312

m1=m[grd==10,grd==10]
sum(m1)/(nrow(m1)*(ncol(m1)-1))

[1] 0.03

m1=m[grd==11,grd==11]
sum(m1)/(nrow(m1)*(ncol(m1)-1))

[1] 0.0615942

m1=m[grd==12,grd==12]
```

```
sum(m1)/(nrow(m1)*(ncol(m1)-1))

[1] 0.09090909

## density of interaction among students from a given grade
## with students from all grades (ignoring gender)
## matrix m1 shown below is not square; it has r rows and
## c columns
## the c columns include the r nodes that determine the
## rows of m1. The number of possible edges in m1 are
## r(r-1) + r(c-r) = r(c-1)

m1=m[grd==7,]
sum(m1)/(nrow(m1)*(ncol(m1)-1))

[1] 0.01209677

m1=m[grd==8,]
sum(m1)/(nrow(m1)*(ncol(m1)-1))

[1] 0.009191176

m1=m[grd==9,]
sum(m1)/(nrow(m1)*(ncol(m1)-1))

[1] 0.007586368

m1=m[grd==10,]
sum(m1)/(nrow(m1)*(ncol(m1)-1))

[1] 0.007058824

m1=m[grd==11,]
sum(m1)/(nrow(m1)*(ncol(m1)-1))

[1] 0.01000817

m1=m[grd==12,]
sum(m1)/(nrow(m1)*(ncol(m1)-1))

[1] 0.01143791

## density of interaction among students from the
## same gender group (ignoring grade)

m1=m[sx=="F",sx=="F"]
sum(m1)/(nrow(m1)*(ncol(m1)-1))

[1] 0.01690373
```

```
m1=m[sx=="M",sx=="M"]
sum(m1)/(nrow(m1)*(ncol(m1)-1))

[1] 0.008984726

## density of interaction among students from a given gender
## group with students of either gender (ignoring grade)

m1=m[sx=="F",]
sum(m1)/(nrow(m1)*(ncol(m1)-1))

[1] 0.01163597

m1=m[sx=="M",]
sum(m1)/(nrow(m1)*(ncol(m1)-1))

[1] 0.00790788
```

Furthermore, densities can be stratified according to gender. For example, the network among seventh graders of the same gender is four times "denser" for females (0.0823) than it is for males (0.0228). Also the network between seventh graders and same gender students of all grades is about four times "denser" for females (0.0291) than it is for males (0.0060). The gender differences become considerably weaker for twelfth graders. The density of the network between twelfth graders and same gender students of all grades for females (0.0117) and males (0.0114) are similar.

```
## density of interaction among students from the
## same grade, for given gender

## female seventh graders
m1=m[sx=="F",sx=="F"]
grd1=grd[sx=="F"]
m2=m1[grd1==7,grd1==7]
sum(m2)/(nrow(m2)*(ncol(m2)-1))

[1] 0.08235294

## male seventh graders
m1=m[sx=="M",sx=="M"]
grd1=grd[sx=="M"]
m2=m1[grd1==7,grd1==7]
sum(m2)/(nrow(m2)*(ncol(m2)-1))

[1] 0.02279202
```

```
## female twelfth graders
m1=m[sx=="F",sx=="F"]
grd1=grd[sx=="F"]
m2=m1[grd1==12,grd1==12]
sum(m2)/(nrow(m2)*(ncol(m2)-1))

[1] 0.04761905

## male twelfth graders
m1=m[sx=="M",sx=="M"]
grd1=grd[sx=="M"]
m2=m1[grd1==12,grd1==12]
sum(m2)/(nrow(m2)*(ncol(m2)-1))

[1] 0.1

## density of interaction among students from a given grade
## with students from all grades, for given gender

## female seventh graders
m1=m[sx=="F",sx=="F"]
grd1=grd[sx=="F"]
m2=m1[grd1==7,]
sum(m2)/(nrow(m2)*(ncol(m2)-1))

[1] 0.02915452

## male seventh graders
m1=m[sx=="M",sx=="M"]
grd1=grd[sx=="M"]
m2=m1[grd1==7,]
sum(m2)/(nrow(m2)*(ncol(m2)-1))

[1] 0.005996473

## female twelfth graders
m1=m[sx=="F",sx=="F"]
grd1=grd[sx=="F"]
m2=m1[grd1==12,]
sum(m2)/(nrow(m2)*(ncol(m2)-1))

[1] 0.01166181

## male twelfth graders
m1=m[sx=="M",sx=="M"]
grd1=grd[sx=="M"]
m2=m1[grd1==12,]
sum(m2)/(nrow(m2)*(ncol(m2)-1))
```

```
[1] 0.01142857
```

A further comment: There are many different ways of plotting the connections of a network, and plots of the very same adjacency matrix may look quite different depending on how nodes are located on the graph. So, even a simple visual display of a network structure makes some assumptions. We illustrate this by visualizing the same network as drawn according to three commonly adopted design principles (Fruchterman and Reingold, the design that is usually recommended; Kamada-Kawai; and the circle arrangement, which at least in this case, is not a very informative design criterion).

```
## Plotting options. Not that easy. Pictures look differently
## Principles of Fruchterman/Reingold:
##     Distribute vertices evenly in the frame
##     Minimize the number of edge crossings
##     Make edge lengths uniform
##     Reflect inherent symmetry
##     Conform to the frame

set.seed(654)       ## to get reproducible graphs
plot(faux.mesa.high,mode="fruchtermanreingold",
+    label=lab,vertex.sides=vs,vertex.rot=45,
+    vertex.cex=2.5,vertex.col=col[grd-6],edge.lwd=2,
+    cex.main=3,displayisolates=FALSE)
legend("bottomright",legend=7:12,fill=col,cex=0.75)
```

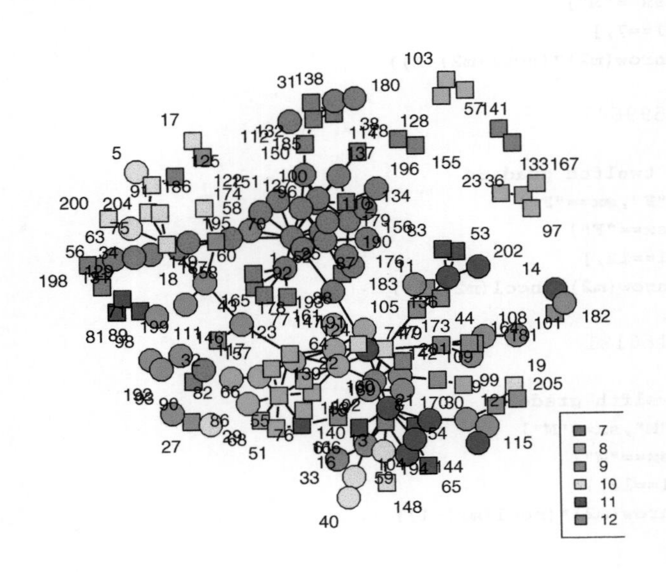

```
set.seed(654)      ## to get reproducible graphs
plot(faux.mesa.high,mode="kamadakawai",label=lab,
+     vertex.sides=vs,vertex.rot=45,vertex.cex=2.5,
+     vertex.col=col[grd-6],edge.lwd=2,cex.main=3,
+     displayisolates=FALSE)
legend("bottomright",legend=7:12,fill=col,cex=0.75)
```

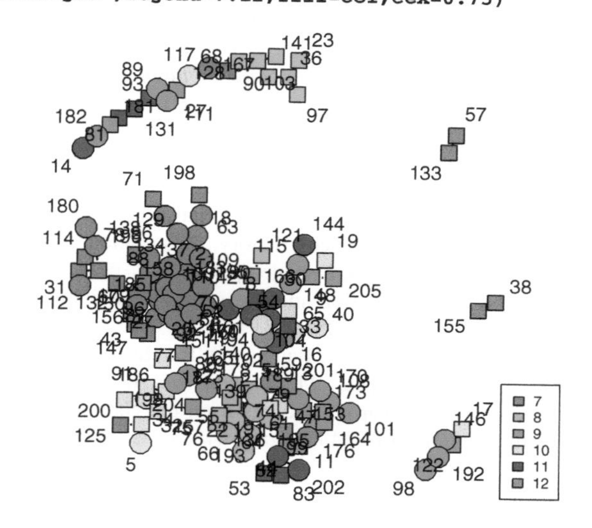


```
set.seed(654)      ## to get reproducible graphs
plot(faux.mesa.high,mode="circle",label=lab,
+     vertex.sides=vs,vertex.rot=45,vertex.cex=2.5,
+     vertex.col=col[grd-6],edge.lwd=2,cex.main=3,
+     displayisolates=FALSE)
legend("bottomright",legend=7:12,fill=col,cex=0.75)
```

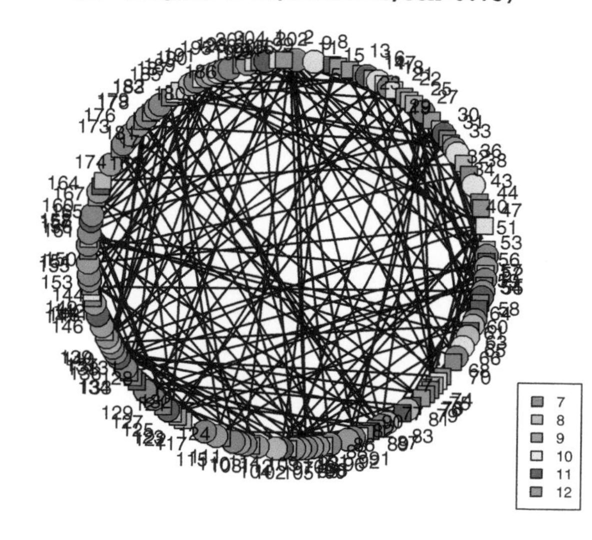

The R program that has created this output can be found on the web page that accompanies this book.

REFERENCES

Breiger, R. and Pattison, P.: Cumulated social roles: The duality of persons and their algebras. *Social Networks*, Vol. 8 (1986), 215–256.

Burt, R.S.: Structural holes and good ideas. *American Journal of Sociology*, Vol. 110 (2004), No 2, 349–399.

Handcock, M.S., Hunter, D.R., Butts, C.T., Goodreau, S.M., and Morris, M.: Statnet: Software tools for the representation, visualization, analysis and simulation of network data. *Journal of Statistical Software*, Vol. 24 (2008) No. 1, 1–11.

Padgett, J.F.: Marriage and elite structure in Renaissance Florence, 1282–1500. Paper delivered to the Social Science History Association, 1994. Available at http://www4.ncsu.edu/~gremaud/MA432/padgett.pdf.

Padgett, J.F. and Ansel, C.K.: Robust action and rise of the Medici 1400–1434. *American Journal of Sociology*, Vol. 98 (1993), 1259–1319.

Resnick, M.D., Bearman, P.S., Blum, R.W., Bauman, K.E., Harris, K.M., Jones, J., Tabor, J., Beuhring, T., Sieving, R.E., Shew, M., Ireland, M., Bearinger, L.H., and Udry, J.R.: Protecting adolescents from harm. Findings from the National Longitudinal Study on Adolescent Health. *Journal of the American Medical Association*, Vol. 278 (1997), 823–832.

Wasserman, S. and Faust, K.: *Social Network Analysis: Methods and Applications*. Cambridge, England: Cambridge University Press, 1994.

APPENDIX A

Exercises

In this appendix I list exercises that should help (i) the reader to master the material and (ii) the instructor using this book to assess student learning. The problems under Exercise 1 should be assigned immediately after having studied each chapter. Exercises 2 through 4 address the analyses of several large data sets: eight data sets from Jank (2011), three data sets from Williams (2011), and several data sets from the annual Data Mining and Knowledge Discovery competitions organized by the ACM Special Interest Group on Knowledge Discovery and Data Mining. The data sets can be used to practice the material of Chapter 2 (on obtaining relevant graphical displays and numerical summaries) and the modeling tools that are discussed in the subsequent chapters. Regression, regression trees, and LASSO methods predict continuous outcome measures, and the quality of the predictions is assessed on evaluation data sets or through cross-validation. Logistic and multinomial logistic regression, classification trees, naïve Bayesian methods, nearest neighbor methods, and discriminant analysis classify observations on categorical outcome variables. Clustering methods divide units into homogeneous groups, and tools such as principal components analysis help reduce the dimensionality the data. Two smaller data sets in Exercises 5 and 6 illustrate logistic regression and classification trees. Exercise 7 contains eight examples on regression, logistic regression, multinomial logistic regression, discriminant analysis, and regression and classification trees. While solutions for these exercises are provided, I suggest that readers reanalyze these eight data sets and explore and evaluate alternative methods. Also, I have found it useful to assign to groups of students relatively unstructured projects that require each group to identify a problem of interest and collect data that is relevant to the problem's solution. Students learn best if they see the practical relevance of the studied material. Larger-scale projects where groups design and carry out their own studies go beyond simple exercises and are designed to challenge students. Written term papers on these projects and subsequent oral presentations of the results can also be used for student evaluation.

Data Mining and Business Analytics with R, First Edition. Johannes Ledolter.

EXERCISE 1

After reading each chapter, duplicate the numerical results given in the chapter's illustrations by executing the R programs that are given on the book's webpage. Investigate discrepancies in case you find differences.

Exercises for students with good computer background. Improve the R programs that are listed on the book's webpage and make them more efficient.

Exercises for students with strong statistics background. Elaborate on the methods that are discussed in the chapter. Search the literature for other interesting examples. Use the R software templates to analyze these examples. Comment on differences in the numerical results if you find that your analyses differ from the published results.

EXERCISE 2

Consider the following data sets taken from the text by Jank (2011).

Data1. The **HousePrices.csv** data set includes prices and characteristics of $n = 128$ houses in a major US metropolitan area. The variables include Price (sale price in dollars), SqFt (size in square feet), Bedrooms (number of), Bathrooms (number of), Offers (number of offers the house has received while on the market), Brick (whether it is brick construction; Yes/No), and Neighborhood (East/North/West). The objective is to explain the sale price of a house as a function of its characteristics.

Data2. The **DirectMarketing.csv** data set includes data from a direct marketer who sells his products only via direct mail. He sends catalogs with product characteristics to customers who then order directly from the catalogs. The marketer has developed customer records to learn what makes some customers spend more than others. The data set includes $n = 1000$ customers and the following variables: Age (of customer; old/middle/young); Gender (male/female); OwnHome (whether customer owns home; yes/no); Married (single/married); Location (far/close; in terms of distance to the nearest brick and mortar store that sells similar products); Salary (yearly salary of customer; in dollars); Children (number of children; 0–3); History (of previous purchase volume; low/medium/high/NA; NA means that this customer has not yet purchased); Catalogs (number of catalogs sent); and AmountSpent (in dollars). The objective is to explain AmountSpent in terms of the provided customer characteristics.

Data3. The **GenderDiscrimination.csv** data set includes Gender (male/female), Experience (work experience, in years), and Salary (annual salary in dollars) of $n = 208$ individuals. The objective is to learn whether the data indicate systematic compensation discrimination against female employees.

Data4. The **LoanData.csv** data set lists the outcome of $n = 5611$ loans. This data set comes from the consumer-to-consumer (C2C) lending market where borrowers post loan listings and lenders invest in those loans by bidding on the borrower's loan rates. The data variables include the Status of the loan (ultimate outcome; whether the loan is current, late, or in default), the Credit.Grade of the loan (categorical; from the best rating AA to the worst one, HC for heavy risk), Amount of loan (in dollars), Age of loan (in months), the Borrower.Rate, and the Debt.To.Income.Ratio. Here the objective is to distinguish among good and bad loans, that is, to classify loans into good (current), late, and default loans.

Data5. The **FinancialIndicators.csv** data set lists indicators of the financial health of $n = 7112$ companies listed at various stock exchanges. The objective here is to explore the relevance of accounting information for explaining and predicting stock returns. Financial indicators include profitability ratios (such as gross margins and profit margins), liquidity ratios (such as the operating cost flow ratio), activity ratios (such as the stock turnover ratio), and debt and market ratios. Analyzing this data set, one appreciates the fact that (i) available financial indicators are closely related (multicollinear) and (ii) it is not particularly easy to obtain strong relationships between stock price (or stock price changes) and the accounting information.

Objectives: Analyze the data sets D1–D5 using the appropriate methods. You may want to use regression and regression trees for data sets D1, D2, and D3 and classification methods such as logistic regression and classification trees for data set D4. For data set D5, you may want to consider dimension reduction techniques such as principal components to describe the state of health of a stock company. You may want to relate the return on a stock to the accounting summaries or their first few principal components.

Before modeling the information, use data summaries and graphical displays such as histograms and scatter plots to illustrate the information that is contained in the data. Stratify histograms and scatter plots whenever possible to achieve displays of the data in more than two dimensions. Evaluate the models that you have fitted to the data. Use cross-classification and consider splitting the data into estimation and test (evaluation) data sets.

```
hp <- read.csv("C:/DataMining/Data/HousePrices.csv")
hp[1:3,]
dm <- read.csv("C:/DataMining/Data/DirectMarketing.csv")
dm[1:3,]
gd <- read.csv("C:/DataMining/Data/GenderDiscrimination.csv")
gd[1:3,]
ld <- read.csv("C:/DataMining/Data/LoanData.csv")
ld[1:3,]
fi <- read.csv("C:/DataMining/Data/FinancialIndicators.csv")
fi[1:3,]
```

EXERCISE 3

Consider the following data sets taken from the text by Williams (2011).

Data6 and Data7. The **weather.csv** data set contains 1 year of daily observations from a single weather station (Canberra); $n = 366$ rows. The **weatherAUS.csv** data set contains $n = 36,881$ daily observations from 45 Australian weather stations. The weather data were obtained from the Australian Commonwealth Bureau of Meteorology. The data has been processed to provide a binary target variable RainTomorrow (whether there is rain during the next day; No/Yes) and a continuous target (risk) variable RISK_MM (the amount of rain recorded during the next day). The data set includes the following variables:

Date: The date of observation (a date object).

Location: The common name of the location of the weather station.

MinTemp: The minimum temperature in degrees centigrade.

MaxTemp: The maximum temperature in degrees centigrade.

Rainfall: The amount of rainfall recorded for the day in millimeters.

Evaporation: Class A pan evaporation (in millimeters) during 24 h (until 9 a.m.).

Sunshine: The number of hours of bright sunshine in the day.

WindGustDir: The direction of the strongest wind gust in the 24 h to midnight.

WindGustSpeed: The speed (in kilometers per hour) of the strongest wind gust in the 24 h to midnight.

WindDir9am: The direction of the wind gust at 9 a.m.

WindDir3pm: The direction of the wind gust at 3 p.m.

WindSpeed9am: Wind speed (in kilometers per hour) averaged over 10 min before 9 a.m.

WindSpeed3pm: Wind speed (in kilometers per hour) averaged over 10 min before 3 p.m.

RelHumid9am: Relative humidity (in percent) at 9 am.

RelHumid3pm: Relative humidity (in percent) at 3 p.m.

Pressure9am: Atmospheric pressure (hpa) reduced to mean sea level at 9 a.m.

Pressure3pm: Atmospheric pressure (hpa) reduced to mean sea level at 3 p.m.

Cloud9am: Fraction of sky obscured by cloud at 9 a.m. This is measured in "oktas," which are a unit of eighths. It records how many eighths of the sky are obscured by cloud. A 0 measure indicates completely clear sky, while an 8 indicates that it is completely overcast.

Cloud3pm: Fraction of sky obscured by cloud at 3 p.m; see Cloud9am for a description of the values.

Temp9am: Temperature (degrees C) at 9 a.m.

Temp3pm: Temperature (degrees C) at 3 p.m.

RainToday: Integer 1 if precipitation (in millimeters) in the 24 h to 9 a.m. exceeds 1 mm, otherwise 0.

RISK_MM: The continuous target variable; the amount of rain recorded during the next day.

RainTomorrow: The binary target variable whether it rains or not during the next day.

Objectives: Analyze the data sets. The objective is to predict tomorrow's rain amount (a continuous target variable) and tomorrow's likelihood of rain (a binary target variable). Use the appropriate tools such as regression and logistic regression, regression and classification trees, discriminant analysis, and naïve Bayesian methods. Compare your model results on holdout samples.

```
weather <- read.csv("C:/DataMining/Data/weather.csv")
weather[1:3,]
weatherAUS <- read.csv("C:/DataMining/Data/weatherAUS.csv")
weatherAUS[1:3,]
```

Data8. The **audit.csv** data set is an artificially constructed data set that contains the characteristics of $n = 2000$ individual tax returns. The data set includes the following variables:

ID: Unique identifier for each person.

Age: Age of person.

Employment: Type of employment.

Education: Highest level of education.

Marital: Current marital status.

Occupation: Type of occupation.

Income: Amount of income declared.

Gender: Gender of person.

Deductions: Total amount of expenses that a person claims in their financial statement.

Hours: Average hours worked on a weekly basis.

RISK_Adjustment: The continuous target variable; this variable records the monetary amount of any adjustment to the person's financial claims as a result of a productive audit. This variable is a measure of the size of the risk associated with the person.

TARGET_Adjusted: The binary target variable for classification modeling (0/1), indicating nonproductive and productive audits, respectively. Productive audits are those that result in an adjustment being made to a client's financial statement.

Objectives: Analyze the data set. Explore the data by preparing useful graphs and tables. Here the objective is to predict the binary (TARGET_Adjusted) and continuous (RISK_Adjustment) target variables. Tools such as regression and logistic regression, and regression and classification trees should be tried. Evaluate the models through cross-validation and on holdout samples.

```
audit <- read.csv("C:/DataMining/Data/audit.csv")
audit[1:3,]
```

EXERCISE 4

An annual Data Mining and Knowledge Discovery competition is organized by the ACM Special Interest Group on Knowledge Discovery and Data Mining, the leading professional organization of data miners. This competition is referred to as the *KDD Cup*. The webpage http://www.sigkdd.org/kddcup/ lists the topics of previous competitions, describes the objectives of the posed problems, and provides zipped files of the relevant training and evaluation data sets. The webpage discusses the metrics that are used to evaluate the submissions and provides links to methodologies that turned out particularly useful. The scope of most problems goes beyond the introductory discussion of this book. The posted problems are quite challenging, with many cases and a very large number of variables (the numbers of cases and variables are quite a bit larger than those considered in this text). Nevertheless, some of the problems are relevant and should be accessible for study to groups of students with good computer science background. Looking through the various competitions, I believe that the challenges in 2000 and 1997/1998 are best suited for group study.

The 2000 Challenge deals with clickstream and purchase data for Gazelle, a manufacturer and distributor of hosiery. The 1997/1998 Challenge deals with the response to a direct mailing and how to maximize donations to a charity. Additional possibilities for student projects are the 2009 Challenge that deals with issues of customer relationship management (CRM) and the 2007 Challenge that addresses the Netflix movie ratings.

It is the problem for the 1997/1998 Challenge that is discussed here. The data has been provided by the Paralyzed Veterans of America (PVA), a not-for-profit organization that provides programs and services for US veterans with spinal cord injuries or disease. With an in-house database of over 13 million donors, PVA is also one of the largest direct-mail fund raisers in the country. The study lists the results of a June 1997 fund-raising mailing to millions of PVA donors. The mailing includes a gift "premium" of personalized name and address labels plus an assortment of 10 note cards and envelopes. The training data includes the response to the appeal (an indicator whether a donation has been made, as well as the amount of the donation), numerous variables that describe the recipient as well as demographic information, and prior donation history. The objective of the analysis is to predict the response to the mailing—whether a donation is made and, if so, the amount of the donation. This task involves classification/discrimination (for predicting the occurrence of a donation) and regression (for predicting the amount of the donation). The performance of any proposed method can be evaluated on a test data set that has been withheld from the estimation data set.

Information about the problem is given in the file **cup98doc.txt** that can be found on the webpage http://www.sigkdd.org/kddcup/. The learning data set in file **cup98lrn.zip** contains 95,412 records, with each record consisting of 481 variables. The dictionary file **cup98dic.txt** gives a description of the variables. TARGET_B is a binary indicator that expresses the response to the most recent mailing; TARGET_D represents the donation amount (in dollars). These are the variables that need to be predicted. The variable CONTROLN is a control number that allows you to link the records in the evaluation data set **cup98val.zip** (96,367 cases and 479 variables) to the actual responses for TARGET_B and TARGET_D that are given in the file **cup98VALtargt.csv**. Download and unzip these files and look at their contents. Zipped versions of comma-delimited Excel files (**cup98LRN_csv.zip**, **cup98VAL_csv.zip**) and the file **cup98VALtargt.csv** are available on the book's webpage. Unzip these files to create the Excel files **cup98LRN.csv** and **cup98VAL.csv**, and read them into your R session (as illustrated below). Use the learning data set to find models that predict the target values and evaluate the predictions on the evaluation data sets.

```
## read the data
cup98LRN <- read.csv("C:/DataMining/Data/cup98LRN.csv")
cup98LRN[1:3,]

## read the data
cup98VAL <- read.csv("C:/DataMining/Data/cup98VAL.csv")
cup98VAL[1:3,]

## read the data
cup98VALtargt <- read.csv("C:/DataMining/Data/cup98VALtargt.csv")
cup98VALtargt[1:3,]
```

EXERCISE 5

The following data are taken from Higgins and Koch (1977). The data come from an extensive survey of workers in the cotton industry. The variable of interest is the presence of lung byssinosis. Byssinosis, also called brown lung disease, is a chronic, asthma-like narrowing of the airways resulting from inhaling particles of cotton, flax, hemp, or jute. It has been recognized as an occupational hazard for textile workers. More than 35,000 textile workers, mostly from textile-producing regions of North and South Carolina, have been disabled by byssinosis, and 183 have died between 1979 and 1992.

Numbers of workers suffering from (yes) and not suffering from (no) byssinosis for different categories of workers are given below. The covariates are race (1, white; 2, other); gender (1, male; 2, female); smoking history (1, smoker;

2, nonsmoker); length of employment in the cotton industry (1, less than 10 years; 2, between 10 and 20 years; 3, more than 20 years); and the dustiness of the workplace (1, high; 2, medium; 3, low).

The information can be arranged as a factorial, with the number of affected workers among the total number of workers in each group as the response variable. The 72 groups of the factorial arrangement are formed by all possible level combinations of the five explanatory variables: 3(Dust) × 2(Race) × 2(Sex) × 2(Smoking) × 3(Employment). Seven of the 72 categories are empty. A subset of the data is shown below. The data, arranged as outcomes (1 = YES and 0 = NO) and their frequencies (Weights), are given in the file **byssinosisWeights.csv**.

Yes	No	Number	Dust	Race	Sex	Smoking	Employ Length
3	37	40	1	1	1	1	1
0	74	74	2	1	1	1	1
2	258	260	3	1	1	1	1
25	139	164	1	2	1	1	1
.	.	.	.	.	.	.	.
.	.	.	.	.	.	.	.
2	340	342	3	1	2	2	3
0	0	0	1	2	2	2	3
0	2	2	2	2	2	2	3
0	3	3	3	2	2	2	3

Relate the outcomes (1 = YES and 0 = NO) and their frequencies to the explanatory variables. Treat the explanatory variables as (categorical) factors. Analyze the data. In particular,

- Fit a logistic regression model. Assess its adequacy and interpret the results. Discuss whether (and which of) the covariates have an influence on the presence of byssinosis.
- Consider classification trees to predict the likelihood of contracting byssinosis. Byssinosis is a relatively rare disease, even under the worst conditions. Interpret the output. Identify the conditions under which byssinosis is most likely to occur.

```
## read the data
bys <- read.csv("C:/DataMining/Data/byssinosisWeights.csv")
```

EXERCISE 6

The following data are taken from Brown et al. (1983). The data, collected in Bradford (UK) between 1968 and 1977, are from 13,384 women giving birth to their first child. The data set includes information on toxemic signs exhibited by the mother during pregnancy: hypertension only; proteinurea (i.e., the presence of protein in urine) only; both hypertension and proteinurea; and neither hypertension nor proteinurea. The aim of the study was to learn if the level of smoking and the social class are related to the incidence of toxemic signs and how this might depend on social class. The two covariates are social class (1 through 5) and the number of cigarettes smoked (1, none; 2, 1–19 cigarettes per day; 3, more than 20 cigarettes per day).

Class	Smoking	Both Hypertension and Proteinurea	Proteinurea Only	Hypertension Only	Neither Problem Exhibited	Total
1	1	28	82	21	286	417
1	2	5	24	5	71	105
1	3	1	3	0	13	17
2	1	50	266	34	785	1135
2	2	13	92	17	284	406
2	3	0	15	3	34	52
3	1	278	1101	164	3160	4703
3	2	120	492	142	2300	3054
3	3	16	92	32	383	523
4	1	63	213	52	656	984
4	2	35	129	46	649	859
4	3	7	40	12	163	222
5	1	20	78	23	245	366
5	2	22	74	34	321	451
5	3	7	14	4	65	90

The data, arranged as outcomes (1 = YES and 0 = NO) and their frequencies (weights), are given in the file **toxaemiaWeights.csv**. Analyze the data. Discuss whether (and which of) the covariates have an influence on the presence of toxemic signs. Consider each symptom group separately. Find suitable logistic models and compare their classification results with those from classification trees.

The information can be arranged as a factorial, with the 15 groups of the factorial arrangement formed by all possible level combinations of the two explanatory variables: 5(Class) × 3(Smoking). Use the binary logistic regression function in R, specifying the outcomes (1 = YES and 0 = NO) and their frequencies, and entering the explanatory variables as (categorical) factors.

```
## read the data
tox <- read.csv("C:/DataMining/Data/toxaemiaWeights.csv")
```

EXERCISE 7

The following eight examples illustrate regression, logistic regression, multinomial logistic regression, discriminant analysis, and regression and classification trees.

The objective in these examples is to predict or classify the observations. Reanalyze these eight data sets and explore and evaluate alternative methods.

EXAMPLE 7.1 CLASSIFICATION TREE FOR IDENTIFYING SOYBEAN DISEASE

The soybean data set is taken from the UCI Machine Learning Repository (http://archive.ics.uci.edu/ml/datasets/Soybean+(Large)). The data come from the paper by Michalski and Chilausky (1980).

The data set **soybean15.csv** contains 290 different diseased soybean samples. The type of disease (there are 15 different classes of disease, such as charcoal rot, brown stem rot, and downey mildew) is given in the first column. The next 35 columns contain categorical factors that describe various attributes of the soybean plant and the growing conditions; a nominal scale is assumed for all attributes. The objective is to use the information on these attributes to predict the disease classification. Some attributes include missing values (they are labeled as blanks in the Excel file, and as NA in the R session); 24 of the 290 cases have a missing observation for at least one of the attributes, which leaves 266 observations for the analysis. The list of attributes with their outcomes is shown as follows:

```
 1. date:               april,may,june,july,august,september,october.
 2. plant-stand:        normal,lt-normal.
 3. precip:             lt-norm,norm,gt-norm.
 4. temp:               lt-norm,norm,gt-norm.
 5. hail:               yes,no.
 6. crop-hist:          diff-lst-year,same-lst-yr,same-lst-two-yrs,
                        same-lst-sev-yrs.
 7. area-damaged:       scattered,low-areas,upper-areas,whole-field.
 8. severity:           minor,pot-severe,severe.
 9. seed-tmt:           none,fungicide,other.
10. germition:          90-100%,80-89%,lt-80%.
11. plant-growth:       norm,abnorm.
12. leaves:             norm,abnorm.
13. leafspots-halo:     absent,yellow-halos,no-yellow-halos.
14. leafspots-marg:     w-s-marg,no-w-s-marg,dna.
15. leafspot-size:      lt-1/8,gt-1/8,dna.
16. leaf-shread:        absent,present.
17. leaf-malf:          absent,present.
18. leaf-mild:          absent,upper-surf,lower-surf.
19. stem:               norm,abnorm.
20. lodging:            yes,no.
21. stem-cankers:       absent,below-soil,above-soil,above-sec-nde.
22. canker-lesion:      d,brown,dk-brown-blk,tan.
23. fruiting-bodies:    absent,present.
24. exterl decay:       absent,firm-and-dry,watery.
25. mycelium:           absent,present.
26. int-discolor:       none,brown,black.
27. sclerotia:          absent,present.
28. fruit-pods:         norm,diseased,few-present,dna.
29. fruit spots:        absent,colored,brown-w/blk-specks,distort,dna.
30. seed:               norm,abnorm.
```

```
31. mold-growth:      absent,present.
32. seed-discolor:    absent,present.
33. seed-size:        norm,lt-norm.
34. shriveling:       absent,present.
35. roots:            norm,rotted,galls-cysts.
```

Below we list the R program and we comment on its output. The R program can be found on the webpage that accompanies this book.

```
library(ares)
## needed to determine the proportion of missing observations
library(tree) ## classification trees

## reading the data
soybean15 <- read.csv("C:/DataMining/Data/soybean15.csv")
soybean15[1:3,]

                  disease C1 C2 C3 C4 C5 C6 C7 C8 C9 C10 C11 C12 C13 C14 C15 C16
1 diaporthe-stem-canker  6  0  2  1  0  1  1  1  0   0   1   1   0   2   2   0
2 diaporthe-stem-canker  4  0  2  1  0  2  0  2  1   1   1   1   0   2   2   0
3 diaporthe-stem-canker  3  0  2  1  0  1  0  2  1   2   1   1   0   2   2   0
  C17 C18 C19 C20 C21 C22 C23 C24 C25 C26 C27 C28 C29 C30 C31 C32 C33 C34 C35
1   0   0   1   1   3   1   1   1   0   0   0   0   4   0   0   0   0   0   0
2   0   0   1   0   3   1   1   1   0   0   0   0   4   0   0   0   0   0   0
3   0   0   1   0   3   0   1   1   0   0   0   0   4   0   0   0   0   0   0

## converting the attributes into factors (nominal scale)
## calculating the proportion of missing observations
miss=dim(36)
for (j in 1:36) {
soybean15[,j]=factor(soybean15[,j])
miss[j]=count.na(soybean15[,j])$na/length(soybean15[,j])
}
miss

 [1] 0.00000000 0.00000000 0.00000000 0.00000000 0.00000000 0.08275862
 [7] 0.00000000 0.00000000 0.08275862 0.08275862 0.08275862 0.00000000
[13] 0.00000000 0.04482759 0.04482759 0.04482759 0.04482759 0.04482759
[19] 0.04482759 0.00000000 0.08275862 0.00000000 0.00000000 0.08275862
[25] 0.00000000 0.00000000 0.00000000 0.00000000 0.08275862 0.08275862
[31] 0.08275862 0.08275862 0.08275862 0.08275862 0.08275862 0.00000000

## fifth attribute (presence/absence of hail) has 8.27% missing
## observations

## constructing the classification tree
soytree <- tree(disease ~., data = soybean15, mincut=1)
soytree
summary(soytree)
plot(soytree, col=8)
text(soytree, digits=2)

## cross-validation to prune the tree
set.seed(2)
cvsoy <- cv.tree(soytree, K=10)
cvsoy$size

 [1] 19 18 17 16 15 14 13 12 11  9  8  7  6  5  4  3  2  1
```

```
cvsoy$dev
```

```
 [1]  269.1145  251.0295  248.2929  287.5657  381.4021  463.4245  475.8316
 [8]  487.3538  526.0126  527.9349  542.4499  550.6483  762.4923  762.4923
[15]  862.7270  868.9765 1022.9153 1365.9087
```

```
plot(cvsoy, pch=21, bg=8, type="p", cex=1.5, ylim=c(0,1400))
## shows that the tree has many terminal nodes
```

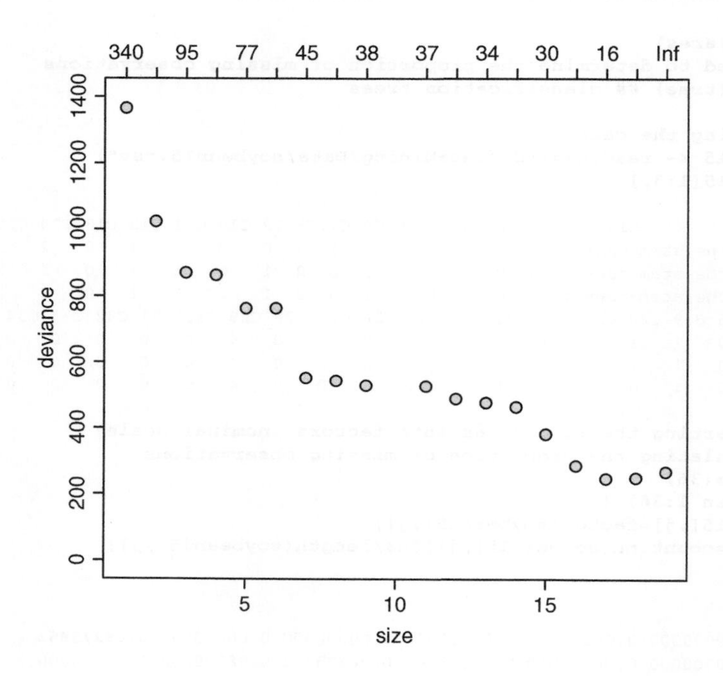


```
soycut <- prune.tree(soytree, best=17)
soycut
```

```
node), split, n, deviance, yval, (yprob/omitted)
      * denotes terminal node

1) root 266 1342.000 alternarialeaf-spot
   2) C15: 0,2 121  548.400 anthracnose
     4) C21: 0 65  231.200 brown-stem-rot
       8) C26: 0 40  110.900 bacterial-blight
        16) C18: 0 30   65.920 bacterial-blight
          32) C22: 0 20   27.730 bacterial-blight
            64) C14: 0 11    6.702 bacterial-blight *
            65) C14: 1 9    0.000 bacterial-pustule *
          33) C22: 3 10    0.000 purple-seed-stain *
        17) C18: 1 10    0.000 powdery-mildew *
```

```
      9) C26: 1,2 25   33.650 brown-stem-rot
       18) C26: 1 15    0.000 brown-stem-rot *
       19) C26: 2 10    0.000 charcoal-rot *
    5) C21: 1,2,3 56  150.200 anthracnose
     10) C28: 0,1 30   38.190 anthracnose
       20) C29: 0,2 20    0.000 anthracnose *
       21) C29: 4 10    0.000 diaporthe-stem-canker *
     11) C28: 3 26   34.650 phytophthora-rot
       22) C12: 0 10    0.000 rhizoctonia-root-rot *
       23) C12: 1 16    0.000 phytophthora-rot *
  3) C15: 1 145  449.700 alternarialeaf-spot
    6) C1: 0,1,2,3 55  134.000 brown-spot
     12) C3: 0,1 14   21.250 phyllosticta-leaf-spot *
     13) C3: 2 41   70.480 brown-spot
       26) C31: 0 36   40.080 brown-spot *
       27) C31: 1 5    0.000 downy-mildew *
    7) C1: 4,5,6 90  221.100 alternarialeaf-spot
     14) C19: 0 54   85.510 alternarialeaf-spot
       28) C31: 0 49   52.190 alternarialeaf-spot *
       29) C31: 1 5    0.000 downy-mildew *
     15) C19: 1 36   58.740 frog-eye-leaf-spot
       30) C28: 0 12   21.300 brown-spot *
       31) C28: 1 24    0.000 frog-eye-leaf-spot *
```

```
summary(soycut)

Classification tree:
snip.tree(tree = soytree, nodes = c(26, 30))
Variables actually used in tree construction:
[1] "C15" "C21" "C26" "C18" "C22" "C14" "C28"
[8] "C29" "C12" "C1"  "C3"  "C31" "C19"
Number of terminal nodes:  17
Residual mean deviance:  0.5684 = 141.5 / 249
Misclassification error rate: 0.1015 = 27 / 266

plot(soycut, col=8)
## below we have omitted the text as it is difficult to read
## terminal node 31 is the one on the far right of the graph
## first split: C15ac (to left) and C15b (to the right)
## second split: C1abcd (to left) and C1efg (to right)
## third split: C19a (to left) and C19b (to right)
## fourth split: C28a (to left) and C28bcd (to right)
```

Here the attributes are factors on a nominal scale. Factor (categorical) attributes are represented with indicator variables. We explained in Chapter 13 how the tree-building approach splits categorical attributes. Not creating factors and treating coded variables as continuous data would split groups according to whether the attribute is above or below a numeric threshold. For coded variables this would be wrong.

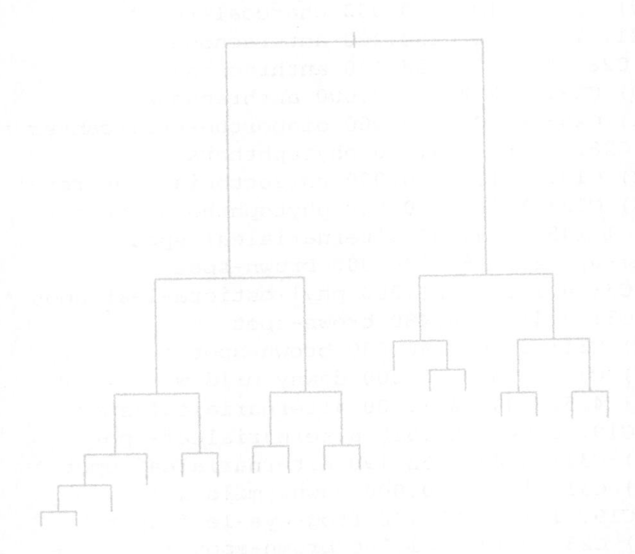

The overall misclassification rate is about 10%. There are 17 terminal nodes. As illustration, consider the last terminal node (listed under 31, and shown as the terminal node to the very right of the plot). It classifies a soybean sample with attributes C15b (leafspot size > 1/8), C1efg (sample taken in August, September, or October), C19b (abnormal stem), and C28bcd (with not normal fruit pod) as suffering from the "frog-eye-leaf-spot" condition.

EXAMPLE 7.2 CLASSIFICATION TREE FOR FITTING CONTACT LENSES

This is a data set for fitting contact lenses (taken from Cendrowska, 1987; the data are available through the UCI Machine Learning Repository). The variable to be predicted is whether a patient should be fitted with hard, soft, or no contact lenses. Four patient attributes (all categorical data on nominal scales) are available to make this choice:

- age {young, pre-presbyopic, presbyopic}
- spectacle-prescription {myope, hypermetrope}
- astigmatism {no, yes}
- tear-prod-rate {reduced, normal}

A listing of all 24 combinations of attributes and the recommended decisions on the fitting of contact lenses is given in the file **ContactLens.csv**. The complete tree is shown later; it leads to a perfect description, with no errors. A somewhat simpler tree, with three terminal nodes and just two attributes (tear production rate and astigmatism), recommends no contact lenses for patients with reduced tear production; soft contacts for patients with normal tear production and no astigmatism; and hard contact lenses for patients with normal tear production and astigmatism present. This tree misclassifies 3 of the 24 possible attribute combinations.

```
library(tree)

## read the data
ContactLens <- read.csv("C:/DataMining/Data/ContactLens.csv")
levels(ContactLens[,1])  ## age
     [1] "pre-presbyopic" "presbyopic"     "young"
levels(ContactLens[,2])  ## spectacle presription
     [1] "hypermetrope" "myope"
levels(ContactLens[,3])  ## astigmatism
     [1] "no"  "yes"
levels(ContactLens[,4])  ## tear production rate
     [1] "normal"  "reduced"
levels(ContactLens[,5])  ## contact lens
     [1] "hard" "none" "soft"

ContactLens

            Age SpectaclePrescrip Astigmatism TearProdRate ContactLens
1         young             myope          no      reduced        none
2         young             myope          no       normal        soft
3         young             myope         yes      reduced        none
4         young             myope         yes       normal        hard
5         young      hypermetrope          no      reduced        none
6         young      hypermetrope          no       normal        soft
7         young      hypermetrope         yes      reduced        none
8         young      hypermetrope         yes       normal        hard
9  pre-presbyopic             myope          no      reduced        none
10 pre-presbyopic             myope          no       normal        soft
11 pre-presbyopic             myope         yes      reduced        none
12 pre-presbyopic             myope         yes       normal        hard
13 pre-presbyopic      hypermetrope          no      reduced        none
14 pre-presbyopic      hypermetrope          no       normal        soft
15 pre-presbyopic      hypermetrope         yes      reduced        none
16 pre-presbyopic      hypermetrope         yes       normal        none
17     presbyopic             myope          no      reduced        none
18     presbyopic             myope          no       normal        none
19     presbyopic             myope         yes      reduced        none
20     presbyopic             myope         yes       normal        hard
21     presbyopic      hypermetrope          no      reduced        none
22     presbyopic      hypermetrope          no       normal        soft
23     presbyopic      hypermetrope         yes      reduced        none
24     presbyopic      hypermetrope         yes       normal        none

## constructing the classification tree that fits the data perfectly
cltree <- tree(ContactLens ~., data = ContactLens, mindev=0,
+     minsize=1)
cltree
```

```
summary(cltree)
plot(cltree, col=8)
text(cltree, digits=2)
```

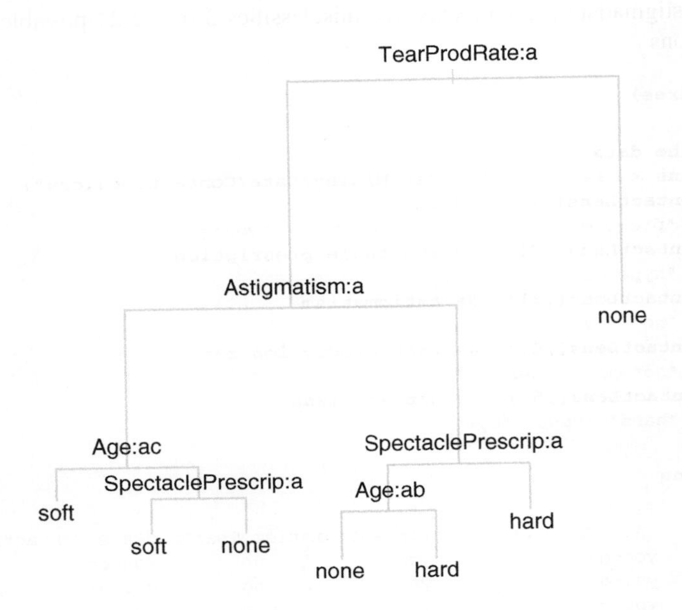


```
## pruning the tree to get a simpler tree
clcut <- prune.tree(cltree, best=3)
clcut

node), split, n, deviance, yval, (yprob)
      * denotes terminal node
1) root 24 44.120 none ( 0.1667 0.6250 0.2083 )
  2) TearProdRate: normal 12 25.860 soft ( 0.3333 0.2500 0.4167 )
    4) Astigmatism: no 6  5.407 soft ( 0.0000 0.1667 0.8333 ) *
    5) Astigmatism: yes 6  7.638 hard ( 0.6667 0.3333 0.0000 ) *
  3) TearProdRate: reduced 12  0.000 none ( 0.0000 1.0000 0.0000 ) *

summary(clcut)

Classification tree:
snip.tree(tree = cltree, nodes = c(4, 5))
Variables actually used in tree construction:
[1] "TearProdRate" "Astigmatism"
Number of terminal nodes:  3
Residual mean deviance:  0.6212 = 13.04 / 21
Misclassification error rate: 0.125 = 3 / 24

plot(clcut, col=8)
text(clcut)
```

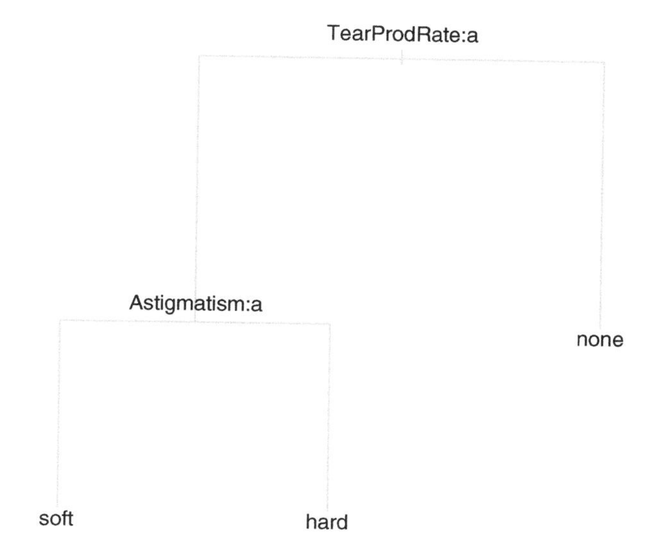

The R program can be found on the webpage that accompanies this book.

EXAMPLE 7.3 DETERMINING THE CREDIT RISK USING A CLASSIFICATION TREE

This is a small data set for assessing the riskiness of certain loans. The riskiness of a loan (low, moderate, and high) in the last column of the data file **credit.csv** is determined from four categorical risk predictors with the following attribute categories:

- Credit history (bad, good, unknown)
- Debt (high, low)
- Collateral (adequate, none)
- Income group (>35, 0–15, 15–35).

Note that a complete factorial arrangement includes $(3)(2)(2)(3) = 36$ attribute combinations; this data set includes only 14.

```
library(tree)

## first we read in the data
credit <- read.csv("C:/DataMining/Data/credit.csv")
credit
```

```
  CreditHist Debt Collateral Income     Risk
1        bad  low       none   0-15     high
2    unknown high       none  15-35     high
3    unknown  low       none  15-35 moderate
```

```
4         bad  low       none  0-15     high
5     unknown  low   adequate   >35      low
6     unknown  low       none   >35      low
7     unknown high       none  0-15     high
8         bad  low   adequate   >35 moderate
9        good  low       none   >35      low
10       good high   adequate   >35      low
11       good high       none  0-15     high
12       good high       none 15-35 moderate
13       good high       none   >35      low
14        bad high       none 15-35     high
```

```
## checking the ordering of the nominal categories
credit[,1]

[1] bad     unknown unknown bad     unknown unknown unknown bad     good
[10] good   good    good    good    bad
Levels: bad good unknown

credit[,2]

[1] low  high low  low  low  low  high low  low  high high high high high
Levels: high low

credit[,3]

[1] none     none     none     none     adequate none     none     adequate
[9] none     adequate none     none     none     none
Levels: adequate none

credit[,4]

[1] 0-15  15-35 15-35 0-15  >35   >35   0-15  >35   >35   >35   0-15  15-35
[13] >35   15-35
Levels: >35 0-15 15-35

credit[,5]

[1] high     high     moderate high     low      low      high     moderate
[9] low      low      high     moderate low      high
Levels: high low moderate

## constructing the classification tree that fits the data perfectly
credittree <- tree(Risk ~., data = credit, mindev=0, minsize=1)
credittree

node), split, n, deviance, yval, (yprob)
      * denotes terminal node

 1) root 14 29.710 high ( 0.4286 0.3571 0.2143 )
   2) Income: >35 6  5.407 low ( 0.0000 0.8333 0.1667 )
     4) CreditHist: bad 1  0.000 moderate ( 0.0000 0.0000 1.0000 ) *
     5) CreditHist: good,unknown 5  0.000 low ( 0.0000 1.0000 0.0000 ) *
   3) Income: 0-15,15-35 8  8.997 high ( 0.7500 0.0000 0.2500 )
     6) Income: 0-15 4  0.000 high ( 1.0000 0.0000 0.0000 ) *
     7) Income: 15-35 4  5.545 high ( 0.5000 0.0000 0.5000 )
      14) CreditHist: bad 1  0.000 high ( 1.0000 0.0000 0.0000 ) *
      15) CreditHist: good,unknown 3  3.819 moderate ( 0.3333 0.0000 .6667)
        30) CreditHist: good 1  0.000 moderate ( 0.0000 0.0000 1.0000 ) *
        31) CreditHist: unknown 2  2.773 high ( 0.5000 0.0000 0.5000 )
```

```
        62) Debt: high 1  0.000 high ( 1.0000 0.0000 0.0000 ) *
        63) Debt: low 1  0.000 moderate ( 0.0000 0.0000 1.0000 ) *
```

summary(credittree)

```
Classification tree:
tree(formula = Risk ~ ., data = credit, mindev = 0, minsize = 2)
Variables actually used in tree construction:
[1] "Income"     "CreditHist" "Debt"
Number of terminal nodes:  7
Residual mean deviance:  0 = 0 / 7
Misclassification error rate: 0 = 0 / 14
```

plot(credittree, col=8)
text(credittree, digits=2)

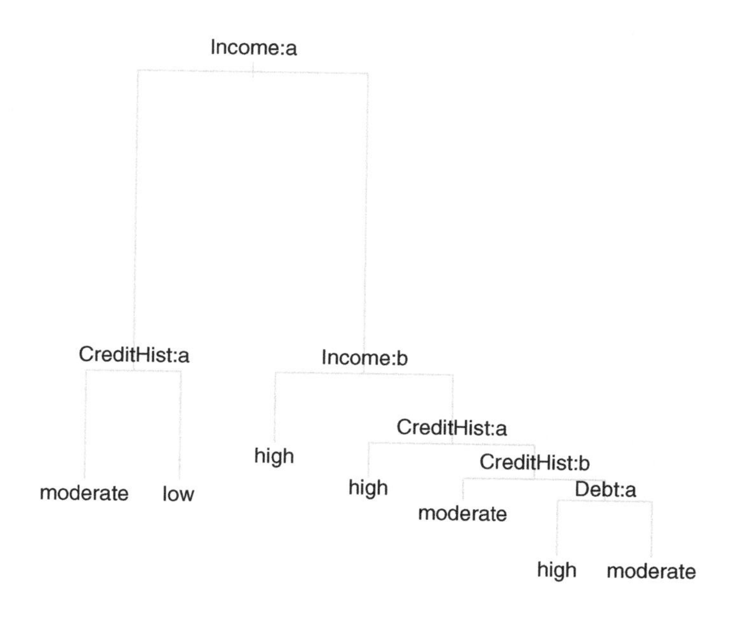

The classification tree explains how to predict the riskiness of a loan. The classification tree has seven terminal nodes and uses three attributes: income, credit history, and the amount of debt. Collateral does not enter the decision tree. Applicants with income above 35K (income group a) and bad credit history (credit history a) carry moderate risk. Applicants with income above 35K (income group a) and good or unknown credit history (credit history not a) carry low risk. Applicants in the low income group (income group b; 0–15) carry high risk. So do applicants from the middle income group (income group c; 15–35) with bad credit history (credit history b). For the medium income group (15–35) with unknown credit history (credit history c), it depends on the prior debt whether the applicant is selected into the high risk group (if debt is high) or into the moderate risk group (if debt is low).

EXAMPLE 7.4 DETERMINING THE PROGRESSION OF LIVER DISEASE USING A CLASSIFICATION TREE

The data in file **hepatitis.csv** is taken from the text by Witten et al. (2011). The file, created by K. Seela, P. Tatavarthy, and S. Tippa, includes liver biopsy results on 306 individuals with confirmed hepatitis transmission. A liver biopsy is usually the most telling test for assessing the nature and severity of liver disease, and biopsy results are expressed in terms of three severity groups (I–III). However, biopsies carry some risk. We would prefer to identify the stage of liver disease from the results of noninvasive tests if this were possible. The objective of this exercise is to use patient characteristics to predict the result of the biopsy (last row of the table shown below). A good predictive model may be able to reduce the number of needed biopsies.

No	Attribute	Type of Data	Values	Description
1	Sex	Categorical	0 = M, 1 = F	Gender
2	DOB	Numeric	Date	Date of birth
3	DOT	Numeric	Date	Date of transmission of disease
4	Route	Categorical	Coc, IV, Tx, N, NRF, Tatt, Sex	Route through which disease was transmitted
5	IV	Categorical	+, −	Intravenous
6	Tx	Categorical	+, −	Blood transfusion
7	Coc	Categorical	+, −	Usage of cocaine
8	Tatt	Categorical	+, −	Presence of tattoo on the body of patient
9	HBV	Categorical	+, −	Presence of hepatitis B virus in patient
10	HIV	Categorical	+, −	Presence of HIV infection
11	EtOH	Categorical	+, −	Alcohol usage by the patient
12	Obes	Categorical	+, −	Whether the patient is obese or not
13	Rx	Categorical	+, −	Treatment, whether patient has been treated
14	Tox	Categorical	+, −	Presence of any toxic elements
15	CLD	Categorical	+, −	Whether the patient has chronic liver disease
16	LFT	Categorical	+, −	Whether or not liver function test was done
17	YWOD	Numeric		DOT − DOB, years without the disease
18	Age	Numeric		Current age of the patient
19	Bx	Categorical	I, II, III	Biopsy result, which specifies HepC

For our analysis we use Gender, Age, YWD(years with the disease) = Age−YWOD(years without the disease), and the 12 indicator variables 5–16 to classify the biopsy results in row 19 of the earlier table. Our selection of variables is slightly different from that of Seela, Tatavarthy, and Tippa, and we have corrected several incorrect entries.

We recommend a tree with six terminal nodes and four covariates: Age, the presence of the hepatitis B virus, alcohol usage of the patient, and the presence of a liver function test. But only 56% of the patients are classified correctly with our rule. This is probably too low; it appears that biopsies will be necessary.

```
library(tree)

## data set from Witten
## missing data
hepatitis <- read.csv("C:/DataMining/Data/hepatitis.csv")
hepatitis
## calculating YWD = (Age - YWOD)
hepatitis[,20]=hepatitis[,18]-hepatitis[,17]
colnames(hepatitis)[20]= "YWD"
hepatitis[1:3,]
## cleaning up the data set
hh=hepatitis[,c(-2:-4,-17)]
hh[1:3,]
## create factors for the categorical variables
for (j in 1:13) {
hh[,j]=factor(hh[,j])
}
hh[1:3,]
levels(hh[,6])
levels(hh[,8])
levels(hh[,13])

## constructing the classification tree
heptree <- tree(Bx ~., data = hh)
heptree
summary(heptree)
plot(heptree, col=8)
text(heptree, digits=2)

## cross-validation to prune the tree
set.seed(2)
cvhep <- cv.tree(heptree, K=10)
cvhep$size
cvhep$dev
plot(cvhep, pch=21, bg=8, type="p", cex=1.5, ylim=c(400,750))

hepcut <- prune.tree(heptree, best=6)
hepcut
```

```
1) root 221 463.80 II ( 0.37557 0.42986 0.19457 )
  2) Age < 62.5 204 429.30 I ( 0.40196 0.40196 0.19608 )
    4) HBV: 0 194 401.90 II ( 0.40722 0.41753 0.17526 )
      8) Age < 46.5 66 127.60 I ( 0.54545 0.31818 0.13636 )
        16) EtOH: 0 43  73.02 I ( 0.65116 0.25581 0.09302 ) *
        17) EtOH: 1 23  48.82 II ( 0.34783 0.43478 0.21739 )
          34) LFT: 0 6   0.00 I ( 1.00000 0.00000 0.00000 ) *
          35) LFT: 1 17  31.41 II ( 0.11765 0.58824 0.29412 )*
      9) Age > 46.5 128 266.40 II (0.33594 0.46875 0.19531)*
    5) HBV: 1 10  17.96 III ( 0.30000 0.10000 0.60000 ) *
  3) Age > 62.5 17  23.05 II ( 0.05882 0.76471 0.17647 ) *
```

```
summary(hepcut)
```

```
Classification tree:
snip.tree(tree = heptree, nodes = c(3, 5, 16))
Variables actually used in tree construction:
[1] "Age"   "HBV"   "EtOH" "LFT"
Number of terminal nodes:  6
Residual mean deviance:  1.915 = 411.8 / 215
Misclassification error rate: 0.4434 = 98 / 221
```

```
plot(hepcut, col=8)
text(hepcut)
```

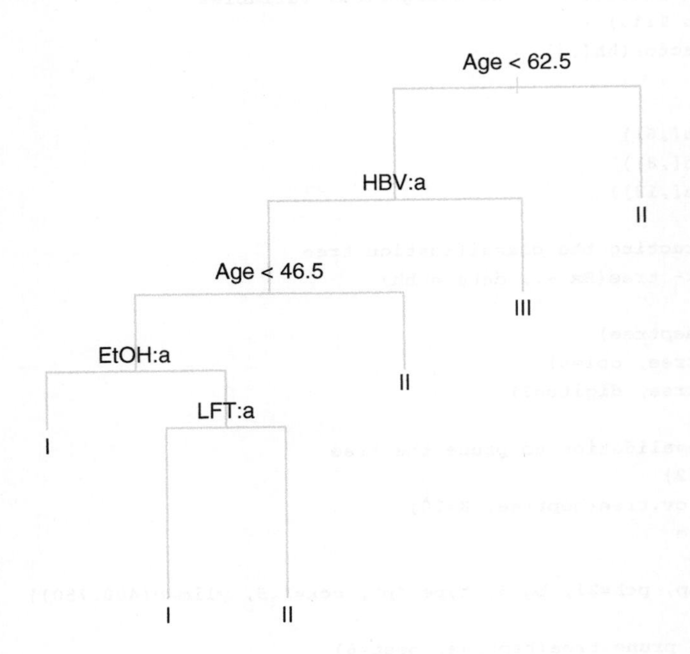

EXAMPLE 7.5 PREDICTING THE OUTCOME OF LABOR NEGOTIATIONS USING A CLASSIFICATION TREE

This is a data set that relates the outcome of labor negotiations (good or bad, as seen by the workers who are covered by these agreements) to attributes of the negotiated contract. The data in file **labor.csv** is available from the UCI Machine Learning Repository. It includes 57 Canadian contracts, 16 attributes, and an overall outcome (0/1) variable. The data set contains many missing variables. For the construction of the classification tree, we have omitted variables with many missing values, and we concentrate our analysis on the following four attributes: duration of the contract, wage increase in the first year of the contract, the number of weekly hours negotiated for, and the quality of the vacation benefit package (categorical on a nominal scale). The objective is to predict the overall quality of the contract.

The first tree with five terminal nodes has one split that leads to identical classifications. Node 2 leads to identical results when splitting on hours (relative to 38.5), and this node and the tree below it can be snipped off. The simplified tree misclassifies 3/47 or 6.4% of the contracts. Contracts with a low first year wage increase are considered bad. Contracts with large first year wage increase and average to generous vacation packages are judged good. It is interesting to learn that contracts with large first year wage increases, but below-average vacation packages and large negotiated hours are perceived as bad contracts. Workers in Canada value their free time.

```
library(tree)

## read the data
labor <- read.csv("C:/DataMining/Data/labor.csv")
labor[1:3,]
## omit variables with lots of missing values
ll=labor[,c(-3:-5,-7:-11,-13:-16)]
ll[1:3,]

levels(ll[,4])  ## vacation benefits

[1] "average"        "below_average" "generous"

levels(ll[,5])  ## response: overall contract quality

[1] "bad"  "good"

## constructing the classification tree
labortree <- tree(Class ~., data = ll)
labortree

node), split, n, deviance, yval, (yprob)
      * denotes terminal node

 1) root 47 63.420 good ( 0.40426 0.59574 )
   2) WageIncY1 < 2.65 14  7.205 bad ( 0.92857 0.07143 )
     4) Hours < 38.5 6  5.407 bad ( 0.83333 0.16667 ) *
```

```
    5) Hours > 38.5 8  0.000 bad ( 1.00000 0.00000 ) *
  3) WageIncY1 > 2.65 33 31.290 good ( 0.18182 0.81818 )
    6) Vacation: below_average 12 16.640 good ( 0.50000 0.50000 )
     12) Hours < 39 6  5.407 good ( 0.16667 0.83333 ) *
     13) Hours > 39 6  5.407 bad ( 0.83333 0.16667 ) *
    7) Vacation: average,generous 21  0.000 good ( 0.00000 1.00000 ) *
```

```
summary(labortree)

Classification tree:
tree(formula = Class ~ ., data = ll)
Variables actually used in tree construction:
[1] "WageIncY1" "Hours"     "Vacation"
Number of terminal nodes:  5
Residual mean deviance:  0.3862 = 16.22 / 42
Misclassification error rate: 0.06383 = 3 / 47

plot(labortree, col=8)
text(labortree, digits=2)
```

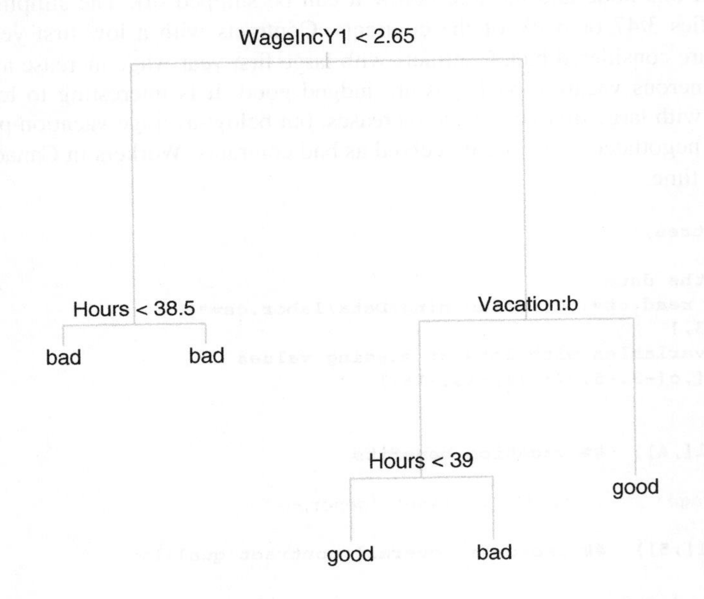


```
p1=snip.tree(labortree,nodes=2)
p1

node), split, n, deviance, yval, (yprob)
      * denotes terminal node

 1) root 47 63.420 good ( 0.40426 0.59574 )
   2) WageIncY1 < 2.65 14  7.205 bad ( 0.92857 0.07143 ) *
   3) WageIncY1 > 2.65 33 31.290 good ( 0.18182 0.81818 )
     6) Vacation: below_average 12 16.640 good ( 0.50000 0.50000 )
      12) Hours < 39 6  5.407 good ( 0.16667 0.83333 ) *
```

```
    13) Hours > 39 6  5.407 bad ( 0.83333 0.16667 ) *
   7) Vacation: average,generous 21  0.000 good ( 0.00000 1.00000 ) *
```

summary(p1)

```
Classification tree:
snip.tree(tree = labortree, nodes = 2)
Variables actually used in tree construction:
[1] "WageIncY1" "Vacation"  "Hours"
Number of terminal nodes:  4
Residual mean deviance:  0.419 = 18.02 / 43
Misclassification error rate: 0.06383 = 3 / 47
```

plot(p1)
text(p1)

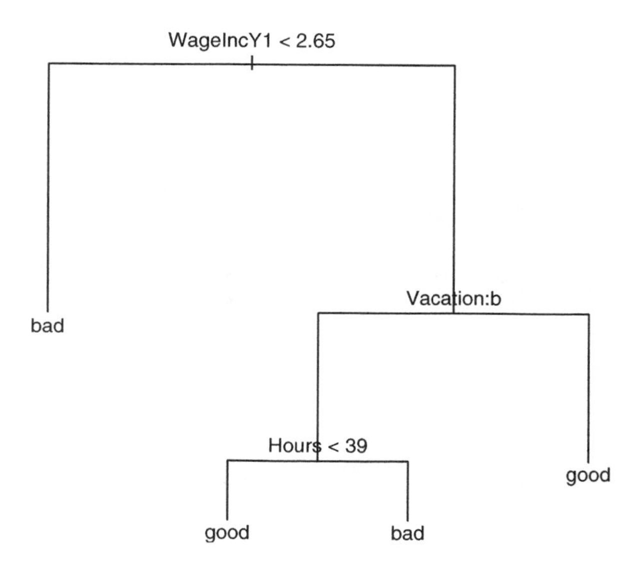

EXAMPLE 7.6 DIABETES AMONG PIMA INDIANS

The data in the file **PimaIndians.csv** are taken from the UCI Machine Learning Repository. The data set consists of 768 females at least 21 years old and of Pima Indian heritage. The response variable is the absence/presence of diabetes (coded 0/1, in the variable Class). The objective is to predict the presence of diabetes from the following eight continuous risk factors:

- Number of pregnancies
- Plasma glucose concentration determined by an oral glucose tolerance test
- Diastolic blood pressure (mmHg)
- Triceps skin fold thickness (mm)
- 2-h serum insulin

- Body mass index [weight in kilograms/(height in meters)^2]
- Diabetes pedigree function
- Age (years)

Several techniques can be used for the purpose of classifying Pima Indians into the diabetes/no diabetes groups, including logistic regression, classification trees, and linear or quadratic discriminant functions. Here we illustrate two techniques: logistic regression and classification trees.

```
## read the data and create plots
PimaIndians <- read.csv("C:/DataMining/Data/PimaIndians.csv")
PimaIndians
plot(PimaIndians)
PI=data.frame(PimaIndians)
```

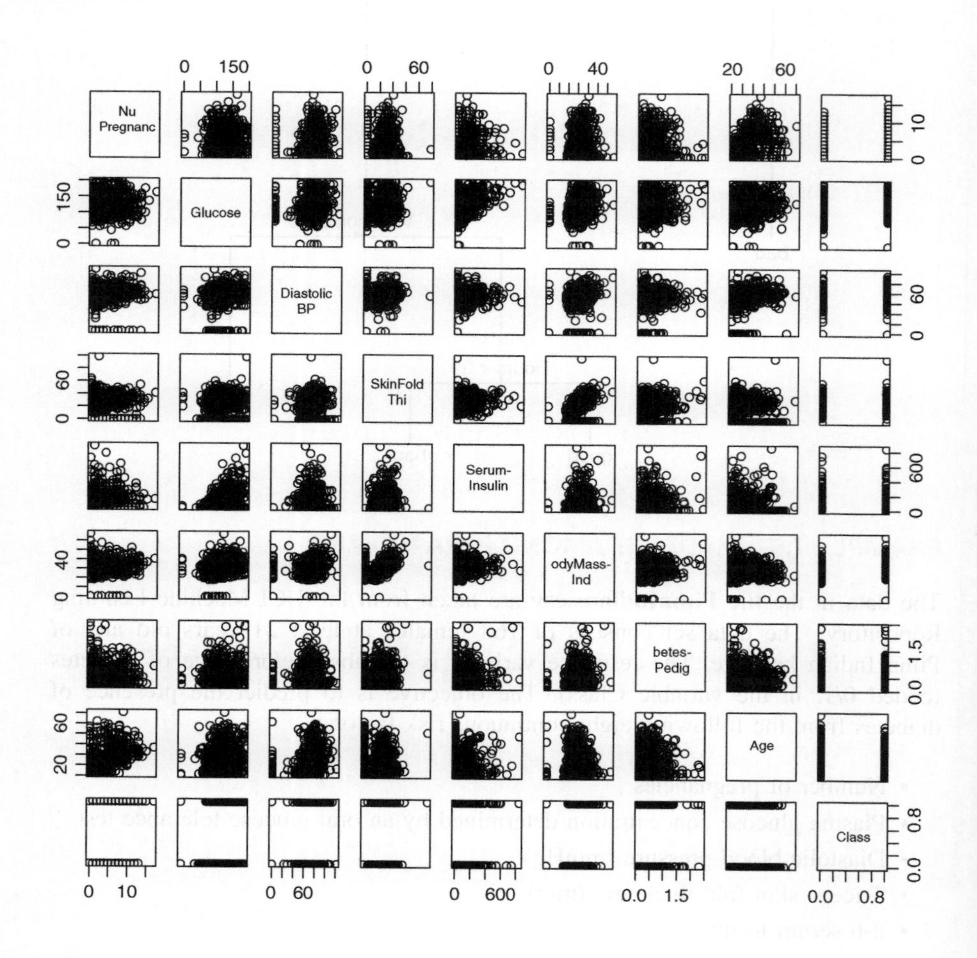

Method 1: Logistic regression The result of the logistic regression fit on all eight covariates (all eight covariates are continuous) is shown below. The model can be simplified as several predictors are not significant. We simplify the model through stepwise backward elimination. Triceps skinfold thickness is the first variable to be omitted. Estimating the simplified model, we find that age can be omitted next. Estimating the model without triceps skinfold thickness and age, we find that serum insulin can be dropped from the model. This leaves a model with five predictors: number of pregnancies, glucose, diastolic blood pressure, body mass index, and diabetes pedigree. As expected, blood glucose turns out to be an important determinant for the diagnosis of diabetes.

We split the data set into two equal halves. We fit the two models (the model with all eight covariates and the simplified model) on the first half and predict the probabilities of the two outcome categories for the subjects in the second half of the data set. We classify a subject into the group with the highest predicted probability. We find that the logistic regression model misclassifies 25% of the subjects. We do the same for the simplified model and find that the difference in the misclassification rates between the full and the simplified models is negligible (24.7% vs 26.0%).

```
## logistic regression model
## mm1: model fitted to all data
mm1=glm(Class~.,family=binomial,data=PI)
mm1
summary(mm1)
```

```
Call:
glm(formula = Class ~ ., family = binomial, data = PI)

Deviance Residuals:
    Min       1Q   Median       3Q      Max
-2.5566  -0.7274  -0.4159   0.7267   2.9297

Coefficients:

                            Estimate Std. Error z value Pr(>|z|)
(Intercept)               -8.4046964  0.7166359 -11.728  < 2e-16 ***
NuPregnancy                0.1231823  0.0320776   3.840 0.000123 ***
Glucose                    0.0351637  0.0037087   9.481  < 2e-16 ***
DiastolicBP               -0.0132955  0.0052336  -2.540 0.011072 *
TricepsSkinFoldThickness   0.0006190  0.0068994   0.090 0.928515
SerumInsulin              -0.0011917  0.0009012  -1.322 0.186065
BodyMassIndex              0.0897010  0.0150876   5.945 2.76e-09 ***
DiabetesPedigree           0.9451797  0.2991475   3.160 0.001580 **
Age                        0.0148690  0.0093348   1.593 0.111192
---
Signif. codes:  0 '***' 0.001 '**' 0.01 '*' 0.05 '.' 0.1 ' ' 1

(Dispersion parameter for binomial family taken to be 1)

    Null deviance: 993.48  on 767  degrees of freedom
Residual deviance: 723.45  on 759  degrees of freedom
AIC: 741.45

Number of Fisher Scoring iterations: 5
```

```
## simplifying the model through backward elimination
RPI=PI[,-4] ## dropping triceps skin fold thickness
mm1=glm(Class~.,family=binomial,data=RPI)
mm1
summary(mm1)
RPI=RPI[,-7]  ## dropping age
mm1=glm(Class~.,family=binomial,data=RPI)
mm1
summary(mm1)
RPI=RPI[,-4]  ## dropping serum insulin
RPI[1:3,]
mm1=glm(Class~.,family=binomial,data=RPI)
mm1
summary(mm1)

Call:
glm(formula = Class ~ ., family = binomial, data = RPI)

Deviance Residuals:
    Min       1Q   Median       3Q      Max
-2.7931  -0.7362  -0.4188   0.7251   2.9555

Coefficients:
                   Estimate Std. Error z value Pr(>|z|)
(Intercept)       -7.954952   0.675823 -11.771  < 2e-16 ***
NuPregnancy        0.153492   0.027835   5.514  3.5e-08 ***
Glucose            0.034658   0.003394  10.213  < 2e-16 ***
DiastolicBP       -0.012007   0.005031  -2.387  0.01700 *
BodyMassIndex      0.084832   0.014125   6.006  1.9e-09 ***
DiabetesPedigree   0.910628   0.294027   3.097  0.00195 **
---
Signif. codes:  0 '***' 0.001 '**' 0.01 '*' 0.05 '.' 0.1 ' ' 1

(Dispersion parameter for binomial family taken to be 1)

    Null deviance: 993.48  on 767  degrees of freedom
Residual deviance: 728.56  on 762  degrees of freedom
AIC: 740.56

Number of Fisher Scoring iterations: 5


## evaluation of the full model
## split the data set into a training (50%) and a test (evaluation)
## set (50%)
set.seed(1)
n=length(PI$Class)
n
n1=floor(n*(0.5))
n1
n2=n-n1
n2
train=sample(1:n,n1)

PI1=data.frame(PI[train,])
PI2=data.frame(PI[-train,])
```

```
## mm2: model fitted on the training data set
mm2=glm(Class~.,family=binomial,data=PI1)
mm2
summary(mm2)

## create predictions for the test (evaluation) data set
gg=predict(mm2,newdata=PI2,type= "response")
gg
hist(gg)
plot(PI$Class[-train]~gg)

## coding as 1 if probability 0.5 or larger
gg1=floor(gg+0.5)
ttt=table(PI$Class[-train],gg1)
ttt

      gg1
      0   1
  0 215  32
  1  63  74

error=(ttt[1,2]+ttt[2,1])/n2
error

[1] 0.2473958

## evaluation of the simplified model
## mm2: model fitted on the training data set
mm2=glm(Class~NuPregnancy+Glucose+DiastolicBP+BodyMassIndex+
+    DiabetesPedigree,family=binomial,data=PI1)
mm2
summary(mm2)

## create predictions for the test (evaluation) data set
gg=predict(mm2,newdata=PI2,type= "response")
gg
hist(gg)
plot(PI$Class[-train]~gg)

## coding as 1 if probability 0.5 or larger
gg1=floor(gg+0.5)
ttt=table(PI$Class[-train],gg1)
ttt

      gg1
      0   1
  0 212  35
  1  65  72

error=(ttt[1,2]+ttt[2,1])/n2
error

[1] 0.2604167
```

Method 2: Classification Trees We start by fitting a fairly large tree to the data. Cross-validation indicates that the resulting tree can be cut back to a smaller size. The tree with seven terminal nodes shown in the following has two splits that lead to identical classifications. Node 7 leads to identical results when splitting on blood glucose (relative to 157.5), and we notice identical classification results in the subtrees below node 2. This indicates that nodes 2 and 7 and their subtrees can be snipped off. Snipping off nodes 2 and 7 leads to our final tree. Women with blood glucose less than 127.5 and women with body mass index less than 29.95 are put in the "no diabetes" class. Women with blood glucose larger than 127.5 and body mass index larger than 29.95 are put in the diabetes risk group. This rule misclassifies $175/768 = 0.23$ of the women. Classifying on glucose alone is slightly worse, with misclassification rate $203/768 = 0.26$.

```
## read the data
PimaIndians <- read.csv("C:/DataMining/Data/PimaIndians.csv")
PimaIndians

## CART analysis
library(tree)
PimaIndians$Class=factor(PimaIndians$Class)
## constructing the classification tree
PItree <- tree(Class ~., data = PimaIndians,mindev=0.01)
PItree

node), split, n, deviance, yval, (yprob)
      * denotes terminal node

 1) root 768 993.50 0 ( 0.65104 0.34896 )
   2) Glucose < 127.5 485 477.00 0 ( 0.80619 0.19381 )
     4) Age < 28.5 271 157.50 0 ( 0.91513 0.08487 )
       8) BodyMassIndex < 30.95 151  21.27 0 ( 0.98675 0.01325 ) *
       9) BodyMassIndex > 30.95 120 111.30 0 ( 0.82500 0.17500 ) *
     5) Age > 28.5 214 272.00 0 ( 0.66822 0.33178 )
      10) BodyMassIndex < 26.35 41  15.98 0 ( 0.95122 0.04878 ) *
      11) BodyMassIndex > 26.35 173 232.70 0 ( 0.60116 0.39884 )
        22) Glucose < 99.5 55  52.16 0 ( 0.81818 0.18182 ) *
        23) Glucose > 99.5 118 163.60 0 ( 0.50000 0.50000 )
          46) DiabetesPedigree < 0.561 84 113.40 0 ( 0.59524 0.40476 ) *
          47) DiabetesPedigree > 0.561 34  39.30 1 ( 0.26471 0.73529 )
            94) NuPregnancy < 6.5 21  28.68 1 ( 0.42857 0.57143 ) *
            95) NuPregnancy > 6.5 13   0.00 1 ( 0.00000 1.00000 ) *
   3) Glucose > 127.5 283 377.30 1 ( 0.38516 0.61484 )
     6) BodyMassIndex < 29.95 76  94.80 0 ( 0.68421 0.31579 )
      12) Glucose < 145.5 41  34.14 0 ( 0.85366 0.14634 ) *
      13) Glucose > 145.5 35  48.49 1 ( 0.48571 0.51429 ) *
     7) BodyMassIndex > 29.95 207 243.60 1 ( 0.27536 0.72464 )
      14) Glucose > 157.5 115 153.90 1 ( 0.39130 0.60870 ) *
      15) Glucose > 157.5 92  71.25 1 ( 0.13043 0.86957 ) *

summary(PItree)

Classification tree:
tree(formula = Class ~ ., data = PimaIndians)
Variables actually used in tree construction:
[1] "Glucose"          "Age"                 "BodyMassIndex"     "DiabetesPedigree"
```

```
[5] "NuPregnancy"
Number of terminal nodes:  11
Residual mean deviance:  0.8594 = 650.6 / 757
Misclassification error rate: 0.2057 = 158 / 768
```

```
plot(PItree, col=8)
text(PItree, digits=2)

## cross-validation to prune the tree
set.seed(2)
cvPI <- cv.tree(PItree, K=10)
cvPI$size
```

```
[1] 11 10  9  8  7  6  5  4  3  2  1
```

```
cvPI$dev
```

```
 [1] 881.4488 834.1200 830.0404 835.6352 781.2184 795.9745 823.2477 828.7466
 [9] 876.0231 886.6899 996.6544
```

```
plot(cvPI, pch=21, bg=8, type="p", cex=1.5, ylim=c(700,1000))
```

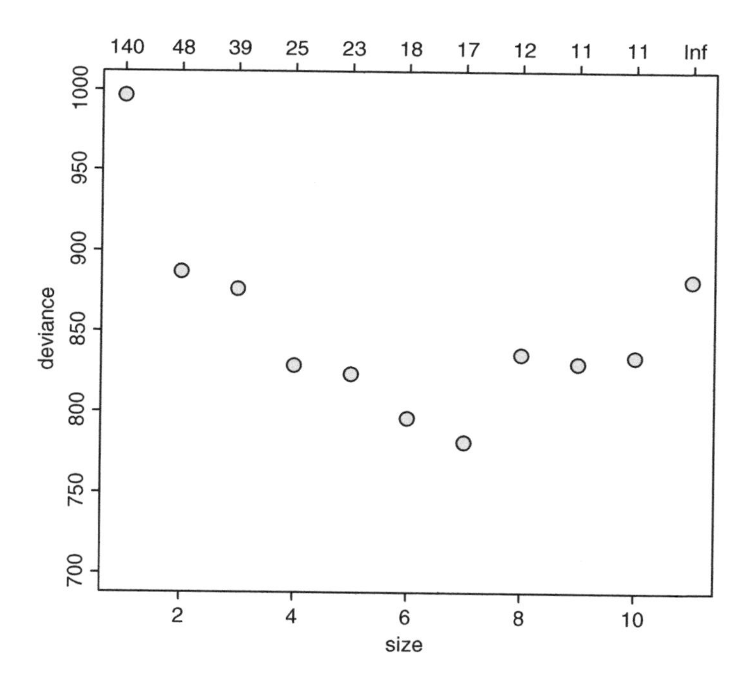


```
PIcut <- prune.tree(PItree, best=7)
PIcut
```

```
node), split, n, deviance, yval, (yprob)
      * denotes terminal node
```

```
1) root 768 993.50 0 ( 0.65104 0.34896 )
  2) Glucose < 127.5 485 477.00 0 ( 0.80619 0.19381 )
    4) Age > 28.5 271 157.50 0 ( 0.91513 0.08487 )
      8) BodyMassIndex < 30.95 151  21.27 0 ( 0.98675 0.01325 ) *
      9) BodyMassIndex > 30.95 120 111.30 0 ( 0.82500 0.17500 ) *
    5) Age > 28.5 214 272.00 0 ( 0.66822 0.33178 )
     10) BodyMassIndex < 26.35 41  15.98 0 ( 0.95122 0.04878 ) *
     11) BodyMassIndex > 26.35 173 232.70 0 ( 0.60116 0.39884 ) *
  3) Glucose > 127.5 283 377.30 1 ( 0.38516 0.61484 )
    6) BodyMassIndex < 29.95 76  94.80 0 ( 0.68421 0.31579 ) *
    7) BodyMassIndex > 29.95 207 243.60 1 ( 0.27536 0.72464 )
     14) Glucose < 157.5 115 153.90 1 ( 0.39130 0.60870 ) *
     15) Glucose > 157.5 92  71.25 1 ( 0.13043 0.86957 ) *
```

```
summary(PIcut)
```

```
Classification tree:
snip.tree(tree = PItree, nodes = c(15, 9, 8, 6, 14, 10, 11))
Variables actually used in tree construction:
[1] "Glucose"        "Age"            "BodyMassIndex"
Number of terminal nodes:  7
Residual mean deviance:  0.9215 = 701.2 / 761
Misclassification error rate: 0.2279 = 175 / 768
```

```
plot(PIcut, col=8)
text(PIcut)
```

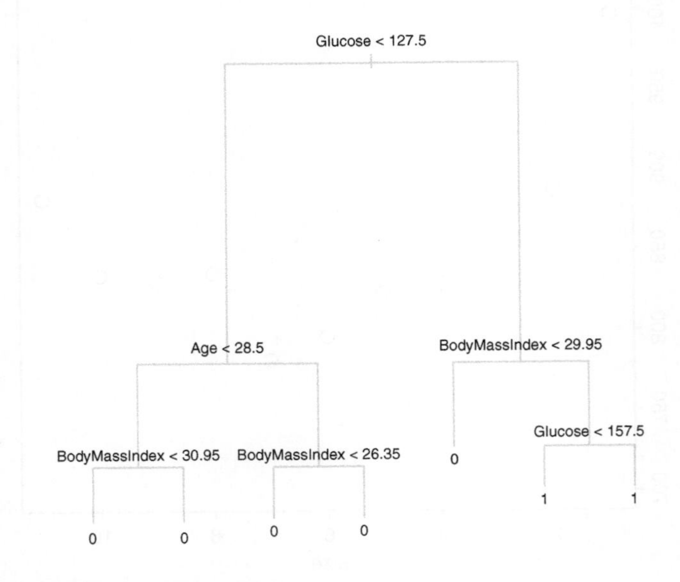


```
P1=snip.tree(PIcut,nodes=c(2,7))
P1
```

```
node), split, n, deviance, yval, (yprob)
      * denotes terminal node

1) root 768 993.5 0 ( 0.6510 0.3490 )
```

```
  2) Glucose < 127.5 485 477.0 0 ( 0.8062 0.1938 ) *
  3) Glucose > 127.5 283 377.3 1 ( 0.3852 0.6148 )
    6) BodyMassIndex < 29.95 76  94.8 0 ( 0.6842 0.3158 ) *
    7) BodyMassIndex > 29.95 207 243.6 1 ( 0.2754 0.7246 ) *
```

```
summary(P1)
```

```
Classification tree:
snip.tree(tree = PIcut, nodes = c(2, 7))
Variables actually used in tree construction:
[1] "Glucose"        "BodyMassIndex"
Number of terminal nodes:  3
Residual mean deviance:  1.066 = 815.4 / 765
Misclassification error rate: 0.2279 = 175 / 768
```

```
plot(P1)
text(P1)
```

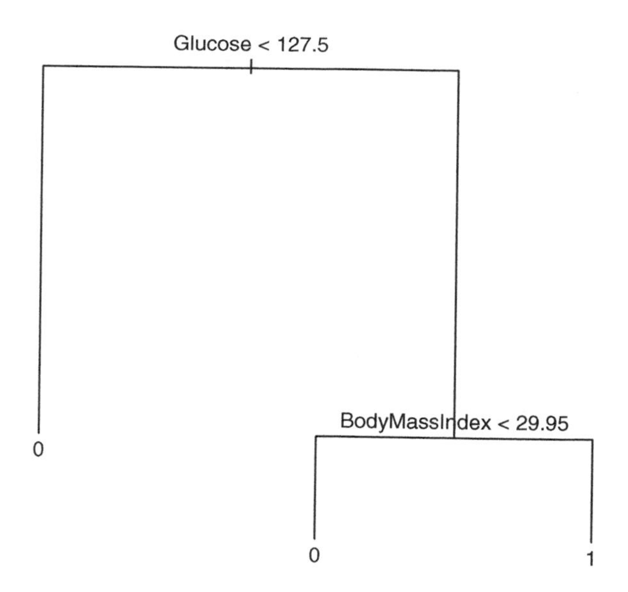

The R programs can be found on the webpage that accompanies this book.

EXAMPLE 7.7 PREDICTING THE CPU PERFORMANCE WITH REGRESSION AND REGRESSION TREES

The data in the file **cpu.csv** are taken from the UCI Machine Learning Repository. The data set consists of 209 different computer configurations, as specified by the following six explanatory variables:

- Cycle Time (MYCT, in nanoseconds),
- Min Main Memory (MMIN, in kilobytes),
- Max Main Memory (MMAX, in kilobytes),
- Cache of Main Memory (CACH, in kilobytes),
- Min Number of Channels,
- Max Number of Channels.

The published and the estimated relative CPU performances [PRP (published relative performance) and ERP (event-related potentials)] are the two response variables that need to be explained. Here we consider PRP as the response.

We use this data set (i) to obtain a prediction equation for the CPU performance in terms of the six numeric interval-scaled attributes and (ii) to construct a regression tree.

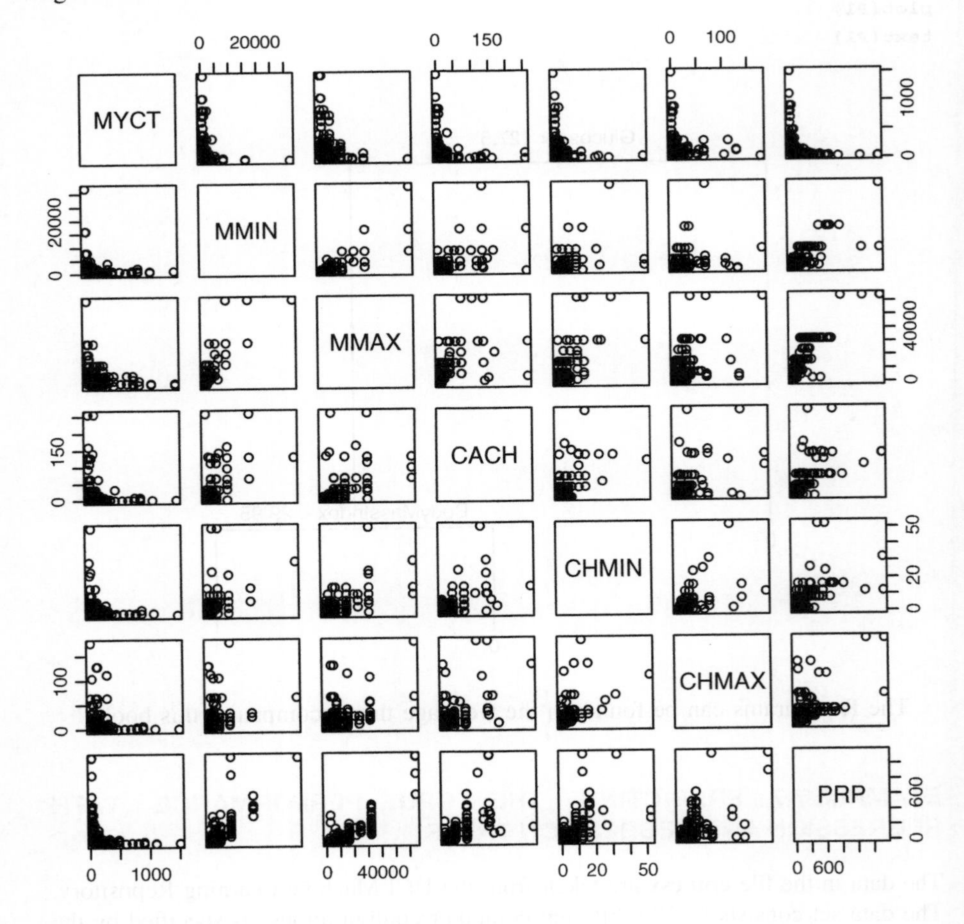

```
## read the data and create a matrix plot
cpu <- read.csv("C:/DataMining/Data/cpu.csv")
cpu
xx=cpu[,c(-1,-9)]
xx[1:3,]

    MYCT  MMIN  MMAX CACH CHMIN CHMAX  PRP
1    125   256  6000  256    16   128  198
2     29  8000 32000   32     8    32  269
3     29  8000 32000   32     8    32  220

plot(xx)


## regression model
regfit=lm(PRP~.,data=xx)
regfit
summary(regfit)

Call:
lm(formula = PRP ~., data = xx)

Residuals:
    Min      1Q  Median      3Q     Max
-195.82  -25.17    5.40   26.52  385.75

Coefficients:
              Estimate Std. Error t value Pr(>|t|)
(Intercept) -5.589e+01  8.045e+00  -6.948 5.00e-11 ***
MYCT         4.885e-02  1.752e-02   2.789   0.0058 **
MMIN         1.529e-02  1.827e-03   8.371 9.42e-15 ***
MMAX         5.571e-03  6.418e-04   8.681 1.32e-15 ***
CACH         6.414e-01  1.396e-01   4.596 7.59e-06 ***
CHMIN       -2.704e-01  8.557e-01  -0.316   0.7524
CHMAX        1.482e+00  2.200e-01   6.737 1.65e-10 ***
---
Signif. codes:  0 '***' 0.001 '**' 0.01 '*' 0.05 '.' 0.1 ' ' 1

Residual standard error: 59.99 on 202 degrees of freedom
Multiple R-squared: 0.8649,     Adjusted R-squared: 0.8609
F-statistic: 215.5 on 6 and 202 DF,  p-value: < 2.2e-16

## cross-validation (leave one out): regression model on all
## six regressors
n=length(cpu$PRP)
diff=dim(n)
percdiff=dim(n)
for (k in 1:n) {
train1=c(1:n)
```

```
train=train1[train1!=k]
m1=lm(PRP~.,data=xx[train,])
pred=predict(m1,newdat=xx[-train,])
obs=xx[-train,7]
diff[k]=obs-pred
percdiff[k]=abs(diff[k])/obs
}
me=mean(diff)
rmse=sqrt(mean(diff**2))
mape=100*(mean(percdiff))
me   # mean error

 [1] -0.3430341

rmse # root mean square error

 [1] 69.64521

mape # mean absolute percent error

 [1] 79.1892


library(tree)

## Construct the regression tree
cputree <- tree(PRP ~., data=xx, mindev=0.1, mincut=1)
cputree <- tree(PRP ~., data= xx, mincut=1)
cputree
summary(cputree)
plot(cputree, col=8)
text(cputree, digits=2)

## Use cross-validation to prune the regression tree
set.seed(2)
cvcpu <- cv.tree(cputree, K=10)
cvcpu$size

 [1] 7 6 5 4 3 2 1

cvcpu$dev

 [1] 1634992 1764535 1748460 2082174 2082174 2619591 5435820

plot(cvcpu, pch=21, bg=8, type="p", cex=1.5,
+     ylim=c(0,6000000))
```

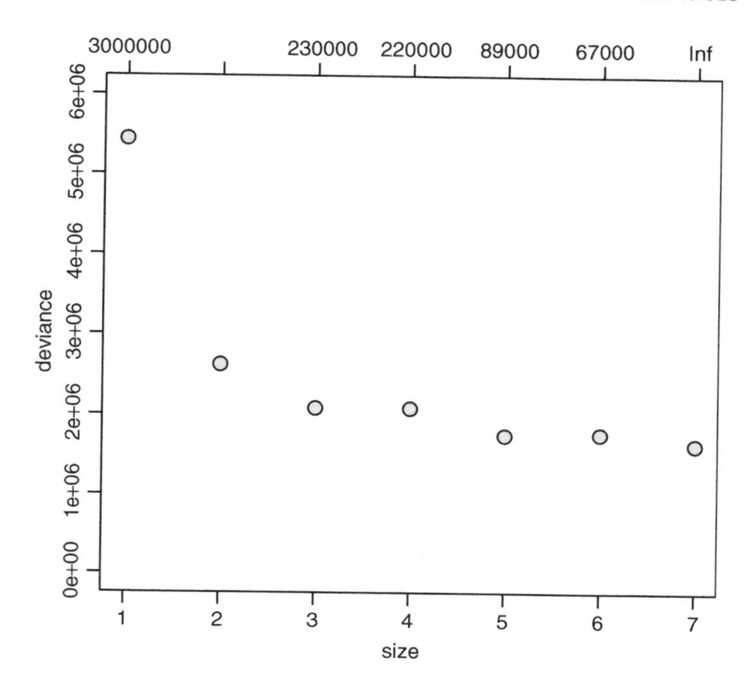


```
cpucut <- prune.tree(cputree, best=7)
cpucut
```

```
node), split, n, deviance, yval
      * denotes terminal node

 1) root 209 5380000 105.60
   2) MMAX < 48000 205 2218000  88.93
     4) MMAX < 22485 178  511000  57.80
       8) CACH < 27 141   97850  39.64 *
       9) CACH > 27 37  189500 127.00
        18) CACH < 96.5 31   65600 105.40 *
        19) CACH > 96.5 6   34840 238.50 *
     5) MMAX > 22485 27  397100 294.10
      10) MMIN < 12000 21  150700 244.60
        20) CHMIN < 7 5    3049 143.60 *
        21) CHMIN > 7 16   80730 276.10 *
      11) MMIN > 12000 6   14190 467.70 *
   3) MMAX > 48000 4  177000 961.20 *
```

```
summary(cpucut)
```

```
Regression tree:
tree(formula = PRP ~., data = xx, mincut = 1)
```

```
Variables actually used in tree construction:
[1] "MMAX"  "CACH"  "MMIN"  "CHMIN"
Number of terminal nodes:  7
Residual mean deviance:  2343 = 473200 / 202
Distribution of residuals:
    Min.   1st Qu.   Median     Mean  3rd Qu.     Max.
-325.200  -19.640   -4.638    0.000   20.360  188.800
```

```
plot(cpucut, col=8)
text(cpucut)
```

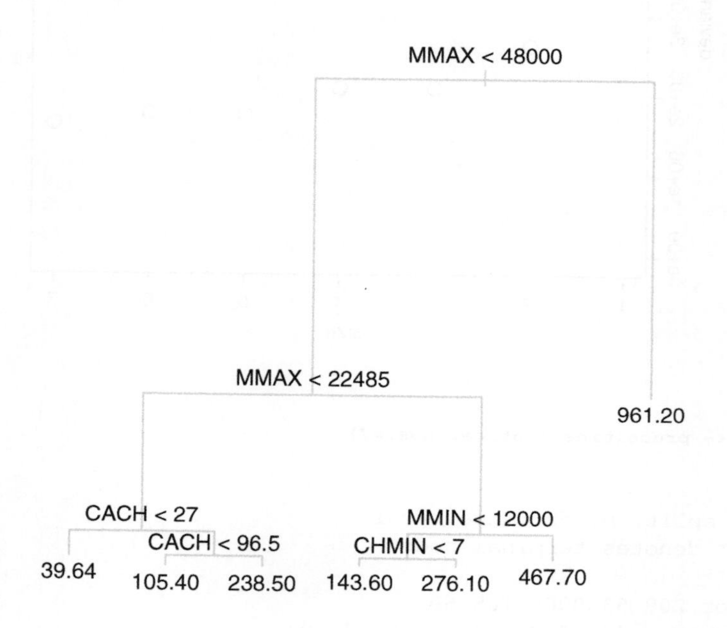

Interpretation of the Results The regression of published performance (PRP) on all six regressor variables explains 86.4% of the variability in PRP and leads to an in-sample root mean square error of 60.0. The leave-one-out cross-validation results in a root mean square error of 69.6, slightly larger than the in-sample estimate. The mean absolute percent error (79%) is certainly not very impressive.

The regression tree performs better, leading to a root mean square error of $\sqrt{2343} = 48.4$. It uses four of the six variables (MMAX, MMIN, CACH, and CHMIN) and results in seven terminal nodes. For example, for a computer with MMAX of less than 22,485 and CACH of less than 27 the model predicts a PRP performance of 39.64 (terminal node under 8). The performance of the third computer on the list, with MMAX $= 32,000$, MMIN $= 8000$, CACH $= 32$, and actual PRP $= 220$, is predicted as 276.10.

EXAMPLE 7.8 INFERRING THE CULTIVAR OF WINE USING CLASSIFICATION TREES, DISCRIMINANT ANALYSIS, AND MULTINOMIAL LOGISTIC REGRESSION

The data in the file **wine.csv** (taken from the UCI Machine Learning Repository) are the results of a chemical analysis of 174 Italian wines from three known cultivars (a cultivar is a group of grapes selected for desirable characteristics that can be maintained by propagation). The chemical analysis determined the quantities of the following 13 different constituents:

- Alcohol
- Malic acid
- Ash
- Alkalinity of ash
- Magnesium
- Total phenols
- Flavanoids
- Nonflavanoid phenols
- Proanthocyanins
- Color intensity
- Hue
- OD280/OD315 of diluted wines
- Proline.

The objective is to classify the 174 wines on the basis of their attributes and to predict their cultivar. We approach this problem from different vantage points: (i) we use classification trees; (ii) we cluster the 174 wines on the basis of their 13 attributes and check whether the resulting clusters agree with the provided information on the cultivars; (iii) we use linear and quadratic discriminant analysis for classifying the information into three groups; and (iv) we use multinomial logistic regression to model the relationship between cultivar and the attributes and to classify the wines. All analyses indicate that this is quite an easy classification problem.

```
## read the data and plots
wine <- read.csv("C:/DataMining/Data/wine.csv")
wine[1:3,]
plot(wine)
```

Method 1: Classification trees

```
## CART
library(tree)
wine$Class=factor(wine$Class)
## constructing the classification tree
Winetree <- tree(Class ~., data = wine)
Winetree
summary(Winetree)
plot(Winetree, col=8)
text(Winetree, digits=2)

## cross-validation to prune the tree
set.seed(1)
cvWine <- cv.tree(Winetree, K=10)
cvWine$size
```

```
[1] 7 6 5 4 3 2 1
```

```
cvWine$dev
```

```
[1] 144.7246 156.8253 157.2120 130.6236 176.0941 271.4292 389.3242
```

```
plot(cvWine, pch=21, bg=8, type="p", cex=1.5, ylim=c(100,400))
```

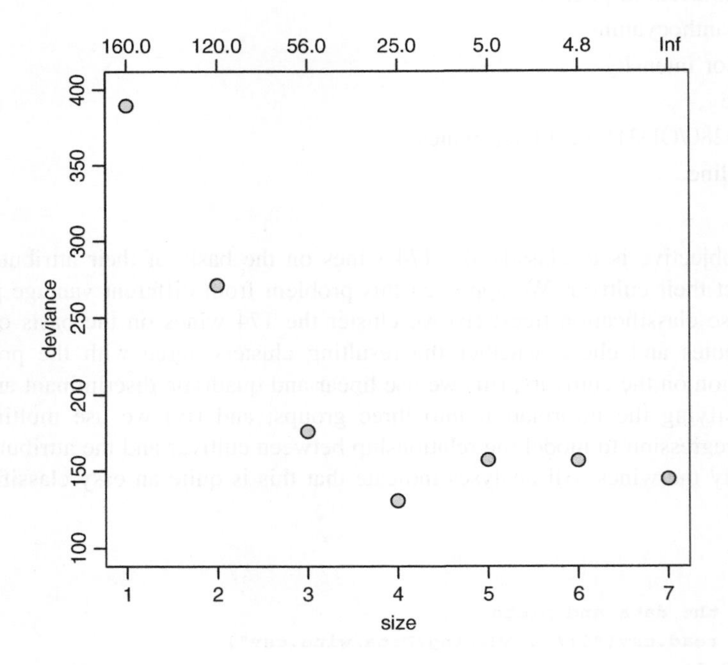


```
Winecut <- prune.tree(Winetree, best=4)
Winecut
```

```
node), split, n, deviance, yval, (yprob)
      * denotes terminal node

1) root 178 386.600 2 ( 0.33146 0.39888 0.26966 )
  2) Flavanoids < 1.575 62  66.240 3 ( 0.00000 0.22581 0.77419 )
    4) Color < 3.825 13   0.000 2 ( 0.00000 1.00000 0.00000 ) *
    5) Color > 3.825 49   9.763 3 ( 0.00000 0.02041 0.97959 ) *
  3) Flavanoids > 1.575 116 160.800 1 ( 0.50862 0.49138 0.00000 )
    6) Proline < 724.5 54   9.959 2 ( 0.01852 0.98148 0.00000 ) *
    7) Proline > 724.5 62  29.660 1 ( 0.93548 0.06452 0.00000 ) *

summary(Winecut)

Classification tree:
snip.tree(tree = Winetree, nodes = c(5, 6, 7))
Variables actually used in tree construction:
[1] "Flavanoids" "Color"      "Proline"
Number of terminal nodes:  4
Residual mean deviance:  0.2838 = 49.39 / 174
Misclassification error rate: 0.03371 = 6 / 178

plot(Winecut, col=8)
text(Winecut)
```

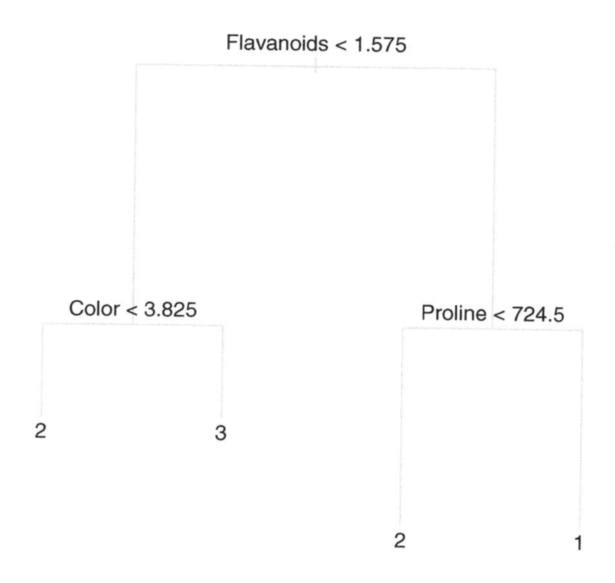

Interpretation of the Results A fairly simple model with just three attributes (flavanoids, proline, and color) and four terminal nodes is quite successful in classifying the cultivar. Wines with flavanoids greater than 1.575 and proline greater than 724.5 are classified as coming from cultivar 1 (node 7). Wines with flavanoids less than 1.575 and color greater than 3.825 are classified as coming from cultivar 3. Wines with flavanoids less than 1.575 and color less than 3.825, and wines with flavanoids greater than 1.575 and proline less than 724.5 are classified as coming from cultivar 2. This simple rule misclassifies 6 of 178 wines, for a misclassification rate of 3.4%.

Method 2: Clustering We standardize the attributes as their units are considerably different. Applying the k-means clustering algorithm to the 13-dimensional attribute vector leads to an excellent prediction of the cultivars. Only 6 out of 176 wines are misclassified. Note that clustering is an unsupervised learning technique. The composition of the groups that we get from clustering is very close to the true grouping of the cultivars. But because clustering is unsupervised learning, clustering cannot tell us which cultivars the groupings represent.

```
## Clustering

## standardizing the attributes as units considerably different
wines=matrix(nrow=length(wine[,1]),ncol=length(wine[1,]))
for (j in 2:14) {
wines[,j]=(wine[,j]-mean(wine[,j]))/sd(wine[,j])
}
wines[,1]=wine[,1]
winesr=wines[,-1]
winesr[1:3,]
## kmeans clustering with 13 standardized attributes
grpwines <- kmeans(winesr, centers=3, nstart=20)
grpwines

K-means clustering with 3 clusters of sizes 51, 65, 62

Cluster means:
        [,1]       [,2]       [,3]       [,4]        [,5]        [,6]
1  0.1644436  0.8690954  0.1863726  0.5228924 -0.07526047 -0.97657548
2 -0.9234669 -0.3929331 -0.4931257  0.1701220 -0.49032869 -0.07576891
3  0.8328826 -0.3029551  0.3636801 -0.6084749  0.57596208  0.88274724
         [,7]        [,8]        [,9]       [,10]      [,11]      [,12]
1 -1.21182921  0.72402116 -0.77751312  0.9388902 -1.1615122 -1.2887761
2  0.02075402 -0.03343924  0.05810161 -0.8993770  0.4605046  0.2700025
3  0.97506900 -0.56050853  0.57865427  0.1705823  0.4726504  0.7770551
       [,13]
1 -0.4059428
2 -0.7517257
3  1.1220202

Clustering vector:
  [1] 3 3 3 3 3 3 3 3 3 3 3 3 3 3 3 3 3 3 3 3 3 3 3 3 3 3 3 3 3 3 3 3 3 3 3 3 3
 [38] 3 3 3 3 3 3 3 3 3 3 3 3 3 3 3 3 3 3 3 3 3 3 2 2 1 2 2 2 2 2 2 2 2 2 2 2 3
 [75] 2 2 2 2 2 2 2 2 2 2 1 2 2 2 2 2 2 2 2 2 2 3 2 2 2 2 2 2 2 2 2 2 2 2 2 2 2
[112] 2 2 2 2 2 2 2 1 2 2 3 2 2 2 2 2 2 2 2 1 1 1 1 1 1 1 1 1 1 1 1 1 1 1 1 1 1
[149] 1 1 1 1 1 1 1 1 1 1 1 1 1 1 1 1 1 1 1 1 1 1 1 1 1 1 1 1 1 1 1 1

Within cluster sum of squares by cluster:
[1] 326.3537 558.6971 385.6983
 (between_SS / total_SS =  44.8 %)

Available components:

[1] "cluster"      "centers"      "totss"        "withinss"     "tot.withinss"
[6] "betweenss"    "size"
```

```
grpwines$cluster ## displaying clustering results
## a bold face entry indicates a classification error

  [1] 3 3 3 3 3 3 3 3 3 3 3 3 3 3 3 3 3 3 3 3 3 3 3 3 3 3 3 3 3 3 3 3 3 3 3 3 3
 [38] 3 3 3 3 3 3 3 3 3 3 3 3 3 3 3 3 3 3 3 3 3 3 2 2 1 2 2 2 2 2 2 2 2 2 2 2 3
 [75] 2 2 2 2 2 2 2 2 2 1 2 2 2 2 2 2 2 2 2 2 2 3 2 2 2 2 2 2 2 2 2 2 2 2 2 2 2
[112] 2 2 2 2 2 2 2 1 2 2 3 2 2 2 2 2 2 2 2 1 1 1 1 1 1 1 1 1 1 1 1 1 1 1 1 1 1
[149] 1 1 1 1 1 1 1 1 1 1 1 1 1 1 1 1 1 1 1 1 1 1 1 1 1 1 1 1 1 1

wine$Class  ## actual classes

  [1] 1 1 1 1 1 1 1 1 1 1 1 1 1 1 1 1 1 1 1 1 1 1 1 1 1 1 1 1 1 1 1 1 1 1 1 1 1
 [38] 1 1 1 1 1 1 1 1 1 1 1 1 1 1 1 1 1 1 1 1 1 1 2 2 2 2 2 2 2 2 2 2 2 2 2 2 2
 [75] 2 2 2 2 2 2 2 2 2 2 2 2 2 2 2 2 2 2 2 2 2 2 2 2 2 2 2 2 2 2 2 2 2 2 2 2 2
[112] 2 2 2 2 2 2 2 2 2 2 2 2 2 2 2 2 2 2 2 3 3 3 3 3 3 3 3 3 3 3 3 3 3 3 3 3 3
[149] 3 3 3 3 3 3 3 3 3 3 3 3 3 3 3 3 3 3 3 3 3 3 3 3 3 3 3 3 3 3
Levels: 1 2 3

## 6 mistakes made among 178 wines
```

Method 3: Discriminant analysis Here we use linear and quadratic discriminant analysis to classify the data into three groups. The linear discriminant analysis makes no classification error, while the quadratic discriminant analysis misclassifies one item. Cross-validation (where we leave out one item in the construction of the classifier and then apply the rule to the item that has been left out) leads to very similar results (two errors for linear, and one error for quadratic discriminant analysis).

```
## Discriminant analysis (linear/quadratic)

library(MASS)
## linear discriminant analysis using the standardized
## attributes
wines[1:3,]
ws=data.frame(wines)
ws[1:3,]
zlin=lda(X1~.,ws,prior=c(1,1,1)/3)
zlin
## quadratic discriminant analysis
zqua=qda(X1~.,ws,prior=c(1,1,1)/3)
zqua

n=dim(ws)[1]
errorlin=1-(sum(ws$X1==predict(zlin,ws)$class)/n)
errorlin

[1] 0

errorqua=1-(sum(ws$X1==predict(zqua,ws)$class)/n)
errorqua
```

```
[1] 0.005617978
```

```
neval=1
corlin=dim(n)
corqua=dim(n)

## leave one out evaluation
for (k in 1:n) {
train1=c(1:n)
train=train1[train1!=k]
## linear discriminant analysis
zlin=lda(X1~.,ws[train,],prior=c(1,1,1)/3)
corlin[k]=ws$X1[-train]==predict(zlin,ws[-train,])$class

## quadratic discriminant analysis
zqua=qda(X1~.,ws[train,],prior=c(1,1,1)/3)
corqua[k]=ws$X1[-train]==predict(zqua,ws[-train,])$class

}
merrlin=1-mean(corlin)
merrlin
```

```
[1] 0.01123596
```

```
merrqua=1-mean(corqua)
merrqua
```

```
[1] 0.005617978
```

Method 4: Multinomial logistic regression Finally, we estimate a multinomial logistic regression model on the standardized attributes and use the predicted class probabilities to classify the wines into three groups. The multinomial logistic model with all 13 attributes is certainly overspecified, and the large standard errors of the estimated coefficients make an interpretation of the model impossible. However, a rule that assigns a wine to the cultivar that has the largest predicted probability leads to a perfect in-sample classification. Of course, a more reliable evaluation would use cross-validation and split the data set into estimation and evaluation data sets. We suggest that you do this as an exercise.

```
## Multinomial logistic regression
## using VGAM

library(VGAM)
ws=data.frame(wines)
gg <- vglm(X1 ~ .,multinomial,data=ws)
summary(gg)
```

```
predict(gg) ## log-odds relative to last group
round(fitted(gg),2)  ## probabilities
cbind(round(fitted(gg),2),ws$X1)
## perfect classification
```

APPENDIX B

References

Abraham, B. and Ledolter, J.: *Statistical Methods for Forecasting*. New York: John Wiley & Sons, Inc., 1983.

Abraham, B. and Ledolter, J.: *Introduction to Regression Modeling*. Belmont, CA: Duxbury Press, 2006.

Adler, J.: *R In a Nutshell: A Desktop Quick Reference*. Sebastopol, CA: O'Reilly Media, 2009.

Agresti, A.: *Categorical Data Analysis*. Second edition, New York: John Wiley & Sons Inc., 2002.

Asuncion, A. and Newman, D.J.: UCI Machine Learning Repository. Irvine, CA: University of California, School of Information and Computer Science, 2007. Available at http://www.ics.uci.edu/~mlearn/MLRepository.html. Accessed 2013 Jan 16.

Benjamini, Y. and Hochberg, Y.: Controlling the False Discovery Rate: A Practical and Powerful Approach to Multiple Testing. *Journal of the Royal Statistical Society: Series B*, Vol. 57 (1995), No 1, 289–300.

Benjamini, Y. and Yekutieli, D.: The control of the false discovery rate in multiple testing under dependency. *Annals of Statistics*, Vol. 29 (2001), No 4, 1165–1188.

Biggs, D., DeVille, B., and Suen, E.: A method of choosing multiway partitions for classification and decision trees. *Journal of Applied Statistics*, Vol. 18 (1991), 49–62.

Box, G. E. P., Jenkins, G. M. and Reinsel, G. C.: *Time Series Analysis, Forecasting and Control*. Third edition. Englewood Cliffs, NJ: Prentice Hall, 1994.

Breiger R. and Pattison P.: Cumulated social roles: the duality of persons and their algebras. *Social Networks*, Vol. 8 (1986), 215–256.

Breiman, L., Friedman, J., Ohlsen, R., and Stone, C.: *Classification and Regression Trees*. Pacific Grove: Wadsworth, 1984.

Brinkman, N.D.: Ethanol fuel – a single-cylinder engine study of efficiency and exhaust emissions, *SAE Transactions*, Vol. 90 (1981).

Brown, P.J., Stone, J. and Ord-Smith, C.: Toximic signs during pregnancy. *Applied Statistics*, Vol. 32 (1983), 69–72.

Burt, R.S.: Structural holes and good ideas, *American Journal of Sociology*, Vol. 110 (2004), No 2, 349–399.

Celma, O.: *Music Recommendation and Discovery*. New York: Springer, 2010.

Cendrowska, J.: PRISM: an algorithm for inducing modular rules. *International Journal of Man–Machine Studies*, Vol. 27 (1987), 349–370.

Data Mining and Business Analytics with R, First Edition. Johannes Ledolter.

Cleveland, W.S. and Devlin, S.J.: Locally weighted regression: an approach to regression analysis by local fitting. *Journal of the American Statistical Association*, Vol. 83 (1988), 596–610.

Craven, P. and Wahba, G.: Smoothing noisy data with spline functions. *Numerische Mathematik*, Vol. 31 (1979), 377–403.

Efron, B., Johnstone, I., Hastie, T. and Tibshirani, R.: (2004). Least angle regression. *Annals of Statistics*, Vol. 32 (2004), No 2, 407–499.

Epanechnikov, V.A.: Nonparametric estimates of a multivariate probability density. *Theory of Probability and its Applications*, Vol. 14 (1969), 153–158.

Fan, J. and Gijbels, I.: *Local Polynomial Modelling and Its Applications*. London: Chapman and Hall, 1996.

Frank, I.E. and Friedman, J.H.: A statistical view of some chemometrics regression tools. *Technometrics*, Vol. 35 (1993), 109–135.

Gentzkow, M. and Shapiro, J.: What drives media slant? Evidence from U.S. daily newspapers. *Econometrica*, Vol. 78 (2010), 35–71.

Granger, C.W.J. and Bates, J.: The combination of forecasts. *Operations Research Quarterly*, Vol. 20 (1969), 451–468.

Hahsler, M., Gruen, B., and Hornik, K.: arules – a computational environment for mining association rules and frequent item sets. *Journal of Statistical Software*, Vol. 14 (2005), 1–25.

Hand, D.J., Daly, F., Lunn, A.D., McConway, K.J., and Ostrowski, E.: *A Handbook of Small Data Sets*. London: Chapman & Hall, 1994.

Handcock, M.S., Hunter, D.R., Butts, C.T., Goodreau, S.M., Morris, M.: statnet: software tools for the representation, visualization, analysis and simulation of network data. *Journal of Statistical Software*, Vol. 24 (2008), No 1, 1–11.

Hartigan, J.A.: *Clustering Algorithms*. New York: John Wiley & Sons, Inc., 1975.

Hartigan, J.A. and Wong, M.A.: A K-means clustering algorithm. *Applied Statistics*, Vol. 28 (1979), 100–108.

Hastie, T., Tibshirani, R., and Friedman, J.: *The Elements of Statistical Learning: Data Mining, Inference and Prediction*. Second edition. New York: Springer, 2009.

Hawkins, D.M.: The problem of overfitting. *Journal of Chemical Information and Computer Sciences*, Vol. 44 (2004), 1–12.

Higgins, J.E. and Koch, G.G.: Variable selection and generalized chi-square analysis of categorical data applied to a large cross-sectional occupational health survey. *International Statistical Review*, Vol. 45 (1977), 51–62.

Jank, W.: *Business Analytics for Managers*. New York: Springer, 2011.

Johnson, R.A. and Wichern, D.W.: *Applied Multivariate Statistical Analysis*. Second edition. Englewood Cliffs, NJ: Prentice Hall, 1988.

Karatzoglou, A., Meyer, D., and Hornik, K.: Support vector machines in R. *Journal of Statistical Software*, Vol. 15 (2006), No 1, 1–28.

Kass, G.V.: An exploratory technique for investigating large quantities of categorical data. *Applied Statistics*, Vol. 29 (1980), 119–127.

KDD (Knowledge Discovery and Data Mining) Cup. Available at http://www.sigkdd.org/kddcup/. Accessed 2013 Jan 16.

Ledolter, J. and Abraham, B.: Parsimony and its importance in time series forecasting. *Technometrics*, Vol. 23 (1981), 411–414.

Ledolter, J. and Burrill, C.: *Statistical Quality Control: Strategies and Tools for Continual Improvement*. New York: John Wiley & Sons, Inc., 1999.

Ledolter, J. and Swersey, A.: Testing 1-2-3: *Experimental Design with Applications in Marketing and Service Operations*. Stanford, CA: Stanford University Press, 2007.

Loader, C.: *Local Regression and Likelihood*. New York: Springer, 1999.

Michalski, R.S. and Chilausky, R.L.: Learning by being told and learning from examples: an experimental comparison of the two methods of knowledge acquisition in the context of developing an expert system for soybean disease diagnosis. *International Journal of Policy Analysis and Information Systems*, Vol. 4 (1980), No 2, 125–161.

Montgomery, A.L.: Creating micro-marketing pricing strategies using supermarket scanner data. *Marketing Science*, Vol. 16 (1987), 315–337.

Padgett, J.F.: Marriage and Elite Structure in Renaissance Florence, 1282–1500. Paper delivered to the Social Science History Association, 1994. Available at http://www4.ncsu.edu/~gremaud/MA432/padgett.pdf. Accessed 2013 Jan 16.

Padgett, J.F. and Ansel, C.K.: Robust Action and Rise of the Medici 1400–1434. *American Journal of Sociology*, Vol. 98 (1993), 1259–1319.

Resnick, M.D., Bearman, P.S., Blum, R.W., Bauman, K.E., Harris, K.M., Jones, J., Tabor, J., Beuhring, T., Sieving, R.E., Shew, M., Ireland, M., Bearinger L.H., Udry, J.R.: Protecting adolescents from harm. Findings from the National Longitudinal Study on Adolescent Health, *Journal of the American Medical Association*, Vol. 278 (1979), 823–832.

Shmueli, G., Patel, N.R., and Bruce, P.C.: *Data Mining for Business Intelligence*. Second edition. New York: John Wiley & Sons, Inc., 2010.

Stamey, T., Kabalin, J., McNeal, J., Johnstone, I., Freiha, F., Redwine, E. and Yang, N.: Prostate specific antigen in the diagnosis and treatment of adenocarcinoma of the prostate, ii: radical prostatectomy treated patients. *Journal of Urology*, Vol. 141 (1989), 1076–1083.

Stock, J.H. and Watson, M.W.: Forecasting using principal components from a large number of predictors. *Journal of the American Statistical Association*, Vol. 97 (2002a), 1167–1179.

Stock, J.H. and Watson, M.W.: Macroeconomic forecasting using diffusion indexes. *Journal of Business & Economic Statistics*, Vol. 20 (2002b), 147–162.

Taddy, M.: Multinomial inverse regression for text analysis. 2012a. Available at http://arxiv.org/abs/1012.2098. Accessed 2013 Jan 16. To appear in *Journal of American Statistical Association*, Vol. 108 (2013).

Taddy, M.: *textir* (R package), 2012b.

Tibshirani, R.: Regression shrinkage and selection via the lasso. *Journal of the Royal Statistical Society: Series B*, Vol. 58 (1996), No 1, 267–288.

UCI Machine Learning Repository. Available at http://archive.ics.uci.edu/ml. Accessed 16 Jan 2013.

Wasserman, S. and Faust, K.: *Social Network Analysis: Methods and Applications*. Cambridge, UK: Cambridge University Press, 1994.

Weber, A.: *Agrarpolitik im Spannungsfeld der Internationalen Ernaehrungspolitik*, Kiel: Institut fuer Agrarpolitik und Marktlehre, 1973.

Williams, G.: *Data Mining with Rattle and R: the Art of Excavating Data for Knowledge Discovery*, New York: Springer, 2011.

Witten, I.H., Frank, E., and Hall, M.A.: *Data Mining: Practical Machine Learning Tools and Techniques*. Third edition. Burlington, MA: Morgan Kaufmann, 2011.

Wold, H.: Soft modeling by latent variables: the nonlinear iterative partial least squares approach. In *Perspectives in Probability and Statistics*, Papers in Honour of M.S. Bartlett; Gani, J., Ed.; Sheffield: Applied Probability Trust, 1975.

INDEX

Printed in the USA/Agawam, MA
July 7, 2021

777558.138